The Logic
of Social Systems

A Unified, Deductive, System-Based
Approach to Social Science

Alfred Kuhn

Foreword by
Kenneth E. Boulding

THE
LOGIC
OF
SOCIAL
SYSTEMS

Jossey-Bass Publishers
San Francisco • Washington • London • 1976

THE LOGIC OF SOCIAL SYSTEMS
A Unified, Deductive, System-Based Approach to Social Science
by Alfred Kuhn

Copyright © 1974 by: Jossey-Bass, Inc., Publishers
615 Montgomery Street
San Francisco, California 94111
&
Jossey-Bass Limited
44 Hatton Garden
London EC1N 8ER

Library of Congress Catalogue Card Number LC 73-20965

International Standard Book Number ISBN 0-87589-221-3

Manufactured in the United States of America

JACKET DESIGN BY WILLI BAUM

FIRST EDITION
First printing: May 1974
Second printing: June 1976

Code 7413

The Jossey-Bass
Behavioral Science Series

*To Kenneth E. Boulding,
one of that sparse species
of truly creative minds*

Foreword

I became convinced at least twenty-five years ago that all the social sciences were studying essentially the same thing, which is the social system, even though they were studying it from different points of view and with different vocabularies. I became convinced of this through studying the labor movement, which it was clear to me could not be understood without invoking the skills not only of the economist but also of the sociologist, the political scientist, and even the moral philosopher and theologian. I was never, however, able to formulate to my own satisfaction the "same thing" that all the social sciences were studying, although I have written a good deal on this subject, as evidenced in the fourth volume of my *Collected Papers*. I have never been able to develop a complete, logical body of theory covering the whole social system in a way that satisfies me, though I think I have many bits and pieces of it.

For many years I felt that I was a voice crying in the wilderness, so it was with particular pleasure that I came across another voice in the same wilderness when I read Alfred Kuhn's *The Study of Society: A Unified Approach* (1963). We soon established contact; and while we may constitute the smallest invisible college of the twentieth century, at least we have been able to give each other mutual consolation and support.

The Logic of Social Systems carries the work of developing an integrated social science an important step further. It is not easy to read, even though it is written in a clear style appropriate to the subject matter and is enlivened with many delightful illustrations and flashes of wit. It is, however, a serious attempt to develop a unified logic of the social system, and it demands constant attention and thoughtful reflection from the reader. It is by no means an elementary textbook, although one can visualize a whole generation of elementary textbooks which will be based on it in the future. I am convinced that this is the way the social sciences must ultimately be taught.

The field of this work might be described as that of empirical logic. Logic is the study of the inescapable necessities of the real world, which includes, of course, the necessities of thought. When fantasy floats through the sky on the wings of imagination, logic draws us back to earth and imprisons us behind its possibility boundaries. It is a no-sayer full of "Thou shall not's." It is critic rather than creator; it involves selection rather than mutation; and we cannot expect it, there-fore, to be either agreeable or popular. Its only claim on us is that it is necessary, and this is a large claim. We cannot bargain with it. Its "effective preference," to use Dr. Kuhn's language, overshadows any preference for fantasy that we may put forth.

Even logic, however, as Alfred Kuhn sees so clearly, has a mushy side to it. There are the iron boundaries of formal logic and mathematics, which are based essentially on the great identities: 2 plus 2 are 4, or are at least equal to 3 plus 1; the circumference of any circle is π times the diameter; growth must be equal to what is added less what is subtracted; and, possibly, $e = mc^2$. After the great necessities come the great stabilities: the gravitational constant, the speed of light, the freezing of water at $0°$ Centigrade, the growth of a kitten into a cat, the aging process, the inexorable march of time. These are empirical stabilities rather than logical necessities, but their probability is so high that for all practical purposes it is 1. I am extremely confident that the sun will rise tomorrow, even though a single indubitable and com-pletely authenticated case of levitation would throw all physics into an ashcan from which it might never be retrieved.

As we move into social systems, the necessities perhaps be-come fewer, the stabilities less stable, and the probabilities significantly less than 1. Nevertheless, there are limits to social systems. They are not pure fantasy; they have a logic, they have necessities, and these are essentially the subject matter of this book. No organization can func-tion without DSE, that is, a detector-selector-effector apparatus, any more than life seems to be able to function without DNA, the carrier of its code of operation and its blueprints for growth and decay. Social systems consist of elements, each of which is in some sense an organization, which are related through communications and through transactions. It is hard to find any other form of relationship. Com-munications involves learning and change; transactions involve redis-tributions of stocks, whether these are money, commodities, prestige, status, authority, or political power. Organizations include both communications and transactions among their component parts, but they are guided by an overall selective system and images of the future on the part of their sponsors. Social systems are a queer medley of ecological structures, which have their own logic but do not have any overall control or sponsorship, and organizational structures, which comprise biological organisms (including humans) or man-made material structures like machines and computers and which are of greater or less degree of formality. All operate within the framework of empirical, logical limits which are the main subject matter of this work.

If I had been writing this book myself, and I hasten to say that I would have had neither the patience nor the nerve to produce a document of such vigor, complexity, and scope, I think I would have produced something like a mirror image beginning at the end of this volume and going back to the front. I would have started, I think, with the large ecological necessities of society which are developed in Chapter Sixteen and worked my way to the organizational necessities and the detector-selector-effector mechanism. What this means is that a book of this kind should be read twice. It is not a linear structure, where one begins at the beginning, goes on to the end, and then stops. Rather, it is a network of interconnected relationships where every part depends on every other. Such a work is inevitably hard to read, but the persistent reader will find it all the more rewarding.

I think this is a work which will have influence for many years to come. Others will attempt in the future, I am sure, to do the same thing better. I wish them well; they may even succeed. Nobody, however, will in the future be able to construct a logic of the social system for the first time, for this book has done it. It is a landmark and a watershed after which one hopes the social sciences will never be quite the same again.

I am not altogether sure that this is a scientific revolution in the sense of that associated with Alfred Kuhn's distinguished namesake, T. S. Kuhn. I am sure it will be resisted and perhaps neglected by social scientists who are all too snug within the niches of their own disciplines. But if it is a scientific revolution, it is certainly one whose time has come. The walls that surround our disciplinary Jerichos are crumbling. This work may be the trumpet that will bring them down. At least it is not an uncertain trumpet! Perhaps it is pitched too high to be audible in the restricted range of the narrow specialist. But its sound will make the disciplinary walls unstable, and these walls will at the least develop breaches through which traffic can pass.

KENNETH E. BOULDING

Preface

I hope that *The Logic of Social Systems* moves far enough toward a unified social science to shift the burden of proof from those who believe it is possible to those who insist it is not. Even if the burden does not move that far, I hope the reduction of distance is nevertheless significant. To paraphrase Margaret Mead, this book is an attempt to produce a behavioral science rather than a batch of behavioral sciences.

This volume differs from its predecessor (Kuhn, 1963) in three major respects. First, it is more explicitly system-based and built on system concepts from the outset, whereas the previous book was to some extent pushed into a system mold after much of it had already been written. Although relatively few system concepts are used here, and those are of a basic character, they nevertheless form an integral part of the analytic structure. In fact, I doubt whether any parsimonious, unified, conceptual structure of social science is possible without the use of system concepts. Although this volume is not intended to contribute directly to the development of system analysis as such, I hope it produces some of that result.

Second, the central section of the book is deductive. From models that consist of appropriate definitions and assumptions it deduces numerous propositions about social behavior. First-level deductions are combined to produce second-level deductions, and so on. Items are code-numbered to indicate their logical relationships and facilitate cross-reference. The central body of economic theory has long since achieved this kind of deductive structure. The main aim of this volume is to put the central core of social science on a similar deductive basis; and the book performs a substantial part of this task rather than merely recommending it. Since there is nothing remotely like this approach elsewhere, and since this kind of analytic structure is presumably a major goal of science, I hope the volume provides at the minimum a reasonable point of departure for the deductive venture, whatever the merit of its particular concepts or conclusions as seen by specialists.

Third, this book is more tightly unified than its predecessor. Its central concepts are more clearly identified, more closely related to one another, and considerably wider in their scope of application. Among other things this volume incorporates a wider range of the interactions dealt with by social psychologists and small-group researchers, and it makes explicit some aspects of social stratification, human ecology, and organization theory that were implicit or not covered in the previous work. It also deals with dynamics and social change.

As a prelude to the deductive models of social action proper, the volume introduces a model of social man. Its function parallels that of economic man but is much broader.

The purpose of unification here is not to eliminate or undercut specialties but to provide them a common set of analytic underpinnings. The repair of watches and of threshing machines do not cease to be separate and specialized fields just because both machines utilize the same basic principles of wheels, gears, and levers. But a person's ability to deal with a wide variety of situations will be improved if he understands the general principles before he proceeds to their special applications.

The unifying concepts are outlined in Chapter One. Since these concepts are not themselves new, though there is some newness in their form, the contribution of this presentation lies in identifying the concepts that have unifying potentials, defining them in ways that utilize their unifying properties, and developing those concepts, separately and in combination, so as to perform the unifying function. I cannot explain how or why the items were selected and assembled into their present structure except to describe the process as essentially evolutionary, involving thousands of trial-and-error steps. The book does little to argue the merits of the approach but simply presents it for examination. The system base of the book should tie it to psychological and biological science, as well as to any other field that can be treated in a system framework.

The deductive and unified aspects are related in the following way. Propositions are deduced from models and are then combined in ways that produce further deduced propositions. Particular cases or particular applications are then given names. Depending on the situation, either additional propositions are generated or additional and more specialized models are constructed. This process continues until a substantial scope of social science basics is encompassed. If the process could ever be carried all the way, the result would be a hierarchical structure in which all the analytic terms and propositions of all social science would form a single interlocking conceptual set. Obviously the volume does not complete the job or even argue that it is in principle completable, which it probably is not. The distance this volume does go may be roughly gauged by the fact that the glossary contains approximately 450 terms, all defined directly or indirectly on the basis of the central unifying concepts.

A standard question is Occam's. What is the price of more general concepts as measured by sacrifice of utility in handling particular problems? The eventual answer will have to come from specialists who are willing to familiarize themselves with this approach and apply it to their own materials. My own interim, biased answer is that even the most general concepts have direct applicability to many real situations. That is, I am suggesting that in important respects the concepts presented here largely avoid the conflict between generality and particularity. A mechanical analogy illustrates why.

The concept of *gear* is general, but its generality does not require it to be abstract: the most general-purpose concept of gear is no more abstract than the more limited-purpose concept of reduction gear. It is not an abstraction in the same sense as is mechanical advantage or work. Similarly, the most general analytic concepts used here are no more abstract than are many analytic tools now applied to relatively narrow problems. Like the general concept of gear, they can be applied directly to many real problems. Also, the special analytic tools in this system are either variations or particular configurations of the basics. As to variations, gears can be made with teeth on the rim (the simplest model), on the face, on a beveled edge, or even stretched out in a line as in rack and pinion. To understand the working of a variation we do not ignore the principles of the simple, or "pure," gear but start with statements about the simple model and then modify them to accommodate the variations. As to configurations, a differential is a special combination of shapes and sizes of gears designed to perform a particular function in automotive vehicles. The fact that it is a special instrument does not, however, remove it from the category of geared instruments or make the general concept of gears unnecessary for one who wants to deal with differentials in a nonsuperficial way. To return to Occam's problem, for the person who starts at the level of the whole differential and its function within the drive-train system, the utility of his concept of differential is in no way diminished by virtue of his

learning that it is simply a special configuration of a more general concept, gears. The same person also acquires a sophisticated understanding of differentials, and of their scope and limitations, if he learns what
is inside them and how their components work. For those worried about
reductionism I add my insistence that although the working differential
consists only of gears, it is simultaneously a system in its own right and
is definitely more than just gears.

The analogy is intended as explanation, not demonstration.
But it may clarify why I think the specialist will lose none of the
analytic usefulness of his own tools by learning that they can be viewed
as special cases of more general tools. In fact, he may understand his
special tools better for the experience. Because the specialized concepts
here are built on a common base, the wording of some definitions may
at first seem unfamiliar to the specialist. But once he has learned the
general concepts, I think he will find that the change in wording involves little or no change in basic meaning. Changed wording is inescapable in some cases if we are to have unified social science. Some
alterations of meaning do occur, however, and the specialist will have
to decide for himself whether the alteration is a gain or loss. Sometimes
the problem was merely to decide which of several extant definitions
fit best in the unified structure.

For related reasons I take sharp issue with those who believe
that a language which cuts across disciplines is therefore a metalanguage. Most terms in this volume are already found in the conventional disciplines, and modifying or choosing their definitions for
congruence with the terms of other disciplines does not make them
metalanguage. Even the language of systems is not metalanguage as
used here; it is simply more general. *Precipitation* is not metalanguage
merely because it is more general than rain, drizzle, snow, or sleet; and
system is just as direct and real as organism, machine, or organization.

To illustrate two cases of new wordings where meanings
have not been changed significantly, I define *culture* as communicated,
learned patterns and *authority* as the ability to grant or withhold
sanctions for the performance or nonperformance of instructions. Once
the anthropologist gets accustomed to this definition of culture I think
he will find it quite compatible with his thinking; and the political
scientist, organization theorist, and sociologist will find the same with
respect to authority. But whereas the two terms have thus far been defined independently and on their own, so to speak, in this format one
is a special case of communication and the other of transaction—two
basic concepts of social mechanics, analogous to, say, the gear and the
lever in physical mechanics. I would insist that one cannot really

understand the process (not the content) of culture without some grounding in the science of pattern transmission, which is communication, and that one cannot really understand the nature of authority without understanding those valued-based interactions that I call transactions. Thus the basic concepts used here not only provide a unifying scheme but may also sharpen more often than blunt existing special tools. But again the specialists will have to decide.

The analogy illustrates another point. In physical mechanics we independently developed the wheel, lever, gear, and pulley before we put them together in complex configurations. Thus we understood the components before we dealt with the larger configurations. In societies, by contrast, we had functioning governments, class structures, markets, cultures, and other complex phenomena, and gave them names, long before we analyzed them systematically. The existing disciplines provide a partial analysis of the whole social complex. I am suggesting that one more level of dissection, or reduction, is needed and that this volume outlines and partially develops one way to provide it. There are presumably others; Dahl and Lindblom (1953), Chamberlain (1955), Walton and McKersie (1965), Parsons and Smelser (1956), and other works of Parsons are cases in point. This volume presents a particular approach and does not argue its merits, except to say that no other approach is deductive in the same sense and degree and that none (in my view) ties together as wide a variety of concepts into a unified set.

Once the basics have been reached, however, we can reverse the process. Following the lead of physical mechanics we can discern and amplify the theories of the basics, identify their main variations, note the modifications of the simple theory that are necessary to explain the variations, and then put the basics and their variations together in as many configurations as we have patience or need for. We must also be aware that the reassembled parts will probably not look quite like the initial whole. At this point I am not aware of any aspect of social analysis that is not amenable to this approach, though I know only too well that my deep immersion may blind me to things that do not fit.

All this explanation does not mean there are no analytic ambiguities. If you push down on one side of a wheel and the other side goes up, should you use the model of the wheel or that of the lever to explain what is happening? Nor does it imply that complex situations are necessarily understandable. What predictability does the model of the lever provide when the "arm" of the lever is a slab of mushy ice, the downward pressure on one end of it is a restless polar bear, and the fulcrum is a mound of compact snow? This "mushy ice" problem also illustrates, incidentally, that precision of prediction declines rapidly as complexity increases, whether the subject be physical or social science.

Relatedly, I think the main models in this volume are no more unrealistic than are such assumptions as economic man, the perfect vacuum for the law of falling bodies, the weightless arms and dimensionless fulcrum of the lever, and the zero-viscosity liquids and frictionless pipes of the introductory models of hydraulics. And for those who properly decry attempts by social scientists to mimic the physical sciences I add that such parallelism was not my goal. The preceding analogy occurred to me as a possible means of clarifying the nature of this conceptual structure, not as the basis for it. In this relation among parts I do not think that integration need "become another specialty [rather than] a work-a-day habit of any substantial number of economists, sociologists, and political scientists" (Landauer, 1971).

For reasons implicit above, this volume makes relatively little reference to the works of others. I had to be alert to many extant concepts and findings before formulating my basic models, and I modified them many times by comparing them with the work of others. But once the models were formulated, the elaborations of and deductions from them proceeded largely under their own steam. Occasional discussion of other works is included mainly to indicate similarities or differences, as well as to maintain contact with reality—as reality is currently perceived by specialists.

Chapter Eleven compares several of my deductions with parallel empirical findings. Even a preliminary discussion of such comparisons requires up to three pages per proposition. Since there are perhaps a thousand deduced propositions here, to check them against extant findings is obviously beyond the scope of a single volume. In any event, many of these propositions apparently have no close counterparts elsewhere. The models thus fill the requirement that a new theory generate previously unstated propositions. Furthermore, I believe that many, perhaps most, are empirically testable.

In several instances certain social structures can be deduced (predicted) from the models of interactions. For example, Chapter Ten includes certain deductions about who will be willing and able to transact with whom, and from those deductions class stratification can in turn be deduced.

Although the models lead to new propositions, the main goal of the volume is more efficient knowledge, not new knowledge. If the volume is at all successful, such efficiency will arise from two sources.

First is the deductive structure. There is no point to a thousand empirical findings that managers in twenty-five different industries in forty different countries prefer lower-cost to higher-cost raw materials if this conclusion can be deduced from a profit-maximizing model. Furthermore, to find the deduction incorrect in a particular case is more informative when viewed as a deviation from theoretical expectation than if taken as an isolated datum. Such excellent compilations as those of Berelson and Steiner (1964) and March and Simon (1958) demonstrate that the social and behavioral sciences are replete with low-level empirical findings. It would certainly constitute a major increase in efficiency if many such empirical details could be replaced by deductions from models. Many low-level (specific) generalizations are nevertheless included here to demonstrate that the models can produce them, and this capacity could not be demonstrated without them. The reader interested only in the main thrust of the volume can treat such generalizations as a compilation for reference and simply scan such sections to sense their rationale and place in the larger context.

Second, unifying concepts can themselves increase efficiency. The fact that one can sometimes repeat the same statement in the languages of two to five different disciplines with many specialists unaware that others are saying the same thing means that we can learn something from several different disciplines in one concept-learning exercise if we make explicit the common components. This book attempts to follow the obvious prescription and to forge the common components into a subdiscipline of their own—or superdiscipline, depending on how one views the hierarchy. The glossary demonstrates that a conceptually integrated vocabulary is possible. At this point I will not argue its utility.

The Logic of Social Systems at times spells out intuitively obvious notions, which some readers may regard as trivial. I make no apologies, on the ground that "to be of any scientific use, the obvious must be reformulated into a conceptual system" (Jordan, 1968, p. 15). Only by making intuitive knowledge explicit can the less obvious be systematized (Boulding, 1970, p. 100).

The logical status of various groups of propositions differs considerably. Propositions on decisions and transactions and in the introductory sections on organizations are tightest, and I hope they stand reasonably close inspection. Those on communications are less tight, for reasons that become clear in Chapter Seven. Chapters Three and Four are mainly statements of the model of social man rather than deductions from it, though some deduction is used in forming the model. Chapters Ten and Eleven involve successively large relaxations of the initial models and frequently degenerate to the *ceteris paribus* assumption (shortened for convenience to *cet. par.*) to avoid unwieldy statements of the then-current assumptions. It may eventually be possible to make explicit all assumptions underlying even propositions distant from the initial model, though a computer would be required to do it.

Later chapters drop the deductive approach and merely sketch how this framework might be applied to the subject matter of the conventional disciplines, including history, anthropology, and geography. Implicit in the last two chapters is the idea that although the Two Worlds cannot be merged, they can be bridged. Whether use of the method will induce more cross-sterilization than fertilization only time will tell.

ACKNOWLEDGMENTS

My most conspicuous debt is to Kenneth Boulding, who gave critical comments on the entire, difficult-to-handle manuscript. Beyond that, I have been reading so many of his articles and perceptive little books for so many years, along with listening to his lectures and conversing with him, that I have quite lost track of where his thinking leaves off and mine begins. Without the groundwork he has laid, this book would have been utterly impossible.

Warren Bennis read much of the manuscript and provided important encouragement toward getting it published. Through extensive correspondence and detailed comment on the predecessor volume Carl Landauer helped form this one. Parts of the manuscript have been read and usefully commented on by William Fox, Alex Michalos, Claude Gruen, Michael Marien, Harland W. Whitmore, Dean Moore, Susan Kellar, and Dodd Bogart. Graduate student Robert Beam did yeoman work in tracing primitive terms through the manuscript and the glossary.

In the background perhaps those most responsible for this volume are the many who one way or another responded enthusiastically to my earlier attempt at integration and thereby built the courage so necessary for pushing through the next step—Richard Meier, Myron Tribus, Meno Lovenstein, Robert North, James Miller, Wesley Allinsmith, Win Wenger, Steuart Henderson Britt, Karl Deutsch, George Homans, Irving Morrissett, Lawrence Senesh, David Singer, Bertram Gross, Walter Buckley, Wroe Alderson, Maurice Levine, C. Lowell Harriss, Richard Fagen, William Jeffrey, and Kurt Mayer. A continued dialogue with Charles Hofling about psychology and psychiatry was most useful. Through financial assistance the Charles Phelps Taft Memorial Fund and the Research Council, both of the University of Cincinnati, assisted with the writing and publication, respectively, of the volume. The typing was complicated, and Nancy Tucker and Miriam Tucker with patience and good will did the bulk of it. Nina Kuhn prepared and typed an unusually difficult index, and her patience with the pressures of family and skipped vacations that such writing entails deserves greater appreciation than I can express.

January 1974   ALFRED KUHN

Numbering System

Because of the numerous internal cross-references, many of which must be explicit, key numbers are used. All such key numbers are in boldface and run in sequence as in the Logical Outline, though in much more detail. When a key item is cited elsewhere in the text its number appears in regular type (like this: 3.18.6). As usual, the number to the left of the first decimal point represents a main heading and digits to the right represent successively lower levels of subdivision. The digits are grouped to assist visual identification and short-term memory. Thus 4.13.5132 is to be treated as if it were written 4.1.3.5.1.3.2.

Sometimes zeros are used in a sequence, as when the numbering goes from 2. to 2.0, for introductory or definitional materials. The effect of the zero is the same as a 1, however, and 2.0 comes after and is logically subordinate to 2. In most citations the whole number is used (like this: 3.28.41). But if the citation is to two or more numbers that have the same first three digits, the first is given in full and the others show only the numbers to the right of the second decimal point (like this: 3.62.114, .2, and .46, which means 3.62.114, 3.62.2, and 3.62.46). A citation to all items between 3.62.114 and 3.62.46, inclusive, appears as 3.62.114–.46. *Section* in front of a number refers to all items beginning with those digits: Section 3.2 means the entire section on the transactional analysis; Section 3.26.2 refers to all the material under the heading "Generosity." Absence of the word *Section* means that only the item so numbered is indicated.

The numbers in the Logical Outline identify the main logical sequence and relations of parts. The text itself often includes introductions, summaries, comments, discussions that cut across parts, and other materials that help the exposition but are omitted from the

Logical Outline to avoid obscuring its structure. Thus all items in the outline appear as headings in the text, but not all headings in the text appear in the outline. The amount of detailed numbering is largely pragmatic. Parts to which there are detailed citations must be numbered in detail, sometimes down to individual sentences, though for readability and conservation of space these may be grouped into paragraphs that have key numbers within them. In parts that are not cited in detail, the detailed numbering has been eliminated. Beyond the third digit (to the right of the second point), and except for four- or five-digit numbers that appear in the Logical Outline, a larger number of digits does not necessarily indicate logical subordination, in which case the number merely identifies the location of an item in a sequence. In fact I doubt whether it is feasible, or even possible, to achieve complete consistency of outline at that level, and the benefit would certainly not justify the cost.

Logical Outline

Contents

ORGANIZATIONS: HIGHER-LEVEL SYSTEMS
OF MULTIPLE HUMANS

OVERVIEW

The Logic
of Social Systems

A Unified, Deductive, System-Based
Approach to Social Science

Chapter 1

Introduction
The Main Concepts

We will start by elaborating the intended meaning of the subtitle of this book: "A Unified, Deductive, System-Based Approach to Social Science." *Social science* refers primarily to the three "core," or "nomothetic," disciplines of sociology, economics, and political science, which to a considerable extent seek generalizations independent of time, place, and the particular society. Some discussions in this volume are also relevant for human geography, history, and anthropology, particularly the last. A substantial amount of material is closely related to social psychology, whose relevance to certain aspects of the social sciences I fully accept.

Unified does not mean that a single principle explains all social science or all social phenomena. Nor is it an attempt to "relate" the various disciplines as they now stand. As used here the term has two main dimensions—that certain types of analysis are, in fact, common to various disciplines and that the types are related to one another. As to the first dimension, many concepts and types of analysis keep cropping up in all the core social sciences and sometimes in the others as well. To illustrate, transactions appear widely in sociology in the works of such authors as George Homans, Richard Emerson, Erving Goffman, and V. W. Thibaut and H. H. Kelley. In economics they appear in purchases, bilateral monopoly, collective bargaining, and in sharp form (even if not called transactions) in comparative advantage in international trade. In political science they are seen in power plays, threats, negotiations, wars, and treaties. My point is that there is a basic theory of transactions which is independent of the social system in which it occurs and which, with appropriate substitutions of assumptions, constitutes the underlying theory of all the variations now found scattered among the several disciplines. The notion that one must study economics to learn about "economic" (selfish) transactions and sociology to learn about "social" (generous) ones is naive indeed.

To the extent that the social sciences have a body of concepts and theory in common there is a unity among them. If additional common concepts and theory can be identified and explicated that body of common items, however large or small, can be referred to as having unifying properties or as constituting unified social science— at least in contrast to the present state of affairs. In addition to transactions, the other concepts here selected as having such unifying properties are communication, organization, and system. If some readers object to *unified* in this connection, let it at least be clear how the term is used here.

As to the interrelatedness of concepts implied by *unified,* many terms from the basic social sciences are defined here: communication, government, market, power, organization, leadership, socialization, culture, sovereignty, influence, authority, legitimacy, citizen, civil liberty, boycott and others. But whereas in current social science the terms in one discipline are defined with little or no regard for the terms in any other, and often rather independently within disciplines, all definitions in this work constitute a single interlocking set. All are defined directly or indirectly in terms of the same few basics, which rest in turn on a relatively parsimonious set of primitive terms. Furthermore, every definition in the entire set is made in light of, and with the goal of consistency with, every other definition. However many the errors, and whether or not certain specialists like certain definitions, I think this arrangement qualifies as unified.

Regarding the *deductive* aspects of the volume, although some deductive thinking is used by all social scientists, economics is the only social science whose central theory is essentially deductive in its entirety. Conclusions that the price and quantity of a product are determined by the intersection of supply and demand, for example, or that wages under competition equal the marginal product of labor are not empirical conclusions. They are deduced from a set of definitions and assumptions. In fact, both conclusions are devilishly difficult to test empirically.

This volume contains several hundred propositions about social behavior—as it may occur in politics, families, corporations, competitive markets, small groups, or union-management bargaining. These generalizations (or theories) are not collected from the social science literature. Like the core of economic theory, they are deduced directly or indirectly from sets of definitions and assumptions. Although the models (definitions plus assumptions) have been designed and modified to elicit deductions that seem to explain reality, their validity in their present context does not depend on theoretical or empirical conclusions found elsewhere in the social science literature. Although errors may appear in the analysis, the deductive portions, or central theory, are self-contained and self-sufficient—or at least are intended to be. Their direct validity can be determined solely by examination of the models and how they relate to the conclusions.

Whether the central core of the volume is useful to specialists and helps them explain reality is an important but different question, one they will have to answer after they have become familiar with the models. This deductive core is discussed primarily in Chapters Six through Twelve. The preceding chapters provide the background for the models; the subsequent ones suggest extensions beyond the areas explicitly developed here. The later chapters also compare this approach with other ways of attacking certain problems, compare the deduced conclusions with empirical findings, and discuss certain problems—such as social change—that are not amenable readily, if at all, to the central models.

For this volume *system-based* means several things. First, it does *not* mean a multiplicity of diagrams showing flows and feedbacks, which is an activity more appropriate for the study of relatively particularized systems. Second, it does mean a taxonomy of systems slightly different from those I have found elsewhere. These categories (which mainly distinguish acting from pattern systems, and controlled ones from uncontrolled) seem essential for social science, yet without doing violence to other well-established usages. Third, some system concepts and terminology seem indispensable for certain aspects of the unification process, particularly in relating the intersystem analysis directly to the intrasystem.

Objections are sometimes raised to system analysis on the ground that it is just one more analogy, like the mechanistic analogies that followed Newton and the organismic ones that followed Darwin, and that it too will pass. I believe that the mechanistic and organismic analogies were gropings for the general concept of system, which encompasses both and more. Furthermore, when a system is defined as any set of interrelated or interacting components, it is difficult to imagine a more comprehensive analogy. In any case, I am only suggesting that system is a useful concept for this purpose at this stage of history, not that it is everything and forever. It is not everything even in this volume.

To return to another aspect of terminology, in this volume all terms are not only interlocking and mutually consistent. They are also defined (as nearly as I could make them) precisely and unambiguously. Unlike the natural sciences, which for the most part have developed their own terms, the social sciences are replete with supposedly technical terms that have been used in ordinary language for centuries. Thus one often discovers such terms as power, trust, cohesion, leadership, or legitimacy used in analytic works without being defined, as if on the assumption that everyone, of course, knows what they mean. To illustrate: in sensible, ordinary usage one can say, "You can always trust Joe to lie." Hence to use *trust* in a serious analytic work without defining it carefully is useless or worse. If an author states that "trust makes a group more cohesive" and then fails to define *group* and *cohesion,* I am skeptical about the meaning and the scientific status of the proposition—though operational definitions may be implicit, and in a sense precise, if the researcher describes how he measured trust and cohesion and identifies the constituency and general behaviors of his experimental group. Furthermore, social science writers who do define their terms often use different definitions. Hence it is often difficult to know whether one writer's conclusions about (say) trust or cohesion can be added to someone else's to form a coherent and cumulative body of knowledge. Given definitions that are either different or unstated, it may not even be possible to tell whether similar-sounding propositions agree or disagree.

There is also a tendency among social scientists to "find" the meanings of their terms, after the fashion of the lexicographer, instead of experimenting to discover which definition provides the greatest analytic usefulness and then adopting it with a hard-nosed indifference to everyday usage. I have consistently tried to follow my own advice by using analytic usefulness and consistency with other terms as criteria. I nevertheless think that there are few definitions herein that do not closely reflect the conceptual content of at least one current usage. However, the language sometimes looks unusual because the definition arises from a different context. This question is discussed in further detail near the end of the concluding chapter.

To preview several specifics, I am convinced that the general failure in the behavioral sciences to make a sharp distinction between communication and transaction, and to construct a rigorous, separate theory of each before trying to deal with the many combinations and mixed cases, has contributed much to keeping these sciences as bulky collections of low-level generalizations (see Berelson and Steiner, 1964, p. 660; DiRenzo, 1966, pp. ix and 285ff). Hence I have separated them before rejoining them. I also think it essential to distinguish power from bargaining power, since many clear analytic statements can then be made that are otherwise not possible. Among other things, power becomes the key analytic tool that many have felt it must be, but has elusively failed to be (March, 1966). A sharp distinction between transactional (a matter of social science) and pretransactional (a matter of technology, broadly construed) stages is also essential to the analysis of coercion. Somewhat the same can be said of the difference between controlled and uncontrolled social systems.

There is also the problem of what Kaplan (1964, pp. 70f) calls "premature closure." At this point in history I think the behavioral sciences are afflicted more with dangers of senile openness. But that is a matter the reader can evaluate better after the discussion in the last chapter.

GENERAL PROBLEMS OF SCIENCE

Science is knowledge. Knowledge takes the form of coded patterns in men's heads—mental constructs or images (Boulding, 1956). If the science is about reality, the patterns in men's heads bear some relation to the patterns outside: connections between patterns made inside the head, through thinking, to some degree parallel the connections between patterns outside the head. The closer the parallel, the more empirically useful and valid the science. If we think of connected patterns as constituting a larger pattern, such parallelism enables us to know (predict or expect) the whole of some larger real pattern from observing only part of it. Having predicted an unobserved from an observed portion of a pattern, if we then compare the reality with the prediction, the closer the correspondence the better the science.

The logic is the same for objects and events. And for events it is the same whether the prediction is made before or after the actual occurrence. The logic is also the same for the events of everyday life as for those of formal science. In this fundamental sense it is also the same for empirical as for analytic science. The common element is this: to know one part of a pattern, by observation or assumption, enables us to know something about another part not currently observed or also assumed. Other things equal, the more information that can be reliably deduced or inferred from a given piece, the greater the efficiency of the knowledge structure and the greater the information content of that piece of information.

Information content is in turn related to improbability of type of pattern, taken in conjunction with uniformity among various instances of that pattern. In empirical science, for example, if it is known that a certain shape of toenail is always found in one species of animal, but only in that species (is relatively improbable), then to observe the toenail alone on a specimen permits a deduction that its possessor displays all the other patterns that reliably belong to that species. But if that same shape of toenail is known to belong to ten equally numerous species (is relatively more probable), then either (a) the same deduction has only one chance in ten of being right or (b) only those traits common to all ten species can reliably be inferred from the toenail. Uniformity from specimen to specimen of the pattern is necessary, of course, if an inference is to be made from one specimen to another. The principle of improbability can also apply, however, to only a single specimen. If there is only one Mona Lisa, then once the whole pattern is known, to observe one part of it permits an inference about the unobserved parts.

The crucial aspect of improbability lies in the ability to isolate one pattern from all other patterns that might be similar to it, since without such isolation it is not possible to infer one part of the pattern from another. In deductive science this isolation is achieved by *defining* a pattern as unique, in contrast to observing it to be. In geometry, for example, one type of line is defined as straight throughout its entire length, thereby uniquely isolating it from a host of more probable lines that are unstraight or are straight through only part of their length. From a given part of a straight line it is therefore possible to infer something about other parts of the same line. And since all straight lines are precisely alike in every detail, knowledge discerned or assumed about one can confidently be assumed to be true of all others, and the same can be said of configurations made by combining such lines. The simplicity of the straight line also enables us to make precise statements about it and about combinations of lines—a matter I discuss in more detail later.

Other things equal, the greater the number of patterns that can be inferred directly or indirectly from a given pattern, the greater its generality. For example, the ability to infer certain aspects of the patterns of falling bodies, weight, and planetary orbits from the mental pattern of *gravity* makes that pattern more general than if patterns of falling bodies alone could be inferred. And, other things equal, the fewer the mental patterns needed to infer or construct a given number of other patterns, the greater the parsimony of the science.

To some extent generality and parsimony are obverse sides of the same coin, since the larger the number of real phenomena dealt with by each mental construct, the smaller the number of constructs needed to deal with a given number of phenomena—though I shall not debate here whether this relation always or necessarily holds. Parsimony, of course, refers to the number of *types* of patterns dealt with, not the number of instances. An all-brick wall is conceptually more parsimonious than one of brick, stone, and block, even if it contains more pieces.

In these senses this volume seeks to define mental constructs of social behavior with greater generality and parsimony than those now available while retaining the direct applicability of narrow empirical findings. Toward this end—and strictly in the spirit of experiment—a few central concepts are named and defined. Besides the definition, a model of each concept is described—typically in the form of a set of assumptions—so designed that a structure of logical inferences can be deduced from the model. That is, one part of a pattern can be deduced from knowledge of another part.

The initial model and the immediate deductions from it are good approximations of certain aspects of reality for the simple reason that alternatives were discarded until some were found that showed this trait. By analogy, the deduction from the economic model that a firm will buy its raw materials from the lower-priced (rather than the higher-priced) supplier has its real counterpart in untold thousands of actual decisions, even though the profit-maximizing model may need to be modified for other purposes. Other aspects of reality are approximated here by two main devices. The first is to introduce additional assumptions, thus narrowing the model to special cases or—having explored the implications of the initial model—to introduce substitute assumptions and explore their implications. After each main model has been examined in isolation, the second device is to combine them— that is, to examine the consequences of having two or more operate simultaneously. This second device is a logical extension of the first in that it adds new assumptions that are distinctive enough to justify separate mention.

Many of the resulting configurations are named and defined in the same spirit that angles, rectangles, and other configurations of straight lines are given names. Given the generally bad odor of neologisms, existing terms are used. But whereas many terms as now used are not significantly related (integrative processes, power, market, attraction, authority, freedom, culture), in this volume each is defined as a particular instance or configuration of the few central concepts. The objective is a single interlocking conceptual set.

The result provides a set of analytic underpinnings common to the three major social science disciplines of sociology, economics, and political science (and possibly history, human geography, and anthropology) and a translation into an integrated conceptual structure of significant portions of the existing theoretical (as contrasted to descriptive) content of those disciplines. Least attention is given here to economics since much of it is already couched in a deductive framework whose relation to the present one will be readily evident.

As will be seen, the variety and complexity of the permutations and combinations of the main models and their modifications are seemingly endless. Some configurations and their real counterparts are of intense practical interest, and specialists are needed to deal with them. The point is thus that the separate disciplines are different variations on the same main themes, not that there are no differences among them. It is nevertheless suggested that some disciplinary boundaries be redrawn and that a common language and fundamental conceptual structure be adopted.

The volume contains altogether nearly a thousand propositions about behavior deduced directly or indirectly from a few elementary models, using terms defined as needed. There is thus a large body of constructs held together by logical relations (Margenau, 1966, pp. 30–32). Rules of identification are discussed only tangentially in a section of Chapter Eleven that compares certain of these deduced propositions with empirical ones. I doubt, however, whether the sophisticated social or social-psychological scientist would have much difficulty identifying aspects of reality that correspond with most of the analytic constructs. In any case empirical tests of many propositions should not be too difficult to apply in principle.

All terms have been carefully defined. Such vague terms as morale, cohesion, attraction, and sense of identity are sharply redefined or else the phenomena they deal with are approached through a different conceptual structure. As noted in the Preface, the goal is not new knowledge but simplification of the old. In any event the extent to which the models produce new propositions can be better assessed later. In short, the book is an experimental proposal of a limited list of social scientific concepts designed to be more parsimonious and general than current ones while remaining as directly applicable to reality. This book, of course, can only begin the task, mostly by outlining and describing the conceptual set and by elaborating some of its ramifications. Broad testing of the model must be done by specialists: thus the fundamental concepts and their interactions are spelled out in detail in code-numbered propositions. Their possible extension to other areas is sketched later in discursive form, as are associated methodological problems.

CONCEPTUAL SET

Like systems in general, this volume is something more than the sum of its parts. In fact its wholeness is intended to be its major contribution. Conversely, the significance of some parts can be understood only in the context of the whole. Although the reader may not sense the function of a part until he has finished the book, the remainder of this chapter presents at least a crude map of the whole that the reader may consult from time to time.

The conceptual set is oriented around system analysis. But rather than attempting to adopt, or even scan, the breadth of that amorphous field, the book takes a few central system concepts, adapts them to the social context, and elaborates them. The most basic distinction is between controlled (cybernetic) systems and uncontrolled systems, which among social phenomena appear as formal and informal organization, respectively. For intrasystem analysis the model proposes that any controlled adaptive system, including the human being, must utilize *information, preferences* or *values,* and *behavioral responses,* which are handled respectively by *detector, selector,* and *effector* functions of the system. The interaction of the three to determine the nature and direction of behavior is considered under *decisions.* The magnitude of the effect produced, particularly as related to the system's goals or preferences, is examined under the heading of *power.*

The three intrasystem functions have intersystem parallels. The detector deals with information within the system, and *communication* is a transfer of information between systems. The selector deals with values within the system, and *transaction* is a transfer of value between systems. When precise analytic questions arise we can say that communication and transaction are interactions analyzed with respect to their information and value content, respectively. Within the system (that is, at the single-system level), the effector produces a behavioral result, and *organization* is the joint production of an effect by two or more human systems. And whereas the essence of organization is its higher-level systemness, the initial analysis of communication and transaction attends solely to the interaction as such while ignoring its higher-level system aspects. We later view the constraints on such interactions when the parties are subsystems of an organization. Nevertheless, communication and transaction are the only two *inter*system interactions, strictly speaking, recognized in this framework.

To illustrate these six concepts, let us imagine a woman shopping in a small supermarket managed by its owner-proprietor. She perceives Spanish and Greek olives in cans (detector, providing information). She prefers Greek to Spanish (selector, reflecting her values). She puts a can of them into her basket (effector, carrying out a selected behavior). On another occasion the cans are dented and the labels missing, so she asks the manager which is which. The transfer of information (involving both detectors) is a communication. At the checkout counter, after arguing with the manager for a discount on the damaged cans, she gives money and receives olives. This transaction is based on the values (selector states) attached to both olives and money by both the woman and the proprietor. She next holds open her shopping bag while a clerk puts in the groceries—a jointly produced effect at a simple level of organization. Thus the conceptual set for the intersystem (social level) analysis is directly parallel to and logically connected with the intrasystem (psychological level) analysis.

We will start with pure models of detector, selector, and effector, and then of communication, transaction, and organization, before dealing with their combinations. In their simplest form, the three intrasystem concepts, the three intersystem concepts, and the relationships between them are diagramed in Figure 1. This conceptual arrangement perhaps reflects a primitive distinction between matter-energy and information levels of action. For the pure cases at the information level we are interested in pattern without regard to substance; at the matter-energy level we are interested in substance without necessarily attending to its pattern. Accepting as it does that substance and pattern are both aspects of matter-energy and hence may seem monistic, this view nevertheless parallels the older mind-matter dualism but seems analytically far more fertile. Applied to social interactions, at the elemental level communication is an exchange of information and transaction is an exchange of matter-energy. Although that distinction is probably valid for all interactions of nonhuman systems, even for the introductory model it seems better to define the interactions more broadly for humans, whose interactions differ from nonhuman ones in two important ways. First, communications between humans is predominantly sign-based, or semantic, involving coded information at the conceptual level. Communications between nonhuman systems, possibly excepting man-made machines and a few animals, do not involve signs in syntactic relations. Second, transactions between human

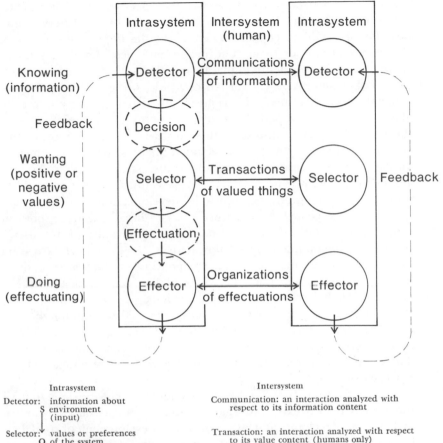

Figure 1. Intrasystem-intersystem axis of controlled systems.

systems are based on the values of the things exchanged whereas those between nonhuman systems are not. Although there is a sense in which certain osmotic exchanges in plants, for example, reflect the "value" of a substance to a plant, it will be clear that the subsequent value-based analysis of human transactions could not be applied to plants— a sensible reason for making the distinction.

   The next main distinction is between cross-sectional and developmental analysis. The first deals mainly with the interactions between two systems, given the states of each. The second deals with changes in those states over time and the effect of those changes on subsequent behavior. To return to the shopper, in time the woman's taste for olives might change and so might the importance to her of the money to pay for them. The proprietor's need for money might change, as well as his desire to placate customers. Hence a later replication of the same overt circumstances would bring a different transactional result. For much the same reason that astronomy is precise about the orbits and periodicity of planets but imprecise about why there are nine planets instead of three or three thousand, most of this volume deals directly with cross-sectional analysis. But developmental analysis is nevertheless important and is discussed in Chapter Sixteen and elsewhere with reasons why the cross-sectional is indispensable to understanding, or even describing, the developmental.

The proposed framework covers a broad spectrum, both of subject matter and of types of analysis. For example, to examine a system as a functioning unit is *holist* analysis. To look downward from that system to the subsystems that compose it is *reductionist,* both in looking at parts rather than the whole and in examining the subject matter through "lower" disciplines. To look upward from a system to the larger system of which it is a subsystem, and to examine the role it plays in that larger system, is *functionalist.* Holist, reductionist, and functionalist analysis are thus different views from the same level. There is no point in arguing which is "better": all are necessary for a balanced view.

All three types can be cross-sectional when we assume that all three levels already exist. But there is also an upward and a downward look on the developmental axis. Here the development of a higher-level system by newly coordinated interactions of existing systems is *emergence*—though once a certain level of system exists, increasing differentiation within it may also be viewed as emergent. The breakdown of higher-level systems so that only the components remain is *decay.* Absence of either movement is *stability, equilibrium,* or simple continuance. Change can also occur without being emergent or decadent, as in stochastic sequences. System analysis thus provides a coherent patterning of otherwise discrepant views. The two trios—holist, reductionist, functionalist; stability, decay, emergence—are the straight, downward, and upward looks from a certain level of system on the cross-sectional and developmental axes, respectively, as diagramed in Figure 2.

Besides offering a framework for knitting together these various views, the hierarchical structure of systems provides a convenient simplifying vehicle. The behavior of the individual human is diagnosed with the intrasystem concepts of detector, selector, and effector functions. When two or more persons interact, but without forming an organization, the interactions are diagnosed through communicational and transactional analysis. If they form an organization, like a family or business, the interactions within it continue to be analyzed as transactions and communications. But to the extent that the organization acts as a unit its behavior is analyzed with respect to *its* detector, selector, and effector processes. If several persons must reach agreement on that unitary decision, they do so in turn through communication and transaction, possibly supplemented by a dominant coalition (see Section 4.12.4).

Whenever an organization interacts as a unit with any person or other organization, those interactions are again communications and transactions. In short, when a controlled system is viewed as a unit, from individual to world government, intrasystem analysis is used.

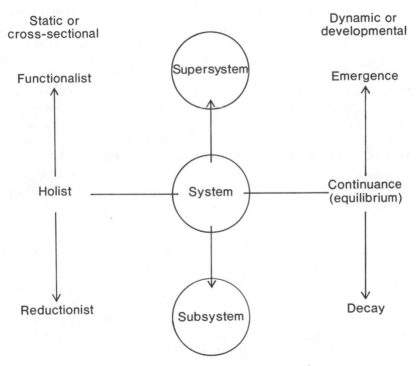

Figure 2. Subsystem and supersystem axis.

When two systems interact, intersystem analysis is used. Additional levels increase the complexity of analysis but do not demand new analytic tools.

Finally, there is the difference between *idiographic* and *nomothetic* analysis. The former deals with the uniqueness of the single case in all its individual glory—the problem of clinician or humanist. The nomothetic deals with the characteristics that different cases have in common; it abstracts from reality the traits of multiple instances. Strictly speaking there can be no science of the idiographic since it is not until uniformities are abstracted from real cases that generalizations can be stated. In that sense all science is a distortion of reality. Even if one views himself as the most idiographic of humanists, however, as soon as he recognizes similarities of pattern between different persons or situations he has to that extent become a scientist and is using a nomothetic approach.

Though individuals may specialize in one approach or another, this volume suggests that a social scientist should have enough experience with each approach to get the feel of it: the controlled system and the uncontrolled; the intrasystem and the intersystem; the cross-sectional and the developmental; the straight, the upward, and the downward look in each; and the idiographic and the nomothetic. These approaches, it is proposed, are the social scientist's basic tools.

Since excellent case studies of personality, historical situations, regional areas, tribal societies, small groups, social problems, and the like abound in literary as well as in scientific works, there is no need to sample them here. The relatively limited attention paid to the idiographic thus implies no discount of its importance; it merely reflects its nature. That the developmental process is accorded relatively small space represents a methodological suspicion (an analytic tool somewhat weaker than a methodological assumption) that the developmental process is itself essentially idiographic—as amplified in Chapter Sixteen.

"Well, the problem is really very complex!" These words, responded to with sage nods, conclude the study of many a real social problem. A central point of this volume is that the inescapable complexity of reality does not necessarily require complex analytic tools—although it may. The proposed relation between this relatively small kit of tools and complex reality can be illustrated in two diagrams. If we accept the customary groupings of social science disciplines, the relationship is illustrated in its simplest form by Figure 3. In the circle to the left are the unified tools, the delineation of which here constitutes basic social science. Their elaboration into particular configurations constitutes the subject matter of the specialties, indicated by arrows radiating to the right. A social problem occupies the circle to the right. For a good understanding of a real problem it is probable that all the social science specialties, perhaps supplemented by natural science, are required. The line going directly from the left-hand to the right-hand circle indicates that the basic tools can also be applied directly to many real problems. Certain conceptual or analytic problems might also appear in the right-hand circle. I call the relation between the left-hand circle and those in the center the unified or integrated approach. It

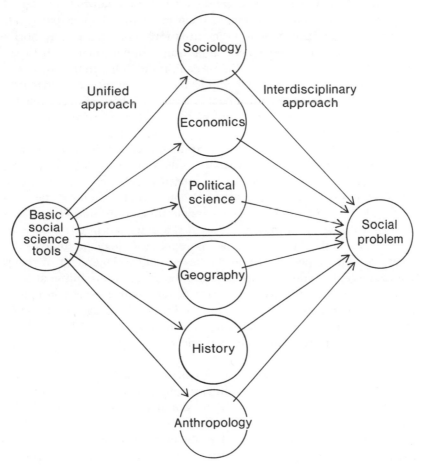

Figure 3. Unified versus interdisciplinary approach.

starts with a compact set of concepts, states their definitions, properties, and propositions about them, and then elaborates them into the configurations of the separate disciplines. The relationship on the right side is widely known as the interdisciplinary approach, which among other things applies concepts from several disciplines to a particular problem.

An expanded view that does not necessarily take the existing disciplines as given appears in Figure 4. The basic tools are again on the left, at A; a problem is again on the right, at E. The basic intrasystem and intersystem concepts and propositions of A are the most general tools—the transaction, for example. By certain limiting assumptions these highly general tools can be made into somewhat specialized tools, at B. Such limitation makes them more useful for a particular purpose but less applicable to other situations. For example, authority is a B-level concept, seen here as a transactional relation between different levels of a hierarchical organization. The limitations that define this transaction are not applicable to many other transactions for which the initial general model is useful. The benefit received for that cost is deeper understanding of that particular relationship—as implied by Occam's law.

At C are social science disciplines, possibly the accepted ones of Figure 3 although I would visualize some restructuring. A discipline consists of a substantial complex of concepts and their interrelations, which concepts in this case consist of those at both A and B, perhaps arranged into still more explicit configurations. All B-level concepts are defined in terms of those at A; all at C are defined in terms of those at B, A, or both. Although some concepts at C may be analytically emergent with respect to A and B, they are not independent of them. For example, I would consider organization theory a discipline at level C, within which authority is only one of the many B-level concepts it uses along with those at the A-level. Continuing the progression, the C-level disciplines are more elaborate and more specialized than B-level concepts, which are more elaborate and specialized than those at the A level.

To skip to level E, every real problem is indefinitely complex; the analytic problem is to abstract from it aspects that are comprehensible and relevant. To do this one would select appropriate A-, B-, and C-level concepts and group them in ways that seemed to constitute a good description of the E-level problem. Such a description would appear at level D and would constitute a simulation of the problem. In later language (2.10b and 3.16.3), the simulation model involves the

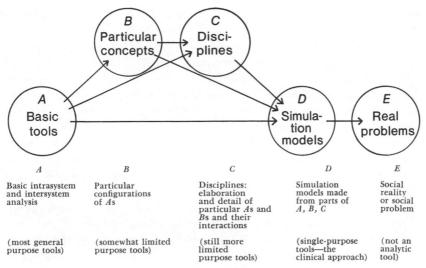

Figure 4. General-purpose and special purpose tools: a continuum.

question of which current concepts can be intersected to describe the problem at hand satisfactorily. Strictly speaking, such a simulation model is a single-purpose tool—like a clinician's description of a personality—that cannot be applied to another problem without modification.

The general logic is the same for Figures 3 and 4; the latter merely contains two additional stages. There is nothing sacred about the number of stages, and the best number may depend on the problem. However, if a hierarchical structure of knowledge is most efficient (Simon, 1965; and 2.10f below), there may be an optimum number of levels. A fully mature science would show an unbroken sequence from the most general to the most specific models. And if the science is integrated, all concepts and language at the right would either consist of or be ultimately reducible to language and concepts from level *A*. Furthermore, *A*-level concepts would often appear directly in the simulation itself. By analogy, the function of a certain resistor in a complex electronic circuit can be understood only in light of its role in the whole arrangement. No matter how complex the whole, however, the resistor never ceases to be a resistor or to follow the principles of resistors. A gear in the most complex machine never follows different rules than a gear in isolation. However much some social scientists may be geared to resist, I propose that this logical relation can and should prevail in the conceptual apparatus of social science. This volume deals primarily with levels *A* and *B*, partially with level *C*, and only in broad terms with levels beyond.

It is pointless to guess how many concepts, disciplines, simulation models, or problems it might be useful to identify. Figure 4 evades the question with only one circle at each level. The single line between *D* and *E* does not mean there can be only one simulation for one problem: there could be many different ones to reflect different degrees of sophistication or types of interest. To anticipate a later point (2.10b): the single line means that the investigator never deals with a real problem as such but only with his image of it, which we here call the simulation. If he is a strict disciplinarian his model comes solely from his own discipline whereas a multidisciplinary team could use a model based on several disciplines, whether or not unified concepts are used. In applying a science to real situations the analyst should learn something about reality and something about the science. All engineering, social or other, is an empirical test of the science it applies, though experiments designed explicitly for testing normally do the job more efficiently. To paraphrase Fisk (1967, p. 19), experience is a badly structured scientific experiment.

It does not seem necessary in the 1970s to reiterate that system analysis has performed the important unifying function of allowing us to apply causal analysis to purposive systems and purposive behavior. Buckley (1967, pp 52f) indicates briefly why this is so and refers to other, more detailed discussions. That contribution is an important reason why system analysis seems to be an indispensable base for rigorous analysis of human systems.

A fundamental question about any social science is whether the units of analysis are actors or actions—see, for example, Harris (1964, p. 182). The present volume uses both. The generalizations, or science, of social analysis concern interactions of systems, be they persons or organizations. However, every interaction, like every action, reflects the nature and states of the systems (actors) that are interacting. It strikes me as nonsensical to talk about behavior without reference to the behaving entity or about the entity without reference to what it does. In a sense that is more than merely verbal, a system is what it does and does what it is. The term *system state* for this purpose includes long-, short-, and intermediate-term conditions; and it comprises conditions as simple as a state of pleasure or pain and as complex as a whole program for instinctive or learned behavior.

In conclusion, the pages that follow suggest a scheme for categorizing systems, system states, and system behaviors and for explaining the relationships among them—with particular emphasis on human systems. Broad categories are defined and models of each are specified, thus providing a scheme of classification. Numerous propositions about behavior are deduced from these models, followed by other propositions about relationships among the categories—thus providing explanation. Some patterns of behavior thus explained are given names and definitions (further categorization), after which interactions among them are deduced (further explanation). To amplify slightly the wording of Homans (1961, pp. 10f), the framework provides a dictionary, a grammar, and many sentences. And as it moves from left to right across Figure 4 it deals first with the most general and then with the successively more specific.

Chapter 2

1.

Systems in General

1.1 INTRODUCTION

Because the term *system* is often bandied about so loosely that it means nothing because it means everything, and because this volume is system-based, this chapter discusses some aspects of the concept of system, a taxonomy of systems, and other relevant terms. The distinction between acting and pattern systems (in 1.201 below) is crucial, and a few generalizations about systems later in the chapter apply to acting systems only.

System analysis has two distinct advantages. First, the notion of system is widely applicable to many kinds of things. Second, generalizations developed from observing one kind of system are often discovered to be valid for other, sometimes discrepant, kinds of systems. In each case the advantage lies in having a concept or generalization that applies to multiple situations and hence that requires a smaller conceptual set than if each situation had to be dealt with independently. This volume as well as the term *general system* encompasses both advantages, though that term will not be used further here.

1.2 DEFINITIONS ABOUT SYSTEMS

The following pages define thirty-five terms that are related to the concepts and analysis of systems. Although all these terms, like all others defined in the volume, appear in the glossary, they are included here where they can be seen in relation to one another and where the definitions can be discussed. This rather large number is necessary to base the whole analysis on terms as fundamental as possible and to avoid ambiguity.

1.20   *Definitions*

a. An *element* is any identifiable entity—concrete or abstract, object or event, individual or collective—such as an electron, a person, a a point, a line, a word, a book, greenness, swimming, the industrial revolution, the universe, economics, the church, existentialism, or all bassoon concerti in F-sharp minor.

b. A *pattern* is any relationship of two or more elements. The elements can be related repeatedly, directly, and in the real world, like protons and electrons, politics and elections, greenness and grass. Or they can be related infrequently, indirectly, or in the imagination alone, like electrons and elections, alchemy and gold, unicorns and satyrs.

c. An *object* is a pattern as it exists at a given moment in time.

d. An *event* is a change in pattern over time.

e. *Matter-energy* is a shortened form for matter and/or energy. (See ii below.)

f. A *system,* in the broadest sense, is any pattern whose elements are related in a sufficiently regular way to justify attention. In this volume the term is used solely in connection with more explicit types of systems indicated below.

g. An *acting (action* or *behaving) system* is a pattern, two or more elements of which interact.

h. A *component* is any interacting element in an acting system.

i. An *interaction of A and B* is a situation in which some change in *A,* through a movement of matter-energy or information, induces some change in *B* or the reverse. If the movement goes indirectly through *C* it nevertheless remains the change in *A,* not *C,* that induces the change in *B.* If *A* and *B* are themselves systems, an interaction of *A* and *B* requires that some output from *A* (matter-energy or information) become an input to *B* or the reverse. (See also 3.10b.)

j. *Mutual interaction* is a situation in which a change induced in *B* by *A* in turn induces some change in *A.* Under certain circumstances the failure of *A*'s action to change *B,* when perceived by *A,* may be construed as an interaction. The meanings of interaction and mutuality are treated in detail in Section 3. The study of mutual interactions between *A* and *B* is perhaps the most distinctive single feature of system analysis, as contrasted to one-way causal effects of *A* on *B*.

k. A *pattern (nonacting or nonbehaving) system* is a pattern, two or more elements of which are mutually consistent or interdependent within some criteria of some action system(s). Not all patterns, of course, are pattern *systems*.

l. *Consistency* or *interdependence:* Elements *A* and *B* are said to be consistent or interdependent for this purpose if some change in element *A* is perceived by some action system as requiring some change in *A, B,* or both to restore a "proper" relation between them.

Discussion of this definition is in order. In an acting system *A* acts on *B* by some movement of matter-energy or information, the nature of the change in *B* depending on the nature both of *B* and of the action. If *A* and *B* are themselves systems, an output of *A* becomes an input to *B*. That is, *A* does something to *B*. But in the pattern system, although elements *A* and *B* are consistent or interdependent, neither *does* anything to the other; there is no movement of matter-energy or information. Language, for example, is a pattern system. If I write "I does" there is a syntactic inconsistency between the two elements. My editor or typist, both acting systems, may change it to "I do." But neither linguistic element itself changes the other. Similarly, a society's beliefs about God may be related to its beliefs about man, and a change in the first may induce a change in the second. But one belief does not itself change the other; the people (acting systems) who hold the beliefs make the change to maintain consistency. If the two beliefs were logic-tight, so that one could change without inducing change in the other, there would be no point in bounding a belief system to include both.

Pattern systems can be real as well as abstract or symbolic. The way in which humans or birds relate their structures both to their goals and to the physical traits of building materials is a pattern system, as is the way a society relates the constituency of its legislature to the groupings of its people—and both are real. In the strict case of pattern system the relationships are definitely selected by an identifiable acting system, whether by instinct, conscious choice, or otherwise. At the opposite end of the scale are purely fortuitous patterned outcomes of the interactions of numerous acting systems, each pursuing its separate goals, as in ecological systems and informal organizations (to be dealt with in later chapters).

My inclination would be to define the term so that nothing would be called a system that is not amenable to system analysis. This category would include only those things defined here as acting systems, which is essentially Miller's approach (J. G. Miller, 1965, 1972). The difficulty, and the reason I do not thus narrow it, is that the term is already so commonly applied to such notions as language systems, logical systems, and mathematical systems that it will not be dethroned easily. Note, for example, that language and logic are actions or thoughts, not actors, and are apparently not subject to the generalizations about systems presented in this volume.

The problem is in part terminological since some acting systems are named for the function they perform, the material they process, or the analysis applied to them, even though there may be no clear-cut physical entity involved. The human circulatory system ties together a variety of physical entities that perform part of the circulatory function. The oxygen or nitrogen systems in nature process those two elements, and their naming with respect to those items is somewhat as if we called General Motors an automobile system because of the thing it makes. The hydraulic and electrical systems of automobiles are named for the analysis required to diagnose their behavior or the forces involved. And although a "threat system" seems to be named for actions, not actors, it can nevertheless be regarded as an acting system if we focus on the entities, such as national governments, that interact through threats.

Levi-Strauss (1963) argues that culture can be expressed quite literally as syntax. I would not quibble over referring to the relationships among the subpatterns of a culture as syntactic. Were I to accept the term, however, I would generalize it in line with the present definitions. We would then say that the components of an action system are tied by action relations while those of a pattern system are tied by syntactic ones. To continue with present definitions, *syntactic* would then be equated with *consistent* as here defined, and *syntactic system* would be the same as *pattern system.*

Regarding language and some aspects of rituals and other behaviors, a syntactic system in this sense might take almost any form one chose to give it. Like conceptual sets, however, some forms would prove more workable than others, *workable* here including both logical consistency and efficiency. Other forms are limited by laws of nature. Whatever its preferences, no society can build skyscrapers of dry sand or conduct rituals that require the medicine man to dance unsupported in midair.

Theories or models of acting systems are definitely pattern

systems, as here defined. And in the same sense that a person's consistent beliefs constitute a pattern system, so does consistency among his actions. If the way a person hits a baseball is related to the way he drives a car the relationship is a pattern, not an acting system. Only actors, not actions or patterns of actions, are construed to be acting systems (Kuhn, 1971b). This concept of pattern system does not carry Parsons' implication of stability (Mack, 1969, p. 60). We would say only that stable patterns make for stability in a person or society and that unstable ones could make for instability.

The subsequent use of system analysis is confined to acting systems. All acting systems have certain traits in common whereas it seems less likely that any generalizations (possibly excepting some related to hierarchy) can be applied to more than one type of pattern system except by accident. Many actions of systems and patterns of actions will be examined, but not as systems in themselves.

  m. A *real system* is an acting or pattern system whose elements consist of matter-energy. Real systems can be linked by information but cannot consist of it.

  n. An *abstract,* or *analytic, system* is a pattern system whose elements consist of signs or concepts: the concepts of length and width; of id, ego, and superego; of supply and demand. Maps and languages are abstract systems that consist of graphic and verbal signs, respectively. The map itself, of course, is real. But the real paper and ink are not themselves a system. It is the pattern of the ink that constitutes the signs whose relationships are a system. Abstract systems may or may not correspond to real systems, or be intended to.

  o. A *nonsystem* is any element which for purposes of a particular investigation does not change pattern—that is, which does not consist of subelements that change relative to one another. For example, for purposes of computing its position in the solar system the earth can be treated as an element, not a system. And despite its convolutions, no part of an automobile crankshaft changes position or velocity relative to any other part; hence it is a component of the engine system but not a subsystem. For other kinds of investigation changes internal to the earth or crankshaft may be relevant, and for those purposes either may be treated as a system. In short, whether something is viewed as system or as nonsystem is determined by the investigator with reference to his purpose; it is not determined by the nature of the thing investigated.

  p. A *system variable* is any element in an acting system that can take at least two distinguishably different states.

q. A *system state* is the condition of a system variable, such as its temperature, color, rate of flow, magnitude, physical location, chemical composition, degree of excitation, on-or-offness, or amount or type of information possessed.

r. The *boundaries of a system* are logically defined by listing all the components of the system; any elements not listed are construed as falling outside the system. For some real systems, or real models of them, a boundary line or surface may be so located that all elements of a specified sort within it are components and all outside are not. All elements within it can then be construed to be "listed" by the boundary. Whether elements not in the system should be ignored or be listed as part of the environment depends on whether they are relevant to the problem under study.

This definition requires elaboration. A functionally defined system may not be amenable to spatial bounding. The circulatory system, for example, extends into virtually every cubic centimeter of the body. A spatial boundary around it would also have to include the respiratory system, the digestive system, the endocrine system, the nervous system, and others. A similar observation would apply to a system defined by the type of analysis applicable to it. For example, that portion of a city subject to electrical analysis would geographically include virtually the whole city—though it would exclude most of its functions.

Although nature provides reasonably clear boundaries for certain types of systems, such as organisms, atoms, river systems, and the solar system, in the sense we are dealing with them the boundaries of a system are determined solely by the investigator and with reference to a particular problem. For example, although the investigator *may* accept the skin-bounded human as the unit of study he may equally well focus solely on that person's circulatory system or his family system—neither of which is clearly bounded by nature. This does not mean, however, that nature has nothing to do with where the investigator will find it useful to draw his boundaries—with carving nature "at the joints." Nature, not man, made the Missouri but not the Sacramento River flow into the Mississippi. Given nature's impact on what is convenient and fruitful, it is nevertheless always the investigator, not nature, that bounds the particular system to be investigated.

s. A *controlled,* or *cybernetic, system* is any acting system whose components and their interactions maintain at least one system variable within some specified range or return it to within that range if the variable goes beyond it, despite changes in one or more forces that influence the state or level of that variable. The cybernetic system

can also be defined as a goal-oriented system. For some purposes
the two definitions can be combined, as when the system variable
consists of deviations from the path to some goal and the control
consists of returning the system to that path. Nothing about the
definition precludes the goals or "preferred states" from changing,
in which case we have a dynamically pathed cybernetic system.

Let me elaborate on this definition. For reasons to be clarified in 2.06 and 2.07 a controlled system is construed to have detector,
selector, and effector functions that operate for the whole system as a
unit and around which analysis of such systems revolves. Without such
unitary controls the system is classified as uncontrolled. Controlled for
this purpose refers only to inner controls, not to constraints imposed
from outside.

Viewed empirically, there are only two known basic types of
controlled systems: all living things, and controlled systems made by
living things. The latter can be subcategorized into certain man-made
machines (servomechanisms) and certain social systems of living things.

With only one major exception in connection with human
social systems we will maintain a strict dichotomy throughout this
volume between controlled and uncontrolled, without recognizing intermediate degrees of control. That is, controlled means that some
internal controls are present, without reference to how rapid, detailed,
or comprehensive they are. No question arises for biological entities;
if they are living they are controlled, no matter how simple or unreliable the controls. Man-made machines are easy to classify once
one gets used to the categories. A guided missile is controlled; a
ballistic one is not. A steam or a mower engine with a governor is
controlled; an automobile engine whose speed depends on how far the
operator depresses the accelerator is not. We will not deal with social
systems of living things other than humans, though some are apparently
controlled in considerable detail and others are not controlled at all.
By contrast social systems of humans might easily be divided into a
continuum between pure controlled and pure uncontrolled, though for
purposes of this volume we will deal formally with only one intermediate type—as in Chapter Fifteen.

Important issues in distinguishing controlled from uncontrolled systems are encompassed in the Rosenblueth-Wiener versus
Taylor controversy (Buckley, 1968, pp. 221–242). The only sensible
response to this difficulty, it seems to me, is the same as for bounding a
system and for deciding whether or not something *is* a system—namely,
deal with it in the way that seems analytically most fruitful. Explicitly,
if the investigator can isolate subsystem behaviors that seem usefully

construed as detector, selector, and effector functions he will pre-sumably understand the system better if he treats it as controlled. Otherwise not. Further discussion of the distinction between controlled and uncontrolled systems is found in Chapter Fourteen in connection with ecological systems. Controlled obviously does not mean the same as determinate.

t. An *uncontrolled system* is any acting system that does not fill the definition of a controlled system. (See the definition of *dynamic equilibrium* below.)

u. *Input* is any movement of information or matter-energy from the environment across the boundaries and into an acting system. Any action on its surface that affects any part of the system is construed to have "crossed the boundaries" into the system. An input neces-sarily modifies the system in some way.

v. *Output* is any movement of information or matter-energy from any acting system across its boundaries to the environment. Any action of the system's surface on its environment (as a foot making a print in the mud) is construed as a movement "across the boundary." Any output necessarily modifies the environment in some way.

w. *Static equilibrium* is a situation in which the forces acting on or exerted by all components in an acting system have come to a state of rest so that no further change takes place—for example, a pendulum at rest or three baseballs resting in a bowl.

x. *Dynamic, or steady-state, equilibrium* is a situation in which com-ponents of an acting system or their states continue to move or change but in which they are balanced so that at least one variable remains within a specified range—for example, the constant height of a river when the inflow of water just equals the outflow or the constant price and quantity of a good in a market when a con-stant volume of sales just equals the constant volume of production. There is a certain sense in which a pattern system, such as a language, might be said to be in equilibrium. However, since the extant analysis of equilibrium conditions is apparently relevant only to acting systems we will apply the term *equilibrium* solely to acting systems. As Singer (1971) notes, strictly speaking it is the variables of the system, not the system itself, that are in equilibrium.

To elaborate the previous definitions, a system is uncon-trolled if the steady state it reaches is simply a state that the variables happen to have reached—such as the height of a river in its natural state or the position of three balls in a bowl. If significant components of an uncontrolled system are themselves controlled systems (say, the natural ecology of a forest—see Chapter Fourteen), any equilibrium

reached or oscillations around it reflect the interactions of the goals
and capacities of the individuals and species, not any goal of the
system as a unit. If the environmental temperature changes, for example, the equilibrium ratio among species may also change. But the
overall system has no mechanism to restore the previous ratio and will
"accept" whichever one happens to occur. Without planning or zoning,
the ecology of a city is uncontrolled since the city accepts whatever
pattern arises. By contrast a system is controlled if it is maintained at a
specified steady state—such as a stretch of river whose depth is regu-
lated by opening or closing dams, or a city subject to zoning.

   y. *Negative (equilibrating) feedback* is an oppositely paired mutual
      interaction. That is, if *A* varies directly with *B*, then *B* varies in-
      versely with *A*. The result is that if *A* changes from some initial
      state its action back on itself through *B* moves it back toward, even
      if not precisely to, its initial state. ("Toward" includes beyond.)
      Hence negative feedback is also known as deviation-correcting.

   z. *Positive (nonequilibrating) feedback* is a similarly (not oppositely)
      paired mutual interaction in which the relationships between *A*
      and *B* are either direct or inverse in both directions. The result is
      that if *A* changes from some initial state its action back on itself
      through *B* is to move it even farther from that initial state. This
      relation is known variously as vicious circle, self-aggravating,
      deviation-amplifying, explosive (if the initial change was up-
      ward), or shrinking or decaying (if the initial change was down-
      ward).

aa. *Homeostasis* is a condition in which a controlled system maintains
      a steady-state equilibrium of one or more system variables.

bb. A *subsystem* is a system that is itself a component of a larger system.

cc. A *supersystem* (or *suprasystem*) is the larger system of which a given
      system is a component.

dd. A *hierarchy of systems* is any relation between systems in which one
      is a subsystem or supersystem relative to another system.

ee. A *closed system* is a system in which interactions occur only among
      components of the system. There are no inputs from or outputs to
      the environment of either information or matter-energy. A real
      system can be partially or temporarily closed by sealing it from its
      environment. An analytic system can be closed by assuming that
      no influences are felt from the environment.

ff. An *open system* is a system that receives inputs from or releases
      outputs to its environment—that is, it is influenced by and in-
      fluences its environment. All real systems are presumably open at
      some times, in some respects, or to some degree.

gg. A *system parameter* is any trait of a system that is relevant to a particular analysis but does not change during the course of the analysis. For example, the number of veins and arteries of the body is relevant to the amount of rise in blood pressure entailed by a given amount of exertion. But since the number does not change during the investigation it is a parameter, not a variable.

hh. An *environmental parameter* is any trait of a system's environment that is relevant to a particular analysis but does not change during the course of the analysis. For example, the average rainfall is relevant to the kind of economy an area will have. But if average rainfall does not change during the period under study it is a parameter, not a variable, during that period. One may, of course, properly argue that nothing remains unchanged and hence that there are no parameters. The question is not whether things *are* parameters but whether the investigator treats them as such. Parameters, like system boundaries, are determined by the investigator, not by "reality," and both are subject to errors of investigative judgment. Some parameters remain remarkably stable, however, such as the length of the year and the absence of asexual reproduction in humans.

ii. *Information* versus *matter-energy:* The distinction between matter-energy and information was introduced in Chapter One (and see J. G. Miller, 1965) and is related to, though not identical with, the distinction between acting and pattern systems. I will defer a more precise definition of information until later (see 2.10a and 2.04), but note several points here.

All information is necessarily carried in some form of matter-energy; hence to distinguish them might not seem tenable. By identifying pure cases we can say that with information we are interested in the pattern without regard to its substance and with matter-energy we are interested in the substance without regard to its pattern—if we may speak of the substance of energy. Suppose a 60-cycle current comes into my house. If my interest is solely with 60-cycleness I do not care whether the current is 10 or 1000 volts, 10 or 1000 amperes, or even whether the 60 cycles occur in electrical current, sound waves, or a hummingbird's wings. But if I want to run my 100-volt AC/DC razor, then I am interested that the substance be electricity and that it be under 110 volts pressure. But since my razor also operates on direct current I do not care whether the current carries the 60-cycle pattern, and I certainly am not interested in 60 cycles of sound waves or hummingbird wings.

Similarly, if I want to know whether a certain kind of

animal excrement will fertilize my garden the problem is one of matter-energy. But if I want to deduce from its shape whether a rabbit or a raccoon has been at my plants the problem is one of information. In the same sense that the boundaries of a system are determined by the interest of the investigator, not by nature, so is the distinction between information and matter-energy. The distinction may often be made on the pragmatic basis of whether we want to *learn* something from a sample of matter-energy or *do* something with it. This way of phrasing it handles some otherwise difficult cases. If an archaeologist were to find aluminum kettles while excavating a Minoan temple, the fact that aluminum was found is part of a pattern of information about technology from which other facts could presumably be deduced. But those facts could also be deduced from incontestable literary evidence that the Minoans had aluminum, and the archaeologist certainly has no interest in *using* the metal as such. He might be able to extract additional information from the shape of the kettle, and for that purpose a good photograph might be just as useful as the kettle itself—"good" meaning that the kettle shows no patterns not reproduced in the photograph.

In controlled systems the distinction between information and matter-energy largely parallels that between signal inputs and maintenance inputs (Berrien, 1968). In the human, signal inputs are represented by the information received through the sensory nerves whereas maintenance inputs include food, water, and air. In the thermostatic system the former is the input of information about temperature level to the sensor, which is transmitted as information to the furnace by a small electrical current; the latter includes the inputs of fuel and electrical energy needed to operate the furnace and blowers.

1.3   PROPOSITIONS ABOUT ACTING SYSTEMS

The following propositions can be found fairly widely in the literature of system analysis. Others equally general may now be extant, and it may certainly be anticipated that others will be added as the subject receives increasing attention. These generalizations, however, are the most conspicuous and satisfactorily handle most of the system aspects of this volume. All seem valid for all acting systems, controlled or uncontrolled, but for no pattern systems—hence the need to distinguish the two types. Related generalizations are noted where they seem evident; otherwise they are a seriatim list. I am not sure how far these propositions are subject to formal proof, and several are empirically based. Subject to expressed reservations about certain statements, we will simply accept and use the propositions when needed, treating them as assumptions if their logical status is otherwise unclear.

1.31   A system can reach and maintain an equilibrium position of some variable under either of two conditions.   **1.31.1** The first is that it is subject to negative feedback, negative meaning opposite, corrective, or compensating. Assume a tank into which water flows at a given rate and a discharge pipe in the bottom through which water flows out at a rate that varies with the pressure in the tank. Assume also that the discharge pipe is of a size that discharges the total input only under significant pressure. The water level will then rise until the pressure is just enough to discharge the water as fast as it comes in. The equilibrium level is a result of the oppositely paired variations: first, that the outflow from the tank varies directly with the pressure (or height) and the pressure (or height) varies inversely with the outflow. Another way to describe this situation is to say that if the water level deviates in either direction from the equilibrating level an opposing action returns it to or toward the equilibrium. A variation is that equilibrium is possible if both variables vary oppositely with a third. Under certain more limited circumstances an equilibrium may be reached if two variables merely have different rates of change or different slopes when plotted.   **1.31.2** A second condition for equilibrium occurs when there is positive, or self-reinforcing, feedback, if there exists an asymptotic limit on at least one variable. For example, fire is subject to positive feedback: the faster it burns, the hotter it gets; and the hotter it gets, the faster it burns. Temperature and rate of combustion may nevertheless reach an upper limit determined by the rate of availability of fuel or oxygen. Interestingly, a reasonably comprehensive survey of the history of the equilibrium concept in social thought (Russett, 1966) gives only passing mention (pp. 169f) to the place of positive and negative feedback in equilibrium.

1.31.3   Except in a closed system or an open system of unchanging parameters (for example, the inflow of water in the tank is constant, as are the size and resistance of the discharge pipe), some oscillation in the value of the equilibrium variables is inescapable unless the changes in the environment can be predicted and appropriate negative feedback be activated simultaneously with those environmental changes. The amplitude of the oscillation will vary inversely with the speed and sensitivity with which the negative feedback operates.   **1.31.4** A system in stable equilibrium is less subject to entropic change than one not in equilibrium.

1.31.5   It might be said that all continuing systems tend toward an equilibrium. Less anthropomorphically, a system that does not maintain equilibrium will not last long enough in a given form to be worth studying as a system. As an extreme example, a stone in the air is much less likely to be observed than one on the ground. Hence the ability to maintain some degree of steady (or steadily oscillating) state is a prerequisite for the continuance of an acting system. **1.31.6** From 1.31.5 we conclude that equilibrium of at least one variable is the most probable state of a continuing open system. In a broad sense that perhaps cannot yet be defined, the most probable states of both closed and open systems seem to be those that minimize energy differentials within the system (see also 1.32.1).

1.32   All systems operate through differentiation and coordination of their components. Although in most cases the components are different in nature (protons and electrons), they may differ only in physical location (*A* carries one end of a table while *B* carries the other).   **1.32.1** Since a brief nontechnical definition of entropy is apparently impossible (Schrödinger, 1945), we will use a modified version of the second law of thermodynamics and state that all closed systems are subject to entropy, which is the most probable state of a closed system. Because of a close conceptual relation between entropy and loss of information, often asserted as an identity, the law is often extended to state that all closed systems are subject to loss of information. Because of the further close conceptual connection between information and variety (or differentiation) and because of its more direct applicability to social phenomena in that form, we will tentatively broaden the principle still further and suggest that all closed systems are subject to loss of differentiation. By corollary, loss of differentiation is, other things equal, a direct function of the degree of closedness.   **1.32.2** By contrast, increases in differentiation can occur only in open systems subject to positive feedback, under conditions of random variation (or generation) and selective retention. This proposition is dealt with in much more detail later, particularly in Chapter Fourteen.   **1.32.3** By joining 1.32.1 and 1.32.2 we conclude that a system can continue at a given level of differentiation only so long as it is open and either generates differentiation by random variation and selective retention among its previous inputs or receives differentiation as inputs as rapidly as it is lost through entropic or other loss.

1.33   Since all inputs and outputs of systems are classified as information (pattern) or matter-energy, all interactions between systems are similarly classified as movements of information or matter-energy. That is, nothing else can get into or out of the system; hence nothing else can affect the system from the environment and nothing else can enable a system to affect its environment.   **1.33.1** The final or any intermediate state of a closed system is determined solely by the nature and states of the system's components at the moment of closure.   **1.33.2** The final or any intermediate state of an open system is determined by the conditions of the environment and by the initial nature and states of the system's components: a given final state may be reached from different initial conditions and by different routes. Items 1.33.1 and 1.33.2 together are sometimes known as the principle of equifinality (J. G. Miller, 1965, p. 233; von Bertalanffy, 1968, p. 40).

1.34   By combining 1.32.1 and 1.33.1 we can apparently conclude that absence of differentiation among components, or movement toward reduction of differentiation, is the most probable state of a long-closed system.   **1.34.1** According to a modified version of 1.32.3, however, a system can presumably go on indefinitely if all its parts are replaceable as they are injured, decay, or wear out. This, of course, is an important difference between biological systems, many of whose parts are not replaceable, especially in complex systems, and social systems that consist of multiple human beings, all of whom are replaceable.

1.35   For many types of systems it is probably impossible to optimize the main system and all its subsystems; perhaps such total optimization is possible only by chance.

1.36   The basic goal or goal structure of a controlled system is put into it from outside the system.

1.4   **TAXONOMY**

        Several kinds of systems have now been defined. Omitted for later discussion is the distinction between acting systems that consist of multiple humans, which we call organizations, and those that do not. As noted in Chapter One, these two types need to be separated because exchanges of information between humans is largely sign-based (symbolic or semantic) while that between all other systems (excepting some man-made machines) is mainly nonsemantic and isomorphic and because those aspects of exchanges of matter-energy between humans that are amenable to social analysis are value-based while matter-energy exchanges between all other systems are based on considerations other than value. This distinction is somewhat oversimplified, but it will do until we add complications in later chapters.

The following outline of categories of systems (adapted from Kuhn, 1971b, p. 118) clarifies the relationship among the systems mentioned thus far. To help place them in this taxonomy we will also include ecological systems and servomechanisms in the outline without defining them at this point. The term *mechanical* in the outline is used in its broad sense to include the solar system, a geological system, a hydraulic system, or an internal combustion engine. There seem to be no convenient collective names for the types of acting systems that are "not multiple humans," and for social science there is no need to invent them. Hence the outline simply lists subcategories of nonsocial systems without pretending that the list is complete.

I. Nonsystems: no interacting components of interest to the investigator (automobile crankshaft, the earth as an orbiting unit).
II. Systems
    A. Acting Systems (real: matter-energy): through movements of matter-energy or information, components act on other components.
        1. Uncontrolled: no goal; no variables maintained within given range. No specific equilibrium.
            a. Not multiple humans.
                (1) Mechanical systems in broad sense: no components are controlled systems (clockworks, the solar system, a natural river system).
                (2) Ecosystems: some components (namely, organisms) are controlled systems.
            b. Multiple humans: informal organizations.
        2. Controlled: goal; at least one variable maintained within given range. Specific equilibrium.
            a. Not multiple humans.
                (1) Servomechanisms (guided missiles, computer-controlled machines).
                (2) Organisms: including individual human beings.
            b. Multiple humans: formal organizations.
    B. Pattern Systems (nonacting): components consistent within criteria of some controlled acting systems but do not act.
        1. Real: Matter-energy components related in certain ways: the shape of society's buildings is related to its technology; the

speed of a car is related to the shape and surface of the highway; the constituency of a legislature is related to the demographic distribution of the population.

2. Analytic: Concepts or their symbolic representations related in certain ways: a society's beliefs about science and religion; concepts of atom, molecule, and chemical reaction; terms in a sentence or mathematical formula; symbols in a map or diagram.

Chapter 3

2.

Human System
Background

2.01   INTRODUCTION TO WHOLE SYSTEM

Society is a system that consists of interactions of individual humans. Through most of this volume the units of analysis are actions, not actors, with heavy attention to interactions. But since the interactions between two birds, or between a bird and a snake, are very different from an interaction between two humans or between one human and a corporation, a study of interactions between systems must first attend to the nature of the systems themselves. As noted in Chapter One, this analysis of interactions is tied explicitly to the system states of the interacting systems. When we move to the organization we assume a certain interchangeability of the persons in its various posts and substitute an abstracted concept of system traits, the role, for the system itself. In other parts of the volume, as with decisions and some aspects of organization, the system (actor) is the unit of analysis.

In any event we need a reasonably explicit description of the units that compose social systems. For this purpose we follow the economist's method of using a model of economic man, but we expand it to a model of social man. By this, incidentally, we mean the kind of man that goes into a society, not man as a product of the society.

As psychology, the concept of economic man is incredibly

naïve. Throughout the ensuing discussion the reader, particularly if he is a psychologist, should keep in mind that although social man is considerably more detailed than economic man the model is still naïve and grossly abbreviated psychology. Its purpose is to do for social science what economic man does for economics. Although it is hoped that it does not fly in the face of established psychological findings, for it to do so would not necessarily invalidate it for the social scientist. Because much of the social science that follows is deductive it is more important that the model from which the deductions are made be explicit than that it be in accord with the reality described by psychologists.

   To keep the approach broad, each statement or definition in the model is couched in terms of the broadest category of system to which it seems applicable. Any statement valid for all controlled systems automatically applies to all organisms, and any statement about organisms automatically applies to all humans.

   The evolutionary origin of man and some its implications are stated because the later model of emergence is tied to evolution and because certain aspects of the model seem clearer if tied to that past. Although some human traits are in a sense deduced from the evolutionary past, the important point is not whether we can deduce our way into the model but what we can deduce out of it. The evolutionary connection seems useful. But it will not subvert the rest of the book if it is not entirely valid. And since genetic man has not changed appreciably for tens of millennia at least, the criterion of a rugged primitive life, not that of an industrial society, will be our guide.

   This chapter deals with the broad nature of the human system, mainly by filling human content into the system concepts while paying particular attention to genetic as contrasted to acquired traits. This is not to exhume the nature-nurture corpse—a distinctly uninspiring venture. But even if we assume high plasticity of the human clay, what can be made of it nevertheless depends on its texture, tensile strength, firing temperature, and ability to take a glaze. And malleable clay makes very different things than malleable gold. Although individuals differ dramatically, many of man's processes and capacities are genetic and relatively uniform. It therefore seems worth identifying some constants on which a society must be built. Chapter Four provides more details and rationale of the model and then summarizes it.

   Although not all details of this model will be used later, they are included here for two main reasons. First, because this model is fairly well ingrained into my system, it might be useful to identify some of the notions I may unconsciously be reflecting in later chapters.

Second, the parts not immediately necessary might constitute a base on which to build additional materials of interest to specialists, particularly sociologists, social psychologists, and organization theorists.

2.02   *Evolution*

Human beings are controlled systems, presumably designed by the process of biological evolution. We define *evolution* as the operation of some selective criterion among essentially random variations. *Random* here means unplanned or uncontrolled, not a mathematically prescribed situation like that of probability theory. In the theory of evolution the selective criterion has been survival of the species and the variation is mutation of the genes.

The selection may come through external competition, as is the central idea of Darwin, through certain kinds of internal subsystem compatibility (Whyte, 1965), or through some combination of these factors. The actual course of evolutionary development would be affected, but not the main concept, if certain mutations are less probable than others or if man-induced alterations of the environment change the likelihood that certain traits will survive. Society itself is a large alteration that greatly changes the survival payoffs. Now that man understands the principle nothing about it prevents him from being his own selector of desired traits, as by selective breeding of humans. For our purposes, however, we can attend to the traits that had presumably evolved by the late Stone Age.

For complex species to survive indefinitely, each generation must on the average live through the age of procreation and care of offspring. To survive that long requires that individuals behave in ways that successfully adapt to their environment. We define *adaptation* as behavior that in some way changes the relation of the system to its environment, whether by altering itself, the environment, or both; *successful adaptation* is an adaptation that increases the likelihood of achieving some goal. Whether the increase is large or small, immediate or delayed, individual or average, cannot be stated in the definition, although there is a clear-cut criterion of living or dead. At the simplest level *adaptation* means that the human may safely step off a stone but not a cliff and the fish may swim in water of 50 degrees but not 200.

Since the notion is often dreadfully misused it should be clear that not every individual need survive to continue the race, nor need every behavior by the individual be successfully adaptive for him to survive. Individuals are tough enough to stand a lot of antisurvival behavior, like walking into doors or hitting thumbs with hammers, since there is high redundancy and large regenerative capacity in their biological mechanisms. Thus only our serious mistakes of relatively low probability cut us down, and "successful adaptation" can sometimes be surprisingly poor. Having noted these points, we may now recall that evolution *is* ruthless and that its bias toward prosurvival structures and processes is high.

To the extent that inherited traits influence behavior, those traits that produce or tend toward survival-oriented behavior are more likely to survive in the species than are contrary traits. In the individual, traits neutral to survival are by definition not a handicap. But the long-run continuance of *any* trait uses information capacity of the genes. If that capacity is limited, the information content used for neutral traits presumably displaces affirmative ones. Hence traits that are neutral in the short run of the species may well be antisurvival traits and disappear in the long run. Since the evolutionary process is continuous, the stage of any species at any moment must therefore be viewed as transitional, not final. Hence although there is a general presumption that each inherited trait has (or had) survival value the presumption is not conclusive for a particular trait.

2.03   *Definitions about Human Systems*

a. *The biological* (or *maintenance*) *system* is the biological entity of the individual human, bounded by skin, hair, nails, and so on. This is the entity that must be maintained for survival of the individual and that executes the actions which constitute his behavior.

b. *States of the biological system,* in addition to the obviously biological blood, liver, and the like, include the states and changes of state in blood pressure, sweating, endocrine secretions, pupil or capillary dilation, and so forth that accompany the emotions.

c. *The behavioral* (or *control*) *system* is the nervous system—afferent, central, and efferent—and associated organs such as eyes and ears that recode information from the environment into a form that will actuate sensory nerves. This is the system that selects behavioral outputs and guides their execution. Our analysis will focus solely on its information (not matter-energy or maintenance) aspects. It does not include the autonomic nervous system, which controls operations of the biological system. Within certain biologically defined limits (such as limited ability to hold the breath) and with exceptions of almost no significance to social analysis (such as outputs of heat or moisture through the skin), all behavioral outputs of the normal noninfant can be controlled by the behavioral system. The behavioral system includes proprioceptive and related nerves that feed information about the state of the biological system into the central nervous system. In short, the behavioral system includes all inputs of information into the central nervous system, all processing that occurs there, and all outputs of instruction from it. The behavioral system can also control certain inputs and outputs of the biological system, as in eating, breathing, and excretion. Inputs of sensation and tissue damage cannot be controlled directly, though many can be controlled indirectly by approaching or avoiding their sources.

d. *The environment of the human,* unless otherwise indicated, means that of the behavioral system. For an individual this environment includes his own biological system. One reason for this formulation is that the individual learns and makes decisions about his own body by the same processes as for the world outside his skin. Even the purely biological aspects of the emotions must feed into the nervous system before they can affect socially interesting behavior. Some borderline cases, including the psychosomatic, do not readily fit this dichotomy. But for social analysis the model can safely focus on the behavioral unit alone. This formulation has interesting ramifications. This age of transplanted and artificial organs, low-temperature trances, intravenous feeding, and other biological supports is approaching the point where death must be defined as a state of the behavioral system, not the biological.

e. *States of the behavioral system* are construed to be those of detector, selector, and effector. For convenience we will use *state* to cover all aspects of the system's condition, from the most permanent to the most ephemeral, learned and unlearned. The state of my selector at a given moment might include, for example, an inborn aversion to sour tastes, an acquired and reasonably permanent liking for lemon flavor, and the fleeting enjoyment of a lemon drink in my mouth. By analogy the state of a computer at a given moment would include its structure and basic operating characteristics, the program it is processing, the totality of data stored in it, the stage of completion of the program, and a datum stored in a particular location. How broadly the term *state* is to be construed depends on the problem at hand.

2.04   *Adaptive Behavior*

To state what adaptive behavior *is* not only provides a framework for analyzing it but also ties it closely to system analysis. Explicitly, adaptive behavior is *behavior* of a *system* in an *environment*.

To omit any of these three elements is to eliminate the phenomenon, and they are thus its irreducible ingredients.

A controlled (adaptive) system cannot respond to its environment unless it possesses information about it. It is much easier to talk about information than to define it. One need not read far into the literature, however, to run across such words as order, organization, selection, nonrandomness, improbability, negentropy, isomorphism, variety, patternness, and Maxie (Maxwell's demon) used as if they were interchangeable with information or somehow captured its essence. It is also relatively easy to measure the quantity of information and to specify the conditions under which it is gained or lost. For the present purpose I doubt whether a long discussion of meaning or measurement of information would move us very far. Hence I will merely identify my own inclination, which I follow throughout this book, and consider pattern to be synonymous with information, mainly on the grounds that I think information theory and communication theory can sensibly be said to deal with pattern and pattern transfer. I avoid confronting the issue directly, however, by defining only the phrase "to possess or contain information." I do this because we do not normally think of a pattern as information if we can learn only about that pattern by examining it. Only when examining a pattern enables us to make inferences about something else do we think of it as information. Information might therefore be viewed, and tentatively defined, as some aspect of a pattern that enables an observer to make inferences about something other than the pattern itself—or at least other than the observed portion of the pattern. Such inferences, or extractions of information, are a function not only of the pattern and its relation to the forces that produced it but also of the observer's prior knowledge of a given pattern and/or of the forces that create it. In the broadest sense all information consists of relationships among parts of pattern, and the ability to extract information depends on the observer's prior knowledge of the degree of reliability of the relationship, whether of objects or events.

To narrow that approach we will say that a system *contains or possesses information* about its environment when some state or pattern within it is functionally related, as a dependent variable, to some pattern in its environment. For this to happen, matter-energy or information must enter the system as an input from the environment, and our attention in this volume will be confined to information inputs. The definition does not imply that the information acquired has only informational consequences. Once the information has entered the system its presence there is an internal state of the system. Any subsequent behavior of the system regarding this information is a response to that internal state, not to the state of the environment. The system responds to its information, or image, not to the environment as such. The information can, of course, be modified by subsequent inputs or by internal processing.

Different systems will give different responses to the same environmental element (the bee responds differently to the flower than does the human), different individuals of the same type of system may give different responses, and the same individual may give different responses at different times. Thus from one environmental situation different kinds of behavior will be selected, depending on the nature (relatively permanent states) and state (relatively transient states) of the *system* involved. As we shall see, the "environmental situation" is not a simple question of "what's there." The "situation" is itself a function of the image structure of the system involved—not merely the baseball traveling through the air but also the fact that it is headed toward the window of a house of an old man who will call the police and the father of the third baseman and nobody will be allowed to play ball again in the lot for at least a week.

The third ingredient of adaptive behavior is the behavioral response. For the moment we will assume that the system has the capacity to execute the response it selects; we will reserve complications on that score until later. The response is a behavioral output.

To combine the points of 2.04 thus far, environment acts on a system (with certain characteristics), which engages in adaptive behavior. If viewed as movements or actions they can be stated as inputs, internal processing, and outputs, which parallel the stimulus—organism—response of experimental psychology. Viewed in the system model as successive steps, they are the *detector* (which receives information about the environment), the *selector* (which reflects the inner tendencies of the system to respond in one way rather than another), and the *effector* (which executes the behavior thus selected). This definition allows any type or degree of action, including the zero quantity of inaction.

Detector, selector, and effector, which will sometimes be referred to as DSE, are thus not merely a useful categorization of functions developed for system analysis; they are also direct representations of the three irreducible elements of adaptive behavior. These items are found under many other headings in various aspects of science—sometimes all three, sometimes with one or more functions subdivided, and sometimes omitting the effector, presumably for the same reasons

that it is deemphasized in this volume. "Irritability" of simple organisms would correspond to the detector. The reality principle and the
pleasure principle are psychiatric designations for detector and selector.
An older categorization of mental processes was knowing, feeling, and
willing. The first two are direct parallels of detector and selector.
Translated, *willing* would be a sufficiently strong preference for the
selected behavior to activate the effector. At the first-grade level the
terms might be knowing, wanting (positive or negative), and doing.
Motivational theory itself has been divided as to whether a "motive" is
the external stimulus object, the inner hedonic state, or the external
goal object toward which behavior is directed (Cofer and Appley, 1964,
p. 368)—a direct parallel of DSE. For obvious reasons the selector in
the present model incorporates only the inner hedonic state. Cybernetic
analysis sometimes distinguishes the item varied by the environment,
the controlling variable, and the controlled variable—another parallel,
but with its focus on the variables instead of on the system processes
that handle them. In a general parallel with other sciences, economists
have long categorized the main aspects of their field as economic science
(what is), economic ethics (what is wanted), and applied economics (the
behavior that achieves the goals). There is a certain logical sense in
which the main activities in the detector, selector, and effector are all
"selections" of the same form that we will call pattern-matching (2.42).
As will be noted, even if these activities are identical in one basic sense
we nevertheless need to distinguish three successive selections for
analytic purposes.

2.05   In light of the preceding paragraphs we may say that in this
model an adaptive system must possess a detector, selector, and effector,
functionally bounded and defined more explicitly as follows:

a. *The detector* is the function by which a system acquires information
   about its environment. For the human being this means information
   about the environment of the behavioral system.
b. *The selector* is the function of selecting a behavioral response to an
   environmental state. As noted, response includes possible inaction.
   Depending on the complexity of the system and the situation the
   selector may be said to contain or reflect the system's tendency, inner
   logic, goal or goal structure, values or value structure, wants, desires,
   valences, affective tendencies, needs, strivings, and so forth. To contrast the two, the detector is concerned with what is (factual judgment) and the selector with what is preferred (value judgment). If
   we assume that the selector has already been provided information

by the detector and utilizes it in making a selection, the selector may also be thought of as the *decider,* the term used by Miller (1965, 1972).

For humans we often distinguish the scope or importance of different preferences. For example, likes and dislikes (tastes) among food or clothing are distinguished from attitudes toward freedom, education, or religion. In the present model these and intermediate levels of values represent different *content* of the selector. But their overall *logic* is the same: all can be reflected in the final analysis as questions of approach or avoidance behavior.

c. *The effector* is the function of executing the behavior selected by the selector.

2.06   *Main System and Subsystems*

If we view the whole behaving system as the main system, then DSE's are functionally bounded first-level subsystems of it. To distinguish these from another level to be discussed shortly, we will refer to them as the main-level DSE.

Detecting, selecting, and effecting are actions of systems, not acting systems. But an action presupposes an actor, for just as we assume (or have empirically concluded) that there is no action at a distance we similarly conclude that there is no action in vacuo. Although we must thus assume a physiological base for these three subsystems, for this volume we need not even speculate what it is—though it seems virtually certain that the central cortex is actively involved in all three. We need not even speculate whether the relation of the three within the cortex is like that of a vacuum tube, which might simultaneously be a component in three distinct subcircuits of an amplifier. Nevertheless we can make some fairly definite statements about the function, as in Section 2.1.

For nearly all controlled systems other than humans it seems sufficient to deal with each of these three functions as a unit. For many purposes in humans, however, and particularly because we attribute consciousness to humans but are skeptical about it elsewhere, we need something more. Without departing from the model, we can accommodate these analytic needs by positing a set of sub-DSE's for each main-level DSE as follows.

2.06.1   There are three detector components. The *subdetector* acquires information about the state of information in the main detector. If information at the main detector consists of images, that at the subdetector may be said to consist of images of images. The *subselector* registers the main detector's information needs or goals. As we will see in Chapter Four, there are certain ways in which the subselector of the detector may be said to have goals of its own. The *subeffector* acquires, seeks, sorts, discards, or otherwise processes the main detector's information.

2.06.2   There are three selector components. The *subdetector* detects the main selector's motives or goals. This information may be thought of as images of goals, of motives, aspirations, hopes, feelings, likes or dislikes, emotions, or valences. The *subselector* registers or reflects desires that will move the main system toward its goals. Its content may be said to consist of the motives for having goals, or motives about motives. The *subeffector* executes a change or continuance of goal structure.

2.06.3   There are three effector components: the *subdetector* discerns the state of the main effector's skills and competence. This information may be said to consist of images of one's own skills. The *subselector* contains or reflects goals about skills or competence to be acquired by the main effector. The *subeffector* engages in actions (study, practice, rest) to achieve skill and competence goals.

One can speculate about whether operations at the first level of DSE should be considered unconscious behavior and those at their subdetectors conscious. Certainly it does not seem amiss to conceive of consciousness as information about our own internal system states, which is what the subdetectors handle. Such a statement does nothing, of course, to explain what consciousness *is*. But it does show that the difference between conscious and unconscious can be categorized as system levels without adding basic concepts. And certainly phenomena of this kind listed for the sublevel DSE's occur regularly in humans. Anyone who likes can add sub-sublevels of DSE's, whose detectors would form an infinite regress of images.

2.07   OPERATION OF WHOLE SYSTEM

The central feature of a system is that the whole is more than or different from the sum of its parts, which means that a system might be defined indirectly as any situation to which the fallacy of composition might apply. Our concern is therefore with the combined effect of DSE in producing behavior. We will first diagram the components of the behavioral system (Figures 5 and 6) and then indicate the sequence of events that produce behavior.

2.07.1   Response selection at the main system level requires an interaction among DSE functions in which (a) the detector provides information about the state of the system's environment; (b) the selector compares that state with the goal state of the system and selects an appropriate response; and (c) the effector receives instructions from the selector and carries out the instructions.

2.07.2   In continuing systems capable of learning (which includes nearly all animal life), a second or learning stage often follows in which (d) the detector provides feedback information about the state of the environment following the previous action; (e) the selector compares the new state with the system's goals and selects an appropriate, possibly corrective, response; and (f) the effector receives new instructions from the selector and carries them out.[1] Unless the outcome of step (f) leaves everything as it was at step (a), this sequence produces a developmental change in the system, the environment, or the relation between the two.

In an ongoing system, and particularly in man, the cycle may be repeated indefinitely, in which case the fourth step of one behavioral cycle merges indistinguishably with the first step of the next, and so on. Logically it is possible to formulate the sequence so that behaviors of all types are treated as a single stream or so that one type of behavior is a separate stream interrupted by other activities such as eating or sleeping. Neither approach is developed here, but I strongly suspect the latter to be more workable.

---

[1] J. G. Miller (1965, p. 394) cites four stages of adaptive decision-making. The first is to establish a purpose or goal, which would here be the selector function. The second is to analyze information relevant to the decision, which is here treated as detector. The third is to synthesize a solution by selecting the alternative most likely to lead to the purpose or goal, which is referred to here as the interaction or combined effect of detector and selector functions. Whether one states this as a separate third step or as an interaction or culmination of the first two seems to me a matter of descriptive convenience, not of disagreement on fundamentals. Miller's fourth step—to issue a command to carry out the selected action—certainly corresponds to the effector here, with the clear implication that the execution is carried out at a level different from that of the control system itself.

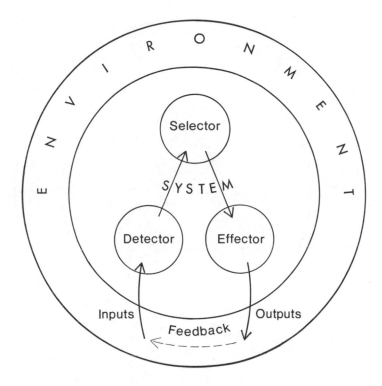

Figure 5. Behavioral system in an environment (main system only).

To analyze complex behavior in complex systems, certainly including humans, it may be necessary to move to the second-level DSE's to deal adequately with the state of the main-level DSE's. In that case the same six steps may be analyzed within each of the first-level DSE's. Although any number of levels of subsystem could be added following this pattern, there is no apparent need here for more than these two.

Nothing in the model precludes multiple or complex actions back and forth between and within DSE subsystems before a response is selected or modifications of DSE's by one another. The model merely requires that all three functions be carried out before behavior can be

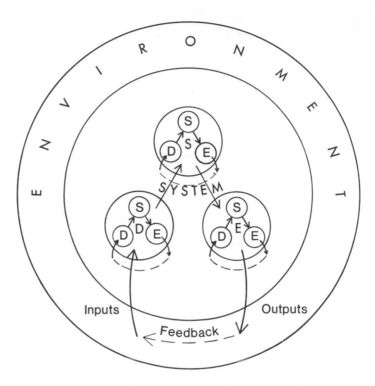

Figure 6. Behavioral system in an environment (including DSE subsystems).

effectuated. And although we discussed consciousness, none of the definitions specifies that the process is either conscious or unconscious.

2.08   *Inherited Versus Learned Behavior*

Some problems and processes concern all three subsystems and can be discussed together before we explore their differences. The questions are: For purposes of our model, which aspects of human behavior are to be considered genetically inherited and which learned? And what is the relationship between the two? Our definition of the behavioral system means that we are taking biological processes as part of the environmental constraints of the behavioral system; they have an important bearing on what it can do but are not variables within it in any direct sense.

2.08.0   Items *a* through *g* that follow are intended to define or clarify terms. Their significance for the model is discussed in the section following the definitions.

a. *Inherited, inborn, and genetically determined* will be used inter-changeably to refer to structures or processes that appear uniformly in every normal human at birth or by maturation independent of experience. The purpose of the definition is to clarify usage of the terms in the model, not to presume that we know precisely what is in fact inborn. Ten toes, not twelve, are genetically normal, and the maturational development of a second set of teeth and pubic hair long after birth is determined by genetics, not learning.

b. A *reflex* is an inborn specific overt response involving muscles to the excitation of specific sets of sensory nerves. Examples are the eye blink, knee jerk, swallowing, coughing, and withdrawing the hand from a flame. For our purposes we also put in this category facial expressions or body postures that reflect certain inner states—like laughing, frowning, tensing. This definition does not preclude learned production or inhibition of the same action, or additional responses to the same stimulus, such as cognition or emotion. As the terms are used below, a reflex is a response to a sensation, not a perception.

c. An *instinct* is here defined as a genetically preprogramed pattern of explicit and relatively detailed overt behavior set off by an explicit environmental situation and essentially invariant with respect to it.[2] Apparently, at least in some cases, a short behavioral pattern is

---

[2] Instinctive performance is subject to many variations, and the term *essentially invariant* can be clarified if viewed as species-specific in the same way structure is. A tomato plant grows tomatoes, not roses or acorns, and has a particular stem and leaf structure. A tomato plant may nevertheless bear large or small fruit, be deep green or sallow, stunted or luxuriant, sprawling or climbing—all depending on water, nutrients, temperature, sunlight, insects, hungry rabbits, and the proclivities of the gardener. It may even die of drought or a careless mower in mid-season. Yet its basic tomatoness never changes. A host of forces can similarly vary the performance of an instinct, including that two or more instincts might be set off concurrently. Yet even though the instinctive nest-building of an oriole may be premature, delayed, aborted, blocked, overstimulated, or confused, the oriole never builds the nest of a robin, hornet, or mouse. To him who knows it the main pattern can nearly always be recognized, and that is what we mean by essentially invariant.

executed instinctively, after which the environmental situation it creates triggers the next step of the pattern.

The instinct generally contrasts with the reflex in two main ways. First, instinct typically requires the inborn ability to recognize a pattern (an inborn perceptual ability, as defined below) rather than the mere excitation of certain neurons—but in the absence of firm knowledge we will not include this in the definition. Second, instinct is usually more complicated than reflex, as illustrated by the difference between blinking the eye and building a beaver dam or even chewing down a tree. Inborn pattern recognition will be construed as an instinct rather than a reflex even if the overt response is relatively simple—as in the greylag gosling's unlearned retreat from the cutout silhouette of a hawk. The Freudian meaning of instinct is considerably different and will be dealt with in connection with urges or drives (2.2 and 2.21).

d. On the assumption that the general meaning of *emotion* is known we will add that the term refers here to the biological state that characterizes a particular emotion (endocrine secretions, pulse changes, sweating) and to the accompanying neural inputs to the behavioral system from those biological states. Such inputs will be construed to be sensations.

e. A variety of *overtly observable states or behaviors* are reliably related by inheritance to inner states: laughing, weeping, tensing, gasping, flushing, sweating. The related inner states can be emotions, sensations, or arousal levels. Like reflexes some of these states may be augmented or inhibited by learning and conscious control.

f. Because other terms that might be used here are vague we will define *sensation (sensory inputs)* neurologically as the actuation of the sensory nerves of any of the five external modalities—sight, hearing, taste, smell, and a conglomerate of skin senses—or of any nerves that convey information about the biological system into the central nervous system, such as kinesthetic senses, hunger pangs, emotional states, fatigue. A sensation consists of *uncoded* information. Whenever a valence (see below) is genetically attached to a sensory stimulation, including those attached to emotions, it will here be considered part of the sensation. As defined here a sensation is solely an internal state of the behavioral system; this definition is independent of the environmental input, if any, that activates it. A sensation could be actuated by spontaneous firing of neurons or by experimentally introduced electrical charges. The definition does not require that the organism be conscious of or "feel" the sensation, though it may.

For example, actuated tactile nerves are normally consciously "felt" while actuated visual nerves are not felt and much activation may go unnoticed, especially on the periphery of vision. In the model all inputs to the behavioral system take the form of sensations.

g. A *valence* is some property of the behavioral system, possibly a collection of neural centers (Olds, 1955), that reinforces approach or avoidance responses and typically (but not necessarily) produces what, for want of a better name, we will call pleasant and unpleasant feelings—the displeasure of a cut or hunger pangs and the pleasure of moderate sweetness or erogenous stimulation. A valence is also internal to the behavioral system.

Since our goal is a model for social analysis, it need not include the qualifications that the psychologist, biologist, or neurologist might want. With this background we now move on to identify the inborn ingredients of DSE.

2.09   *Inborn Aspects of DSE*

2.09.1   *Detector.* The human has a variety of distinct sensations that are correlated, often almost perfectly, on an inborn basis with certain environmental states that actuate them. At the gross level light waves actuate optic nerves and sound waves aural nerves. The chemical constituency of gases acts on olfactory nerves and that of solids and liquids on nerves of taste. Pressures, temperatures, and other environmental conditions act on assorted nerves of the skin. Within the senses we find further specialization. High- and low-pitched sounds are carried by different nerves and different colors of light by different nerves or combinations of nerves. The same is true for different tastes and odors, while in the skin different sets of neurons are sensitive to heat, cold, pressure, and other external situations.

In short, the human inherits sensory mechanisms such that the mere actuation of a neuron or a combination of neurons normally constitutes highly reliable information that a certain kind of thing exists in his environment. Nonactuation is reliable evidence that it is *not* present. With possible exceptions too minor to concern social analysis, light-sensitive neurons are not actuated by sound or the reverse, and temperature-sensitive neurons are not actuated by the sweetness of honey or the reverse. Furthermore the skin senses are topographically specialized, so that a pinprick on the ear actuates different nerves than a pinprick on the finger. Moreover the pressure of a pin actuates a different number of nerves than the pressure of a blunt stick, and intense light or sound actuates more nerve impulses than a weak stimulus. All in all human sensations are considerably ordered with respect to the environment on a purely inborn basis.

Although sensations are highly correlated with certain (relatively simple) environmental states and although a person may be said to "know" that he has sensations, both by inherited mechanisms, there is apparently no inborn connecting link that enables one to know *what* in the environment is actuating his sensations. To speak of awareness, on an inborn basis the human is aware of those internal states that constitute sensations but not of the environmental inputs that initiate them: that warmth on the arm is due to the sun rather than to hot water.

2.09.2   *Selector.* Some sensations and possibly all emotions have inborn valences attached, the main exceptions apparently being sight and sound. Through inborn structures actuation of certain nerves by high temperatures or skin rupture is unpleasant, as are bitter and sour taste and numerous odors. By contrast certain tastes, odors, and skin sensations are pleasant. Subject to certain complications (Section 2.2) some emotions are inheritedly pleasant and others unpleasant. It is important for the information-processing aspects of the model that there are such inborn connections but not that we specify exactly what they are.

As to the exceptions, at low and moderate levels of intensity light entering the eye or sound entering the ear is neither pleasant nor unpleasant in itself. Unlike certain other sensations, and to use language to be defined below (2.10b), inputs of light or sound must be decoded before valences can be invoked. Any inborn preferences we may have for certain kinds of light or sound are not important enough to include in the model. Preference for activity per se appears below (Section 2.41.2).

2.09.3   *Effector.* Humans possess a variety of inborn reflexive responses to particular sensations, such as the reflexive withdrawal from intense heat. To the extent that the sensations are actuated by explicit environmental circumstances—such as heat—and the extent is rather large—the reflexes are tied to those environmental conditions. Humans inherit rather explicit patterns of overt manifestations of some inner states, as with some emotions, sensations, or levels of arousal.

Man, however, has no instincts in this model. The most obvious behavioral requisites for survival of the species are food-getting, eating, and procreating, and if any instincts were left in humans the logic of evolution would presumably include these. Yet even these behaviors must be learned in humans, and to a considerable extent by the apes as well. Should man possess inherited depth perception, which seems distinctly possible, or an inborn withdrawal response to snakes which is dubious, these would be instincts as defined. Neither would be relevant to a model of social man, however. The former would almost certainly be learned at an early age in any event, and although the latter might be important to primitive man it is hardly of great moment to social structures or behavior. Fear of sudden loud noises is a valenced emotion attached to a certain sensation, a kind of inborn connection we have already included in the model. For these reasons no instincts are included. Of prime importance, man also inherits a high-capacity information-processing mechanism in the central cortex, the main contours of whose operation are discussed in connection with learning.

2.09.4   *Summary: First Stage of Model*

The first stage of this model of man, subject to refinements but not basic changes, is that he possesses the inborn aspects of his behavioral system described above, and only those. He has sensations closely geared to important aspects of his environment. But on an inborn basis those sensations produce no explicit behavior other than the reflexes and the outward manifestations of emotions and related states. Beyond the narrow scope of reflexes and some emotions man's inherited behavior is not ordered with respect to environment: it is totally random. Beyond the level of the infant, however, we observe that human responses are not random with respect to environment: they are highly selective. They include such prosurvival acts as sleeping, keeping warm, eating fruits and grain, and mating; they exclude such antisurvival acts as sleeping naked in the snow, eating gravel, walking off cliffs, and totally eschewing sex.

Since man engages in much behavior that is selectively slanted toward survival, but the patterns of which are not inherited, it appears that he inherits mechanisms that *select* survival-oriented behavior even though they do not *prescribe* it—and this model will so assume. Subject to more detail below, the selection mechanism consists of the factors already mentioned, which means that all man's behavior, however complex, is presumed to derive from certain sensations related to externals, a handful of simple inherited responses, certain valences attached to sensations and emotions, and the ability to learn through conditioning and other internal information-processing. Viewed from the context of seemingly incredible complexity, the question then is: How does man get from there to here? To re-phrase the question in a way that points toward an answer: How does man learn to identify and respond to an environmental state by means of sensations and responses when neither has any inborn relation to that state? That is, how can you learn to recognize hot coals or honey from a distance and learn to avoid or approach them without having to wait for the painful burn or pleasant taste in your mouth? The answer will also tell us much about how man can learn to identify and respond to things for which he has no specific inborn sensitivity at all—like tree, wind, uncle, government, symphony—and no inborn valences or responses. In system language, how does he learn responses to things for which he has no inherited detector, selector, or effector states?

It is irrelevant to the model whether the fetus learns prior to birth. Given its location it could hardly learn much about its future external environment, though through proprioceptive nerves it could learn something about its own system. Since the behavioral system is highly active even prior to birth, the point is not that man is largely a blank slate at birth but that anything already written on his slate is mainly about himself, not his environment. His task is to convert behavior that is at first mostly random with respect to environment into behavior increasingly ordered toward it. Our task is to explain how he can do it with the ingredients already specified. However far randomness may be reduced it nevertheless seems safe to say that much always remains, even outside of dreaming. Potter (1966) has suggested that this residual may be the basis of what we call free will.

2.09.5   *Learning Process*

Conditioning is here defined as the main process by which the behavioral system forms noninherited internal patterns that are significantly correlated in some way with patterns in the environment. All such internal patterns are categorized in this model as falling within detector, selector, and effector, and all constitute information about the environment.

In psychology *reinforcement* refers to the gradual strengthening by conditioning of the connection between an environ-mental stimulus and a motivated or reflexive response. For convenience we will extend it to include single nonrepetitive instances of cognitively guided behavior. For example, we will say that a fondness for roast beef reinforces a man's entrance into a restaurant when he sees a magnificent roast through the window, even if he had never seen that restaurant before or previously eaten in one. The fondness may also reinforce fantasizing about roast beef.

Although "classical" conditioning of reflexes was extensively investigated before operant conditioning of rewarding and punishing behaviors, the latter seems the more basic for learning survival-oriented behavior. It is largely irrelevant whether or not a reflex becomes conditioned to a new stimulus—though see the discussion of conditioning of secondary motives from primary (2.22)—but highly important that actions which produce food or avoid pain are reinforced. Hence we explore here the logic of operant conditioning, which connects an action with the feedback of information about its perceived consequences.

If feedback inputs are only randomly related to behavioral outputs no conditioning occurs. For example, for all practical purposes a person breathes all the time. Breathing is therefore associated with everything he does; it is not correlated with a particular action and hence becomes conditioned to no particular feedback. But breathing in smoke or water is highly correlated with unpleasant feedback and quickly reinforces an avoidance response. If pressing a bar and getting food pellets are significantly correlated in the environment a parallel connection can be formed within the organism. But if they are only randomly related in the environment (that is, not related) the conditioning mechanism will not relate them in the organism.

In the usual experiments with animal conditioning, attention is focused on externals—the stimulus, the number of trials, the extent of learning an observable response. Neurological correlates may be assumed, but the experimental results are normally reported without mentioning them. For greater generality of the information model, the present model reverses the emphasis. It takes the basics of conditioning as established and focuses on the change of internal information that must (in a cybernetic-information model) underlie the change in overt behavior. The model avoids the question of *what* happens in the neurons by speaking simply of patterns. When a pattern is repeated as input or output the neural representations of its elements become somehow associated internally by conditioning.

Conditioning often involves overt trial and error.[3] Since this

---

[3] The term *trial and error* is fairly well understood. But the process is strictly a trial, of course, followed by success or failure. As will be seen in Chapter Sixteen, it is a special case of the broad developmental process of random generation and selective retention. That is also why *conditioning* takes in a wider scope of learning processes here than psychologists accord it. This model potentially includes higher-level cognitive processes that elsewhere might be labeled the "self-organizing" behaviors of the brain.

model assumes an inner pattern that controls the overt movements, nothing about the general notion of conditioning excludes mental trial and error preceding the overt variety, even of the type we call creative thinking. Since this is impossible for an experimenter to observe and virtually impossible to report verbally, experimentation is sticky indeed. Without going so far as to specify it for the model I nevertheless think that the "laws" of conditioning probably apply to all uncoded information and to the way pieces of it come to be connected. I feel sure they do not apply to certain connections made between pieces of coded information. Yet I also feel sure that it often takes the same kind of repetition to make a firm connection between two abstract concepts in my head that it takes the rat to connect pressing a bar with getting food. For these reasons *conditioning* is construed more broadly here than is customary in experimental psychology. To describe it as the "main" process of making connections leaves the way open for distinctly different ways of making them, ways that I suspect apply only to coded information.

2.09.6   The number of trials required for conditioning is presumably determined by evolution. There is no reason for an organism to learn a pattern it will encounter only once, since such learning could never be used again. Evolution would therefore presumably not retain one-trial conditioning, and with reduced force the same is true of patterns encountered only a few times. But by the time a pattern occurs ten, fifteen, or twenty times at not-to-distant intervals the likelihood that it is random drops precipitously. The likelihood that the organism will again encounter it rises at about the same rate, and there may then be some reason for learning it. The number of reinforcement trials required for reasonably firm conditioning is thus statistically related to the amount of evidence required to establish that there *is* a pattern in the organism's inputs, and that number is quite "rational" in the logic of evolution.

Extinction follows the same rationale. If the environment changes after a pattern has been learned, it would be a poorly adaptive organism that kept giving a response long after it had stopped being appropriate. The principle of *frequency* in conditioning can therefore be viewed as a genetic adaptation to the probability that a given pattern regularly exists in the environment, that it will be encountered often enough to be worth learning, or both.

The principle of *recency,* or shortness of time lapse between the two elements to be conditioned, reflects similar considerations. Among factors that are crucial to survival, a long delay between action and feedback probably means that the feedback is not in fact reliably related to the action. In other words, relatively fast-paced events are of more concern to survival than slower-paced ones—a not improbable situation for our prehistoric forebears—and if the "effect" does not follow fairly soon after the "cause" there is probably no causal relation. An implicit *post hoc ergo propter hoc* is apparently built solidly into our adaptive processes, and it is adaptively "fallacious" only when adequate frequency is lacking.

The earth nevertheless contains far more reliable patterns than the organism can learn; its problem is which to learn and which to ignore. If its motives are geared to what is important to it (2.2) and if its speed of learning a pattern is positively correlated with the strength of the motive tapped by that pattern, then the organism will learn most quickly, and hence more probably, the patterns most closely related to survival. The principle of *intensity* states this positive relation between motive strength and learning speed and is rationally "selected in" by evolution.

To connect the genetic and the learning aspects, we may say that evolution "reinforces" in the species those mechanisms which reinforce prosurvival but not antisurvival behavior by the individual and that evolution "extinguishes" in the species those mechanisms which do the opposite. Conditioning is the mechanism of reinforcement and extinction in the individual, and evolution has selected into the species those types and rates of conditioning that contribute to survival. Of the three major principles of conditioning experimentally established, *frequency* concerns the question whether there is a reliable pattern and *intensity* whether the pattern matters. *Recency* crosses the two and is concerned with whether there is a pattern among those things that matter. Our purpose here is not to state empirical conclusions but to identify the apparent logical connection between man's behavior-selecting mechanisms and his evolutionary origin.

To continue in a more tenuous vein, we can experiment with the language of developmental change (2.02 and Chapter Sixteen) and describe conditioning as positive feedback in an open system subject to random variation and selective retention. The organism generates random behavior. If the ensuing feedback information is not correlated with certain behavior nothing happens. But if the feedback from the environment is correlated with the behavioral output (a pellet drops every time the rat randomly presses the bar) this ordered pattern in the environment selectively strengthens a related pattern in the organism. Through the learning process order will arise in the organism only if order is first there in the environment, though the complexity of its environment compared to its limited information-processing capacity may force the organism to oversimplify—to act as if the environment were more orderly than it is, one possible result of which is superstition (see 2.11). It may be similarly hypothesized that complexly ordered behavior can be learned only in a highly ordered environment.

A bit of speculation about positive and negative feedback may be useful here. Positive feedback is seen in this situation: the more strongly the rat associates pressing the bar with getting food, the more he presses the bar; and the more he presses the bar, the more strongly he associates the two. The relationship reaches an equilibrium as the correlation inside the rat asymptotically approaches unity. We are speaking here solely of the associative process; the motivational aspects come later.

Assuming that we learn to avoid antisurvival threats more quickly than we learn to approach prosurvival elements (2.41.1), the avoidance response can also be called positive feedback—the more the rat avoids the charged grid, the more freedom from shock he has; and the more freedom from shock he has, the more he avoids the grid. A difficulty with this formulation (and a practical difficulty of avoidance responses) is that the organism must learn that something produces unpleasantness before there is any value in avoiding it. A theoretical difficulty arises in the positive-feedback hypothesis if we merely change the language of the avoidance response because the change then becomes a case of negative feedback—the more the rat touchs the grid, the more displeasure he receives; and the more displeasure he receives, the less he touches the grid. More work is needed to discover whether the positive/negative-feedback approach to learning is fruitful. Its use in a different context appears in Chapter Seventeen.

2.09.7   Through higher-order conditioning, the human is able to form long chains of conditioned relations. To begin with the second order, by forming two independent conditioned connections of A to B and B to C he possesses a conditioned connection between A and C. A long chain might extend from A to Z. It might also have many branches and circle back on itself at points. It is crucial to the formation of complex conceptual and motivational structures to be able to connect A to C by virtue of having independently connected both to a middle item B without having ever experienced A and C in conjunction—and so on into higher-order conditioning.

As these connections become more complicated we approach a theoretical limit at which every output is a response to every previous input (Boulding, 1970, p. 54). In Hullian language, the whole of past learning is a potential intervening variable between a stimulus (input) and the ensuing response (output). In other terms, it is potentially the total apperceptive mass within which a behavior occurs.

Chapter 4

Human System

Detector, Selector, Effector
Subsystems

In the preceding chapter we narrowed our focus from systems in general to the human system, stating certain human traits that the model will construe as inborn, tying the system to its evolutionary background, and specifying the model's central feature—the learning process. This chapter adds details to the detector and selector and effector (DSE) subsystems, with principal emphasis on building complex learned patterns with the inborn ingredients already identified. The section on the detector is relatively long, mainly because an understanding of its details is necessary for the later discussion of human communication. That on the effector is relatively short because this subsystem is relatively uninteresting for social analysis.

2.1   DETECTOR

The detector provides a system with information about its environment, without which it would be unable to adapt (2.07.1). A substantial fraction of this analysis is taken up with definitions and discussions, after which we move into a less formal discussion of the detector processes. Among possible other things the detector handles the processes known as sensation, perception, and cognition. The terms *sensation* and *perception* are given precise definitions here that may not coincide with those a psychologist might use. *Cognition* is not used here because other terms seem to do the job.

2.10   *Definitions about Detector Processes*

a. *Uncoded (nonsemantic) information* is defined as patterns in matter-energy which, by virtue of having been imposed by other matter-energy, may be said to contain information about the latter. This definition must be read in the light of the earlier discussion of information (2.04) and the definition of coded information that follows immediately—the latter because the only sensible definition of uncoded information is to say that it is any information that is not coded. As an example, a tire track in mud is a modulation (modification) of the mud that constitutes information about the shape and former presence of the tire; the modulated radio wave contains information about the sounds spoken into the microphone; the bend of the tree constitutes information about the prevailing winds; the light waves reflected from a vase contain information about the vase, and so in due course do any optic nerves activated by that light. All isomorphic transformations of pattern would be uncoded according to this definition. For reasons to be seen, information can be "extracted" from an uncoded form only by an information-processing mechanism that uses coded information.

b. *Coded (semantic) information* is information that has been separated into distinguishable patterns—that is, categorized or grouped by similarity of pattern. If uncoded information is pattern, coded information is pattern of pattern: metapattern or metainformation. Coded information may or may not be conscious. This definition makes no implication about the number of hierarchical levels of coded information as that term is used below. Coded information is a form of pattern system in which parts are related on the basis of consistency or inconsistency according to the criteria of an acting system, in this case the human.

The crucial distinction is operational in that the pattern of uncoded information is processed as such (isomorphically) whereas the initial pattern of coded information is relevant only until it has enabled an information possessor to identify its class or category, after which the class pattern is processed instead. As the term is defined below, the initial pattern serves only as a cue, not as a pattern in its own right. Referred to as *uncertainty absorption,* the parallel at a different level is that an interpretation of evidence replaces the evidence itself (March and Simon, 1958, p. 165).

When a single class pattern is inadequate for the information processor, he may substitute an *intersection* of two or more class patterns. If light reflected from a ball enters my eyes, for example, once I have categorized the pattern as *ball* I thereafter process that class pattern rather than the pattern of visual sensations. But if I need to discriminate it from some other ball, I may process this one by a new pattern made by the intersection of the concepts *ball* and *green*— assuming that I had already formed the latter. To handle information about a particular green ball, let us assume that I had also formed class patterns representing, respectively, classes of things and particular instances of things. Instead of the visual sensory pattern, my brain would then process the intersection of three existing class patterns: *ball, green,* and *particular instances.*

There is no obvious limit to the number of classes that can thus be tapped. A particular floating large green plastic beach ball is an intersection of seven class patterns and can be adequately processed only by someone who has already formed all seven. That names are attached to these concepts merely reflects the logistics of verbal communication; the overt behavior of preverbal children and nonverbal animals attests that they apparently do the same things. This point is emphasized because it seems indispensable to the later discussion of semantic communication, the first of three main building blocks of social analysis.

   c. *Knowledge* is coded information in a human brain. As Boulding (1970, p. 2) puts it, knowledge is gained by the orderly loss of information, which is essentially the process described here as substituting the coded image for the data. By defining knowledge as patterns in human brains this model also coincides with Boulding's proposition that all knowledge disappears with the death of its holders, even though collections of signs from which knowledge can be extracted remain in libraries. The definition does not presume that other animals do not use substantially similar processes, even if they stop at lower levels of complexity; it merely ignores them as irrelevant to a model of man. However, I suspect that the time is somewhat overdue for experimental psychologists to become more explicit about intervening variables and to try to identify what minimal image structure is logically necessary for experimental rats or pigeons to make the connections they do between stimulus and response.

d. *Sensation (sensory input)* has already been defined. Now we add that sensations, as such, are uncoded information whether or not they have valences attached or are subsequently coded or decoded. However, they unequivocally are *recodings*—isomorphic transformations of external patterns into neural ones. (The terms *coding, decoding, recoding,* and others are dealt with in more detail in the later discussion of communication [3.10.0].) The sensations are equally uncoded in themselves whether or not the receiver already possesses codes for handling them.

As noted, a sensation is a state of the system. Although correlated closely with certain relatively simple states of the environment (2.09.1), a sensation alone may provide remarkably little information about the specific stimulus that actuates it. For example, and to speak of conscious awareness, sensation alone may make a person aware that his finger is hot but give him no indication at all whether he is encountering a candle, a hot plate, a focused beam of sunlight, or a light bulb. Sensations of the optic nerves are even less informative until patterns of light and color are coded; and until they are, such sensations provide little information about what's out there (Kuhn, 1963, pp. 110–113). For our purposes we can say that by inherited mechanisms a person can know that he has a sensation, but not the nature of the environmental phenomenon that set it off. The main problem of the detector is to learn how to identify the latter and the former.

e. *Codes (concepts) of sensations:* To try to express the inexpressible in words, if a collection of pain-sensitive nerves in the finger is actuated by a pinprick, this collective actuation constitutes a simple, uncoded pattern accompanied by the feeling of pain. After repeated experience the individual learns categories of sensations, so that he can recognize and respond differentially to "sharp, narrow, pain-in-the-finger" as distinct from "big, dull, pain-in-the-stomach." More particularly, he forms *concepts* of these things, a concept being the same as a class pattern, and can then deal with the thought, idea, or image of each sensation as distinct from the sensation itself. In this model one could not talk of pain, pleasure, odor, sound, or light unless he had first conceptualized each. Since sensations are themselves internal states, codes of sensations are codes of internal states. Other sensory modalities behave similarly. A single exposure to light reflected from a visual field produces a pattern of

visual sensation. Through experience these patterns are grouped
and categorized in various ways according to certain similarities or
differences in the pattern of sensation. In simple terms these
patterns can be blueness and yellowness, squareness and roundness,
bigness and littleness, thinness and thickness.

f. *Codes (concepts) of externals:* By combining different patterns of
sensations actuated by an element in the environment and typically
by patterns taken from more than one modality, the individual
constructs a pattern which represents that external element. For
example, the infant's nearly simultaneous sensations of milk in the
mouth, assuaged hunger, tactile hardness, a certain visual pattern,
and the sensations of holding and sucking all combine into a single
pattern of his bottle. Additional experiences with it and with other
things, including experiences with the bottle when not in use,
eventually produce the refined pattern of "bottle as a thing outside
of me and existing independently."

In this model the individual initially *has* sensations but
perceives things. Any sensation except the single firing of a single
sensory neuron is itself a pattern since many nerves are actuated, many
of them repeatedly. Although the pattern of one sensation is not quite
the same as that of any other, the repetition of a sensory input (the
pattern of mother's face) is similar in important respects to the previous
exposure to the same input but very different from some other input,
such as the pattern of the doorway. Because the environment is ordered
in many respects, the infant forms a code of his sensations by grouping
similar patterns and distinguishing them from other patterns. He iso-
lates certain patterns of patterns, which we may call second-level
patterns. It probably requires third-level patterns (see 2.11), or patterns
of second-level patterns, and often the combining of second-level
patterns from two or more senses, to form the internal pattern we would
call an image of external reality. There is no limit to the number of
levels that could thus be formed, but three are probably required to
produce such an image. Although the general process is presumably
the same for all individuals, the content, precise sequence, and timing
of the parts are presumably unique to each. There may or may not be
any significance to the parallel suggestion that third-order conditioning
is necessary for "foresightful" behavior (Kuhn, 1963, pp. 137–138).

For cogent reasons spelled out by Simon (1965) it requires
less information capacity to describe, and presumably to recall and
otherwise process, a given amount of information arranged in a
hierarchy of patterns than the same amount treated as a single
composite. The whole pattern is also more durable, for reasons he also

gives. The hierarchical model of pattern formation is adopted here because it parallels the approach of system analysis and because it is efficiently described. Moreover, being itself efficient it seems likely to have been selectively retained by evolution, just as evolution retained the hierarchical arrangement of biological structures. Another argument in its favor is that the different levels of the nervous system virtually cry out for it (in the way the afferent nerves from each type of sensation collect in their own coordination center before going into a grand collection in the central cortex, and the efferent nerves reverse the sequence on the way to the muscles). And, finally, learning of complex motor activities often follows the same pattern—especially, but not exclusively, when done consciously. Regarding the last point, since conscious control of motor activities is probably carried on by the same major area of the brain that forms information concepts it seems reasonable to presume that the basic processes are the same.

g. *Codes versus instances:* We have said (2.04) that a person "possesses information about his environment" when a pattern within his behavioral system is functionally related (dependently) to patterns outside. This relationship takes two main forms. The first is that some relatively permanent patterns in the brain correspond to external patterns the individual has been exposed to repeatedly; these durable patterns may be referred to as concepts, code items, or images. Once learned, such a pattern persists for long periods when the external pattern is not present in the environment, particularly if it is regularly reactivated by recollection. The second is a temporary pattern brought into being by sensations currently received from the environment.

Although we remain uncertain about the neurological basis of this distinction, the main contours of the problem seem satisfactorily represented by a model in which the durable patterns consist of networks of neurons that have become hooked together in the learning process while the temporary ones consist of the selective activation of one or more networks that "represent" the external pattern encountered. The durable pattern will be designated a concept and the temporary one a perception. The importance of this distinction appears below in the difference between concept-learning and perception.

h. *Concept-learning (pattern formation)* is the process of forming relatively durable patterns in the cortex that correspond directly in some way with recurrent patterns of sensations and indirectly with external patterns that actuated the sensations. Unless otherwise specified we will be concerned only with sensations activated from the environment, in which case the sensations are intermediaries between the external and the cortical patterns. Although they are indispensable in that intermediary role, we will take them for granted and deal solely with the relation between the environmental pattern and that in the cortex. The patterns thus developed by concept-learning constitute the individual's *code* of information about his environment; the process of learning concepts constitutes *coding*. At the linguistic level, coding includes learning the signs for patterns that are given names.

i. *Concept formation* is a type of concept-learning in which the individual sorts reality into his own patterns. This is the only type available for the infant, for the scientist developing new theory, and for creativity in general. (See *information concept* below.)

j. *Concept attainment* is a type of concept-learning in which the individual adopts patterns learned from others. Concept attainment presupposes communication of some sort.

k. An *information concept* is an idea, mental picture, or image, as distinguished from a motor concept (see below). *Concept* used alone will mean *information concept*. As used here a concept can be a pattern representing a general class of objects or events (chairs or purchases), a single object or event (this chair or my act of buying it last Tuesday), or a null class (elephant eggs or my piano recital at Carnegie Hall). The content of the concept may be any element, pattern, or system as defined here. This definition includes both consciously and unconsciously held concepts.

l. A *motor concept* is a neural pattern that corresponds to a pattern of overt behavior involving muscles and controls its execution. A pure motor concept can be executed as behavior without conscious thought or control.

m. *Perception (pattern recognition or identification)* is the process by which uncoded sensory inputs from some pattern in the environment selectively activate a coded pattern (concept) or group of patterns already learned and stored in the cortex and lead to an inference (Kuhn, 1963, pp. 14–17, 104–110) that an instance of that pattern currently exists in the environment. The selective recognition or identification of the pattern is an act of *decoding* an information input, or *channel selection*. That is, in this model the fundamental distinction between sensation and perception is that the former involves uncoded information while the latter utilizes coded information. In language to be developed in Chapter Seven, perception involves detection plus decoding.

    As noted (item b above), to perceive a thing may require

the intersection of multiple stored concepts. Even acknowledging the greatly increased scope of perceivable things facilitated by the use of such intersections, there nevertheless remains an important sense in which a person cannot perceive any pattern not already stored in his head—which is a different way of stating a central theme of Gestalt psychology. However, by joining existing concepts into new combinations one can perceive additional things represented by those new combinations.

I would be inclined to translate Polanyi (1966) into the present language by suggesting that his "grasping" of knowledge corresponds to pattern recognition (perception) and that his "shaping" corresponds to concept-learning (pattern formation).

n. *Cues* are current receipts of information via the senses that help the organism identify a pattern in its environment. That is, cues assist perception, decoding, or channel selection—depending on the language one prefers.

o. *Sufficient cues* are cues that permit identification of a pattern with a degree of confidence satisfactory for the purpose at hand. Defined operationally in the language of communication theory, cues activate scanning processes and sufficient cues terminate them by identifying a pattern. If the initial cues are themselves sufficient, scanning is presumably not eliminated but merely proceeds more efficiently. (At this date we cannot say with certainty that there is such a thing as a scanning process in the brain, but an information model seems to dictate that we assume one.) Although sensations occur at or before birth, they are not cues until conceptual code items have been developed.

p. *Abstraction* is the formation of higher-level emergent patterns that consist of similar elements among lower-level patterns. *Rectangular* can be abstracted from door, book, sheet of paper, and brick; *obstinate* can be abstracted from a series of stubborn behaviors. Except that we must all presumably build initially on patterns of sensations, the learning sequence of the individual or society may have quite as much to do with the hierarchical position of an idea as does any inherent logic of the idea itself—though once a whole hierarchy is formed, some ways of structuring it are probably more efficient than others.

Whether or not the neural processes used in forming new concepts by abstraction differ from those used in forming new concepts by intersection, the two are logical opposites in that the former generalizes and the latter particularizes or discriminates. In this model abstraction moves up the hierarchical scale, forming superpatterns by combining parts of existing ones. As noted, this is an emergent process and hence presumably occurs according to the principles of emergence (random generation and selective retention) detailed in Chapter Sixteen. The process seems to be relatively slow and haphazard, as would be expected of any emergent development.

By contrast, intersection moves down the hierarchical scale. Instead of abstracting, it starts with concepts that are already relatively abstracted (this, green, ball) and combines them in such a way as to produce a new concept. Such a concept is the kind from which, when taken with other concepts at the same level, the initial concepts could have been abstracted. In contrast to the cumbersomeness of abstracting, intersections can be made almost instantaneously in perceptions (2.10b and m) and in linguistic communications (3.16.3c and 3.16.44). If these suggested differences in speed are in fact prevalent (and I do not recall ever having seen the problem stated in quite this way), presumably we would have to conclude that the neural processes in abstraction and intersection are different.

2.11   *Discussion and Summary*

The detector is the subsystem of the behavioral system that provides the main system information about its environment. In the present model it operates in the following way. Inborn structures provide sensations that are highly correlated with certain relatively simple aspects of the environment. A sensation consists of a pattern of sensory nerve discharges. These the cortex groups into second-level patterns (or concepts) of sensations. And these in turn are grouped in ways that provide images (concepts) of the external objects or events which set off the sensations. Through this process the system comes to possess higher-level inner patterns that are ordered relative to patterns in the environment.

All behavior of the system consists of responses to its inner states; only as inputs from the environment modify those inner states can the system be said to respond to its environment. Although at the level of reflexes and in some arousal of emotions the system may respond to sensations alone, all more complex behavior is a response to the conceptual patterns, or images, formed hierarchically from abstractions and intersections.

Once these images have been formed they can be tapped by new sensations acting as cues, at which point a person perceives as currently present in his environment an instance of whatever concept those sensations select. The process involves an important multiplication of information. For what the person perceives is not merely the information currently received via his sensations but all the information stored in his concept from his accumulated experience—or as much of it as is worth calling into play. The information theorist would add that the more improbable the pattern (and the greater the "distance" to the most similar one), the smaller need be the cue to identify the pattern with a given degree of confidence..Except at the level of sensations, a stimulus is not a stimulus until it has been identified, and it is the identification, not the sensory inputs, that constitute the stimulus. This distinction is important to behaviorist psychology. Although electric shock or warmth can constitute valenced stimuli without being identified in this sense, a circle cannot constitute a stimulus to a rat until the animal has first formed a concept of it and learned the sufficient cues for identifying it. In all such cases the stimulus consists in part of current sensory inputs and in part of past experiences. For virtually all behavior relevant to social analysis, the stimulus in a situation is perhaps more a reflection of past experience than of current inputs.

In perception part of a pattern is presented and the brain responds by activating the whole pattern already stored there or a part of it. Again, such behavior is a central feature of Gestalt psychology. Among numerous consequences of this pattern-completing facility is that during one passage of a familiar symphony the listener can start enjoying the climax he knows lies ahead. By the same token every perception is in some degree false. No two objects or events are precisely alike, and even the same object may never be observed identically twice—that is, with the same lighting, angle of view, and position relative to other objects. Hence when the stored image is substituted for the pattern of actual inputs (the phenomenon of *persistence*) the environment is misrepresented. But if the behavior learned for the class pattern is appropriate for the instance at hand the distortion is of no consequence, although the saving in information-processing is immense.

In a fraction of a second we can perceive things we never encountered before and which therefore do not tap an existing concept. This implies that we have formed a new concept. It also raises the question of how this is possible when concept formation itself generally proceeds at the same pace as other conditioning—though with distinct traits of its own. How can you quickly perceive a cerise beach ball when you have never seen anything but blue and yellow ones? To categorize without explaining, it seems that once concepts are formed the brain can make new ones from them by intersection far faster than it can make the connections that constitute initial concepts. Without such ability every sentence containing a new idea, even a simple one, would have to be repeated often enough to produce conditioned connections before it could be understood (3.16.3c).

If the action pursuant to a perception does not "work," appropriate responses are to modify the image, try a different one, or abandon the effort. In this model every concept not construed as imaginary is a hypothesis about reality, and every perception based on it—or at least every perception acted on—is an empirical test of it. "Reality-testing" is really image-testing. In case of mismatch the (good?) adaptive detector revises the image, not reality.

Since concept formation is a developmental process that results in successively higher levels of patterns in a hierarchical structure, the language hypothesized to explain such emergence (5.21.2) should at least tentatively be applied to it. Thus described, man's brain randomly forms patterns to represent the environment—the random variation (or generation) half of the emergent process. With them he then perceives and acts on his perceptions. If the perception "works" he retains the pattern on which it was based; if it does not he discards or modifies it—the selective retention half of the process. (See Campbell, 1962 and 1965, for a detailed application of this version of the emergent process to creative thinking.) So long as a person recognizes their status and does not try to act on them, wholly imaginary concepts are not unadaptive and may lead to improved real ones.

In the fullest sense, however, all concepts are imaginary; they differ only in the degree to which feedback from behavior based on concepts coincides with the feedback they predict. As Boulding puts it (in a personal communication), the senses are critics, not authors, though they may have mutational functions. At the self-conscious level of science, theorizing is the random generation and empirical testing is the selective retention. As will be seen in Chapter Seventeen in connection with personality, concept formation may perhaps be usefully viewed as involving a positive-feedback relation to environment whereas perception involves a negative-feedback relation.

Every normal human forms a self-image, partly conscious and partly not. Since all overt behavior might be said to change the relation of oneself to his environment in some way, some image of self is at least potentially involved in every overt act. By comparison any one image about the external environment is involved far less often. By simple frequency of use, the image of self is likely to be learned more deeply and in more detail than the image of an external. The effect of this frequency is greatly amplified by the direct internal information one gets about himself—about his sensations, feelings, and muscular movements—and the feedback information about the match between his intended actions and his actual ones.

Both the content of one's self-image and the process of forming it rely heavily on his observations of how others respond to his actions. We will therefore defer further discussion of self-image until Chapter Seventeen, after the main interpersonal analysis. Meanwhile we have discovered why the image of self is presumably more important to behavior than is any other image, at least once it is well formed.

This model, incidentally, would suggest the following relation between learning in humans and learning in lower animals. For obvious reasons an individual that could develop complex structures of preprogramed motor responses would have a better chance of surviving emergencies than one that had to improvise responses. There is thus high survival value in the ability to form hierarchically arranged effector responses. But a coordinated sequence of behavioral elements would normally move through a sequence of environmental elements, and the behavior could not be coordinated into a single large pattern without a parallel coordinated image of the environment. The evolution of a brain with high-level intellectual capacity is thus not distinct from the evolution of complex motor skills but is an indispensable accompaniment of it. Furthermore, in the present model the neural processes are the same for both. To review an important term, if an individual's conceptual structure, either information or motor, is viewed as a system it is clearly a pattern system, not an acting one, and except for its hierarchical structure is not subject to generalizations about system analysis.

A hierarchical structure of knowledge has important consequences. Although lower-level concepts are the essential blocks from which the higher-level ones are later constructed—which is why we use examples to communicate abstractions—once the latter are available the former can apparently be discarded or deactivated for many purposes. If they are discarded, the hierarchical structure vastly increases the amount of knowledge, though not necessarily of information, that the brain can acquire and process. And although there must be some limit to this quantity, this model provides no basis for locating it. With reference to learning and teaching, the question obviously arises as to the minimum number and variety of lower-level concepts that are necessary as the base for higher-level ones.

Drawing partly from psychology and partly from information theory, system analysis, and logic, and omitting considerable detail, we will use the preceding model of the human detector. When more detail is needed to accommodate conscious and complex processes, as in decision-making, the detector can be divided into its own DSE subparts. Interactions and mutual modifications among detector, selector, and effector come later.

Extrasensory perception is obviously omitted from this model. Whether it occurs or not—a matter on which I take no position—there is apparently no significant evidence that it occurs regularly or reliably enough to influence social behavior sufficiently to require inclusion in a model of social man.

2.2   SELECTOR

Once the detector has discerned the environmental situation the function of the selector is to select one response rather than another, including a possible zero response of inaction. To simplify we will use *approach* and *avoid* in a wide sense, more or less synonymous with *reinforce* and *extinguish* (2.09.5). Drawing money out of the bank, taking a bus downtown, and then purchasing a symphony ticket are all regarded as "approaching" music. And writing a letter urging a senator to support fines on emissions from smokestacks is similarly "avoiding" pollution.

The selector is here associated with such terms as motive, goal, goal object, preferences, likes and dislikes, and value judgments. Other psychological terms appropriate to the selector would be drive, drive reduction, valence, needs, hedonic selections, motivation, emotions, id, instinct (Freudian sense), self-actualization, and activation. As with the detector, we will use psychological terms without depending on psychology for the ingredients of our model. We note also that, among scientists, confidence that genetic man arrived at his present state by an evolutionary process is remarkably close to 1.0. The preceding chapter pointed out that man inherits behavior selectors but no behavior of interest to social science. Since these behavior selectors are the product of evolution we assume that they are designed to enhance the likelihood of survival in the individual and the species.

To date psychologists have not provided an authoritative list of inborn motives or even agreed on what they mean by *motivation* (Cofer and Appley, 1964). To have something more explicit to work with this model distinguishes inborn from learned motives, describes how (in the model) the latter can be built from the former (with the help of experience), and somewhat arbitrarily selects four emotions as essential for basic social science. Although we assume that the behavior selectors are oriented toward survival, to select prosurvival behavior does not require that the organism "want" to survive, that it have a "goal" of surviving, that it understand what survival and death are, or even that it know what actions favor each. It is only necessary that its behavior selectors reinforce approach to those behaviors that favor survival and avoidance of those that do not.

The selector includes selection only, not activation. It might also be called an information, switching, hedonic, or channel-selecting system. In this respect it follows Grinker (1968) and Breger (1968) in rejecting the notion of psychic energy and in focusing instead on the controls that release muscular energy—not because we presume that Grinker and Breger are necessarily right but because that approach makes a more parsimonious model for the social scientist. Without attempting to specify how the selector works in the model we can nevertheless note that the amount of energy released can logically be handled solely as a question of which channel is selected: 6, 110, or 440 volts can be put into a circuit depending solely on which switch is closed. By putting energy in the id and not in the ego, Freud left it in the control system. In this model, however, energy is entirely outside the control system: it is in the maintenance system. Even if one wished to use the concept of psychic energy he could still use it as a question of which sets of neurons are activated; the "feelings" associated with so-called high levels of psychic energy, such as excitement and tension, do not involve significantly higher levels of neural energy than do relaxed feelings.

The role of motivation is quite different for the individual than for the species. To illustrate, at the level of the individual it is sufficient that he avoid injury because it hurts, not because it endangers survival. But in the evolutionary perspective it is essential that injury hurt, so that the individual will avoid it. This is what is meant by the evolutionary selection of a behavior-selecting mechanism, and the reasoning about it is sometimes fuzzy.

For positively needed things the logic is the same but requires a second step. To repeat the first, the individual can eat something solely because he likes it and not because he needs it, whereas for the evolutionary survival of the species he must like it because he needs it. Although perpetual avoidance of pain will not itself harm the organism, perpetual eating will. Hence backup controls in such forms as nausea are required to stop the eating after a certain point. Sexual activity is similarly engaged in because the individual likes it, while in evolutionary terms he likes it so that he will engage in it. In this case the activity stops when a climactic cycle makes immediate continuance impossible rather than unpleasant. For humans who know the relation between intercourse and pregnancy the act can be instrumentally directed toward procreation. But quantitatively this is a minor motive. It is unknown to some human societies and presumably to all other mammals. Incidentally, any teleological flavor to the preceding examples is an accident of discourse. Evolution does not select traits *in order to* assist survival. It is simply that those who are less motivated to seek food, engage in intercourse, or avoid pain are also less likely to survive long enough to help continue the gene pool. In short, evolution has functioned in such a way as to make an individual want what he (or the species) needs. Unfortunately, to understand this relationship provides no information about the mechanism in the individual that performs the function.

2.21   The human system contains obvious selector mechanisms of the sort deduced from evolutionary logic. First, most acts or environmental conditions that clearly contribute to survival give rise to sensations we classify as pleasant: witness the ingestion of food and water when we are biologically ready for them, the discharge of wastes, external temperatures that favor the maintenance of the biologically optimal internal ones, genital stimulation, softness next to the skin, and the taste and smell of many nutrients. Second, most acts or environmental conditions that clearly threaten survival give rise to sensations we classify as unpleasant: witness continued absence of food and water, the buildup of wastes in bladder or bowels, external temperatures beyond a safe range, rupture of the boundaries of the organism, pressures of an intensity that threaten tissue damage, and the taste and smell of many biologically harmful substances.

     During the rugged eons that patterned man's genes, acts that brought pleasure and avoided pain had high survival value—on the average. If the circumstances have now changed, man's genes and learning processes have not. *Unpleasant* is used rather than *pain* because some valences are definitely unpleasant though not painful—frustration, claustrophobia, fear, anxiety, itch, moderately uncomfortable temperatures, and noxious tastes and odors.

2.22   We have empirical evidence of "pleasure" and "displeasure" centers in the brain of man and some other mammals (Olds, 1955). The simplest available theory of motivation is that some sensations have inborn connections to such centers—prosurvival ones to positive-valence centers and antisurvival ones to negative-valence centers. These valence centers would be logical equivalents of reflexes except for two factors. First, whereas in the reflex a sensation activates motor nerves, in this case the sensation activates valence nerves. And second, whereas reflexes activate motor responses, these valence centers activate operant learning processes, reinforcing approach responses to those things that set off the pleasant valences and avoidance responses to those that set off unpleasant valences. As noted (2.08.0f), *sensation* includes inputs from the biological system and its emotional states; hence valence centers activated by biological or emotional states follow the same rationale. In this simplest theory, which we will adopt for the model, all inborn, or primary, motivation consists of activation of these centers. The valence centers are mentioned here to clarify the logic of the model, which requires only the dual assumptions that pleasant sensations reinforce approach and unpleasant ones avoidance and that in the evolutionary logic those reinforcements were on the average successfully adaptive.

Realistically, we do have feelings of pleasantness and unpleasantness, and I am willing to assert against any Watsonian-minded skeptic that introspective reports alone are compelling empirical evidence of their reality. (The most hard-bitten behaviorist takes for granted that the rat will approach food and avoid electric shock. Without valence mechanisms?) For the model it is not necessary to specify these valences; it is only necessary that there be at least one positive and one negative.

The next step is to move from inborn to learned motives. Stated in its simplest form, in parallel with the way reflexive salivation can be conditioned to the new stimulus of a bell, the "reflexive" activation of a positive-valence center by the taste of candy can be conditioned so that it will be activated by the sight of it or by the word *candy*. Again, this statement may be viewed as explanatory analogy. The empirical status of such mechanisms is not directly relevant to the usefulness of the model to social science.

2.23   After noting that this psychedelic age seems to put a premium on sensations and primary motives, we will return to the problem of man's complex motives after the following definitions.

a. A *primary motive (primary reinforcer)* is an inborn valence that reinforces approach or avoidance behavior toward those things that set it off—in nontechnical language an urge, drive, need, or (in the Freudian sense only) instinct. Defining these motives as inborn means that all primary motives are formed and put into the system from outside, in this case by evolutionary selection.

b. A *secondary motive (secondary reinforcer)* is a conscious or unconscious concept, often but not necessarily corresponding to a pattern in the environment, which, by virtue of being conditioned directly or indirectly to a primary motive, is able to reinforce approach or avoidance behavior toward those things that set *it* off. If the primary motive were described as a need (say, for food) the secondary motive would be described as a want (say, for steak), although the distinction between needs and wants is not used formally here. If the concepts whose attached valences constitute secondary motives are hierarchically arranged, the motives may have a parallel arrangement.

c. An *intrinsic secondary motive*—such as the satisfaction gained from listening to music—is one whose reinforcing power is independent of whether achievement of the reinforcing state leads to another reinforcing state.

d. An *instrumental secondary motive* is one whose reinforcing power depends on its leading to another reinforcing state. The satisfaction gained from a concert ticket, for example, depends on its ability to lead to the hearing of music.

2.24   It is evident from the model and definitions that secondary motives in the selector depend on concept formation in the detector. One cannot effect approach behavior toward juicy steak not already tasted or smelled until he has learned to distinguish it perceptually from a brick or a skyscraper—which distinction presupposes formation of the concepts involved. If the thought of steak is pleasing, this means that one's concept of *steak* has acquired a conditioned connection with a positive-valence center, and to say "I like steak" implies consciousness of the effect of this connection. Given man's capacity for complex and interconnected concept formation (2.10f) and his ability to form long chains of conditioning of elements in either concepts or motives (2.09.7), the model assumes that no additional theory, only additional learning, is necessary to account for likes or dislikes of mother, mountain climbing, the Beatles, Picasso, the Koran, or pornography.

Life's vicissitudes virtually guarantee that many repeated experiences, and hence the concepts of them, will become at least partially conditioned to both pleasant and unpleasant feelings, perhaps of several different sorts, through numerous different chains of conditioning. Moreover, each step in each chain may similarly acquire cross-connections to both pleasant and unpleasant feelings of various types. Simplicity of the basic model in no way denies richness of actual patterns. The model thus provides a principle simple enough for the ardent nomotheticist and a potential outcome rich enough for the most idiographic humanist. It simultaneously indicates why behavior can be so unpredictable in one case (the lovers' quarrel may end in embrace or murder) and yet allow such high confidence under certain analytically or actually circumscribed conditions (the candidate will almost certainly try to win the election, not lose it). In short, there is nothing necessarily wrong with the hedonic calculus if one allows for a richness of complex secondary conditioned connections to its primary base.

2.25   It is also evident from the model and definitions that, except at the level of reflexes, to satisfy primary or secondary motives requires the formation of motor concepts in the effector. That is, the selector reinforces behavior when the feedback is pleasant; it does not specify it. Details follow in Section 2.3. To repeat, the selector only selects; it does not provide the information, the energy, or the program to carry out the selected behavior. Several aspects of detector and effector, however, seem to require certain selector aspects if they are to contribute effectively toward survival—a matter for the next chapter.

2.3   EFFECTOR

The best laid plans may come to naught if bady effectuated. Any well run individual or organizational life therefore needs to put as much thought into carrying out behaviors as into selecting them in the first place. But thinking and selecting are functions of detector and selector and the interplay between them. An organization might have a sizeable decision structure within its effector subdivisions, but in the model of the individual all decision processes are assigned to detector and selector. Effector processes are here confined to the muscular execution of already selected behavior, including the learning of motor skills. This section on the effector is short because such matters are of little concern to the social scientist and rather little to the psychologist as well. The social scientist is much interested in how man perceives and values the ingredients in his environment, including other people and his interactions with them, and the behaviors he selects in consequence. But the social scientist is almost totally unconcerned about which muscles are used, and in what sequence, to carry out the selected behavior. The short shrift here given the effector is not meant to downgrade its overall importance. It means rather that socially interesting behavior can rarely be carried out by the effector alone; it requires repeated, if not continuous, direction by detector and selector. More specifically, social scientists are interested in directed behavior, and the direction comes from detector and selector.

To illustrate, in a baseball game the effector encompasses only the motor skills involved in throwing, hitting, catching, tagging, or running. Whether to bunt or hit, to walk or to strike out the batter, to throw the fielded ball to second base or home plate—though "effectuational" in the broad sense—all require new decisions. These decisions often involve complicated information: what inning it is, how many players are on base, how many outs there are, what the score is, who the next batter will be, and so on. They also involve goals: given the present state of affairs, which of the next states is the most desired? In all such situations the problem is again one for detector and selector, not effector. In this model the effector, strictly speaking, deals only with the stage of behavior that can be executed without referring back to detector and selector stages, which is another way of saying that complex effectuations can rarely, if ever, be done by the effector alone. Two previous points nevertheless need mention in this connection.

2.31   First, all controls of muscular execution are informational in the broad sense, consisting in the human of information concepts and perceptions, connections to valence centers, and motor concepts. Hence all external behavior is a replication in some medium outside the behavioral system of some pattern inside it—a symbol of it (3.10.0d). And to amplify slightly the point of the preceding paragraph, in well-coordinated actions, like running or throwing a ball, the overt behavior can be controlled by the motor concept alone with no conscious thought about the nature or sequence of motions. In more complex behavior a sequence of such actions is coordinated with a succession of detections of the environment and comparisons with the selector's goal.

2.32   Second, an organism can give an inborn response to a sensation as by reflex. But except for externals like heat that produce distinctive sensations, the organism cannot give an inborn response to an element in the environment unless it also has an inborn ability to perceive that element—which in this model implies an inborn concept of it. To combine the two observations, an organism cannot give an inborn response to a learned concept, although the imprinting process in some animals produces essentially that result. Since humans seem to have no inborn concepts of externals, possibly excepting spatial depth, they must learn their responses to all externals except those that produce distinct sensations.

To illustrate with a basic kind of behavior, a human has an inborn positive valence for erotic sensations, which valence can presumably reinforce *any* behavior that produces them. Society, of course, may condition the individual to avoid some sources of stimulation, but that fact is irrelevant to the mechanism. The human may also inherit a reflexive pelvic-thrust response to it—though this is apparently not reliably established (Cofer and Appley, 1964). But regarding the selector the human male presumably cannot inherit a specific desire to mount a human female, since he has no inborn mechanism for distinguishing a female from a male or a file cabinet. The model thus apparently implies that all overt sexual behavior of humans is learned. And regarding the effector, since in all known instinctive sexual unions among land-dwelling mammals the male mounts the female from the rear it seems most unlikely that between the anthropoid apes and the human there would occur a sudden instinctive reversal to the customary frontal position.

This brief review of instincts is intended to reemphasize that all human behavior of significance to social analysis results from the prior interaction of detector and selector, not from some inborn impulse. The inborn impulse can only be toward one sensation or away from another; the behavior that produces the sensation is not preprogramed in the effector and must be learned.

Chapter 5

Human System

DSE Interactions and
Summary of Model

2.4   DSE INTERACTIONS: INTRODUCTION

The DSE systems are functionally bounded. All operate through neurons. At their more complex levels they presumably all utilize the same neurons—those of the cortex. However, it is not nonsensical to think of their inborn components as independent. The detector alone receives sensory inputs and organizes them into image structures. The selector alone has valence centers that reinforce approach or avoidance responses. The effector alone is connected to muscles and actuates their responses. Although the coordinated action of all three is necessary for adaptive behavior, coordination does not itself imply interdependence: the selector's negative valence for intense heat is there before the detector learns about hot stoves, and the detector can learn to distinguish glue from pencils without the selector's having any preference between them.

By contrast, learned levels of all three depend on one another. The detector's images can hardly be developed at all without the large motor movements by the effector that constitute exploratory activities and the smaller ones that guide the eyes, turn the head to catch sound, or bring the fingers into contact with something to touch. We have already indicated that learned motives are detector concepts attached to selector valences and that all effector responses beyond those that can be executed without thought required repeated guidance from both detector and selector. To distinguish that kind of interdependence from what follows, note that the three systems can utilize one another without necessarily affecting one another's content. That effector processes are required to move my eyes to see or my fingers to feel does not mean that my effector significantly affects the content of my visual or tactile images. Although secondary motives are attached to concepts, the detector does not itself determine which valences will be attached to which concepts. And even if we assume that the detector does not form images until it is motivated to, the specific content of the images can be quite independent of the nature of the motive that triggered the learning of them: my concept of a transformer may be much the same whether my motive to learn about it was to fix my door bell, run an electric train, satisfy idle curiosity, or impress a prospective employer.

For simplicity I have dealt mostly with a simple sequence through the DSE systems, in the same order that behaviorist psychology puts stimulus, organism, response. Behavior of humans (and probably that of most higher animals) is more complicated. For one thing, a given unit of behavior need not start with an input of information. Often it starts with the selector stage—perhaps some recollection that sets off a desire. Only then may the person start gathering information to help satisfy the desire. (Note that an urge arising in the biological system, as from hunger or fatigue, comes from outside the behavioral system; it must first be detected, even as must the wet of rain.) Since all human behavior can be viewed as an unending sequence of cycles (2.07.1–.2), there is always a certain arbitrariness about where one breaks into it. And not only may the cycle start in any of the DSE systems, but it may also go back and forth in highly unpredictable ways: an urge, a little information gathered, a modification of the goal, a partial effectuation, a reassessment of the situation, a second goal modification, a temporary abandonment, and so on.

Before concluding with a summary of the model this chapter explores two types of interactions or interdependencies that differ from those just described. Under the heading of "Cross-System Motives" we will note certain respects in which each DSE system is independently motivated, and under "DSE: Mutual Modifications" we will discuss some ways the content of each affects the content of the others.

Cross-System Motives

The purpose of this section is to provide a conceptual frame-work that encompasses some of the richness of motivation described by the psychologist while staying within our own model. I will not debate whether to describe the self-motivation aspects of the DSE systems as a subselector of each or as motives that cut across all three systems. Four such cross-system motives will be identified: priorities, exercise, efficiency, and frustration.

2.41.1   *Priorities, Including the Role of Emotions*

The first cross-system motivation concerns priorities. Some things in the environment are more relevant to survival than others, and a survival-oriented system should logically give them priority. Thus the main questions are these: What things are most relevant? And do the organism's mechanisms reflect these priorities?

Many things that cause pain can also destroy the organism in short order. Hence avoidance of pain should produce a prompt response. By contrast, all pleasant sensations (as distinct from relief of unpleasant ones) can be postponed for hours, days, or some indefinitely without biological damage. From this we would deduce that pain-avoiding behaviors take precedence over pleasure-producing ones. (Here, as elsewhere, to say that something has priority gives no clues as to how completely lower-priority items are excluded or for how long. The notion nevertheless seems to have broad usefulness regarding basic selector mechanisms.)

2.41.11   *Detector.* The two senses with highest information capacity, sight and hearing, seem to show definite inborn internal priorities. As to sight, moving objects have greater capacity to injure than do still ones and hence should command prior attention. This expectation is empirically supported: from early infancy the eye reliably fixes first on the moving element in an otherwise static field.

Because (a) virtually all sounds other than those issued by animals are generated by motion, (b) sound diminishes rapidly with distance and ceases with the motion that generates it, (c) small things in nature rarely generate loud sounds, and (d) large things do not move without large force, a loud sound in nature normally means that a large force is operating, close at hand, and now—and that the organism had better attend to it promptly. Unless the force is under control of the organism itself it is much more likely to be dangerous than helpful —pleasant sensations seldom arise from movements that make loud noises. Hence there is logic in responding with both apprehension and attention. This expectation is supported in the apparent inborn attention to and fear of sudden loud noises.

For animals with an inborn ability to distinguish friend from foe by smell, similar projections might be made for that sense. But humans cannot do this, and no such priority can be inferred. The remaining senses are for the most part directly valenced and hence are matters for the selector rather than the detector.

2.41.12   *Selector.* If we accept these priorities we should expect that in cases of conflict the valences which reinforce avoidance will take priority in time or strength over those of approach. Here the role of the emotions as a positive-feedback, deviation-amplifying mechanism seems important. In a general sense the valences already discussed can and do set priorities—that which we like or dislike most intensely receives prior attention. But in matters of intense concern to survival those valences in the selector may need to be supplemented, and so may detector and selector processes. In this connection the emotions will be assigned the following role.

By sensation or perception the organism first identifies that something in its environment is of concern to survival. The behavioral system then sends outputs into the biological system, which undergoes some change that returns positive feedbacks which additionally stimulate the initially activated valences in the behavioral system. The level of valence arousal is thus greatly multiplied. Additional valence centers not subject to arousal from direct environmental stimulation or perception could thus also be brought into play. Valences that support "appropriate" behavior could thus unquestionably dominate conflicting ones. If the same emotional state can also focus the detector's attention and energize the effector, the likelihood is further increased that the appropriate behavior will be effectively accomplished.

It is not the purpose of this volume to get into neurological controversies. Yet this model is clearly related to the James-Lange and the Cannon theories of emotions. If I stall my car on the railroad track just as I hear two longs, a short, and a long blast from a locomotive, the emotion of fear is with me—promptly. Since I have no inborn ability to perceive a stalled car or railroad tracks or to decode that particular sound, much less an inborn reason to fear any or all of them, the situation and its implied consequences must be perceived in the cortex. This arrangement would be in accord with the James-Lange theory—that messages go from the sense receptor first to the cortex and then to the visceral regions. Nevertheless sensations from a rupture of the skin have inborn connections to valence centers as well as access to the cortex. That source of fear does not have to be perceived but could be produced directly from sensations. It does not require prior cortical activity and hence could operate according to the Cannon theory—that incoming information goes first to the thalamus and thence to both cortex and visceral regions. Since fear can be aroused in infants before they have had a chance to form concepts and hence to perceive, and since it can also be aroused by situations that presuppose perception, the present model requires both channels.

For routine performance the valences in the behavioral system are adequate and are even superior in adaptive detail. But for matters of strong concern for survival (which we will not call *emergencies* unless we broaden the term to include sex), massive supplementation may be needed at times. This is the role of emotion adopted for the model.

Approach to positively valenced things is in a sense single-dimensional—to do or to learn what will bring the desired result. Avoidance, however, has a double dimension. In the same sense that we can avoid spoiled food by not eating it or by ejecting it, we can avoid an undesired thing by leaving it alone or by actively getting rid of it. If it happens to be an attacker, as was common when man was evolving, the alternatives are escape or counterattack. Escape in turn could consist of either flight or "freezing," the latter a form of hiding.

Simple avoidance of something that does not attack requires no particular action and hence no emotional assistance. But flight, counterattack, or even freezing may benefit from such assistance. Being attacked usually results in pain, which is associated with fear; and (given learning) expectation of attack or pain would produce the same emotion. If a feared potential attacker is detected in time for successful flight, an emotion that mobilizes for intensive performance will assist survival. If the threat is detected too late for flight—a situation appropriately calling for more intense fear—and if counterattack is not feasible, the best option may be to freeze and perhaps go unnoticed. Despite complications to be mentioned, the above alternatives could be the evolutionary logic for the fact that moderate fear is mobilizing and intense fear paralyzing.

Counterattack can also be assisted by mobilization. But whereas flight is a response to fear of pain, counterattack is more likely to succeed if it temporarily ignores pain. This is the priority that appears in anger. Although both flight and counterattack (aggression) are assisted by emotion, they do not necessarily depend on it; one can do a lot of running and slugging without it.

With this background we can regard the role of emotions as priority controls. Among the three actions successful flight is safest, and to optimize survival it should take temporal priority over attack. Between the corresponding emotions, nonparalyzing fear should take temporal priority over anger. Between paralyzing fear and anger the odds are mixed, but the overt choice is likely to be beyond the control of the subject. Once attack has begun, however, anger should take priority in intensity over fear. In this sense anger would not be a response to fear but a supplementing or displacing response to the same thing that aroused the fear. Anger need not be tied solely to feared things, however, since attack might also be a sensible response to frustration (see 2.41.4) or a necessary means of getting something wanted.

Moderate fear is compatible with patterned, controlled responses; intense fear brings freezing and loss of control. In parallel, moderate anger accompanies controlled attack; rage (intense anger) brings loss of control. In the first case the loss of control produces freezing, which seems appropriate when fear is intense. In the second it brings intense random behavior, which seems appropriate when controlled attack is failing. (See the discussion of reversion to random behavior in connection with frustration in Section 2.41.4.)

To these emotions we will add love and sexual arousal, which differ from the previous ones in that neither is essential to the survival of the individual who experiences them. We will assume that the evolutionary origin of love lies in the necessity that the child be cared for by its parent(s) and that sexual arousal is necessary if the pleasant sensations of genital stimulation are to be carried to their biological conclusion. Although in our society love of mate is considered an important basis for marriage and sexual relations, for the evolutionary criterion it is relevant only to care of the children.

Watson and Morgan (1917) reported that they could find only three innate emotions in young children: love, fear, and rage. If this is true (and contemporary psychologists seem reticent about coming to grips with the role of emotion in motivation), these emotions correspond to the three basic types of overt response suggested above: approach, escape, and attack. It seems safe to add sexual emotion as genetically developed at puberty.

On the basis of the preceding observations, this model posits the inborn emotional ingredients of the selector listed in Table 1, including frustration, which is discussed later. Without trying to be encyclopedic we will nevertheless mention other states construed as emotions to indicate why they are not included here. Excitement is a level of arousal and is unspecific about what induces it or where it will lead. The internal state manifested overtly in weeping may simply be breakdown from overload. It too is unspecific and can be elicited by pain, love, pleasure, anger, frustration, or fear. My suspicion is that the main component of grief is massive frustration, as when a love relation or strongly sought goal is destroyed, with weeping or exhaustion occurring at overload levels. As noted below (2.41.43), frustration itself may be a sort of overload. Laughter seems essentially a relief of mild dissonance or frustration. Our frequent uncertainty whether to laugh or cry is compatible with viewing laughter and grief as responses to different levels of frustration. Other responses to it will be examined shortly.

TABLE 1

Inborn Emotional Ingredients

Valence	Behavior Reinforced	Accompanying Emotional State (If Aroused)
Positive (liking)	Approach	Love
		Sexual passion
	Simple avoidance	None or fear
Negative (disliking)	Withdrawal (flight or hiding)	Fear
	Attack (aggression)	Anger or rage
Negative	Avoidance or removal of dissonance	Frustration

Some so-called emotions are valenced but seem to be mere combinations or modifications of other, more basic ones. Jealousy, for example, appears to be a combination of love, fear, anger, and frustration. Disgust, if it rises to the somatic level, seems to be a partial arousal of nausea. Hate, at least for this model, can be subsumed as intense anger. Some so-called emotional responses are subordinated to others. For example, whether sorrow or joy will be elicited by a death depends on whether the deceased was loved, hated, or feared. Surprise is a discrepancy between expectation and reality; embarrassment is a discrepancy between the reality and what one wanted others to know of him.

Thus whereas certain responses commonly listed as emotions are concerned with state levels, mere configurations of basics, non-specifics, secondary responses, or essentially temporary conditions, by contrast love, fear, and anger augment relatively durable valenced positions (as between persons and groups) and have reasonably pre-dictable consequences. Erotic emotion also attaches a valence to certain relations between individuals. For these reasons we retain in the model the emotions of love, erotic response, fear, and anger. Because circum-stances can be perceived so differently by different persons or by the same person at different times, it seems imperative that emotions be classified by the nature of the internal state rather than by the environ-mental situations that elicit them. The priority we assign to fear is compatible with the fact that fear in various forms (apprehension, anxiety, inadequacy, guilt) is by a large margin the main source of emotional disturbance encountered by psychiatrists. In speaking of apprehension, anxiety, inadequacy, and guilt as fears, we are of course speaking of their emotional, not their conceptual, aspect.

2.41.13   *Effector.* There seem to be no priorities for the effector that are useful for this model. It might be assumed that evolution has in-corporated in us some mechanism for selecting more efficient over less efficient ways of executing behavior. But since efficiency is discussed separately below for all three subsystems, we defer this item.

2.41.2   *Exercise*

The first cross-system motivation concerned priorities. The second concerns exercise. All closed systems are subject to loss of differentiation (1.32). At the biological level this means that the system must continue to receive inputs of relatively structured material while excreting those from which structure has been extracted (Miller, 1965). We will at this point ignore the possibility that the genetically de-termined content of DSE may deteriorate with time (that is, loss differentiation). Instead we will focus on their learned content. The parallel statement here is that learned abilities deteriorate without practice. Even if they do not drop below a certain level of "knack," such as the basic ability to swim or ride a bicycle, some of the finer coordinations disappear. In present language "learned abilities" in-clude conceptual structures and secondary motives as well as motor skills. The rapid loss of orientation among persons experimentally cut off from all externally generated sensations parallels the loss of skill of the concert pianist who does not practice. Having identified loss of information as the underlying theory we will simply say hereafter that all behavioral faculties require regular use and exercise merely to maintain a certain level of competence.

It seems clear that an individual who keeps his faculties in good working order will adapt more successfully than one who does not and that the one who is intrinsically motivated to exercise his faculties will be the more likely to keep them in good order. Because of its survival logic, because of the behavior of many mammals, particularly the play and exploratory activities of the young, and because of the widespread experience that doing and learning can be fun, we will posit a general positive valence for simple use and exercise of behavioral equipment and posit a general negative valence for continued disuse. As a point of reference we will assume that there is an optimal rate of exercise which varies with the individual and his state. We will call exercise of faculties at this optimal level *self-expression* or *self-actualization,* with a positive valence. Operation substantially below or above this level shows a negative valence, with boredom below and frustration above. The positive valence at optimal level and the negative on either side would thus tend to move the organism toward optimal if it deviates in either direction. We will describe the content of these motivations for each subsystem without discussing the question of optimal level for each. The concept of optimal level of exercise seems to accommodate both tension-seeking and tension-reducing activities (to use the psychologist's terms) when the organism is below and above optimal levels, respectively. There is also a certain existentialist flavor to this enjoyment from simple use of the faculties. Moreover the tenor of the overall model is much more Western than Oriental.

2.41.21   *Detector.* In the model the main activities of the detector are pattern formation and pattern recognition (concept-learning and perception), both based on sensations.

a. *Pattern formation* is learning patterns in the environment: exercise of this ability would call for avoidance of an unchanging pattern, approach to moderate change or shift of pattern, and frustration at a dramatically new pattern of unmanageably high information content. Insofar as exploratory and play activities of children are directed toward image-testing (rather than mere muscular effort or skill) this principle would predict manipulative interest in a

moderately new situation, boredom and withdrawal when its
pattern has been reasonably well learned, and frustrated avoidance
of a pattern dramatically different from those already known. Such
a motive could also be described as a desire for new experience or
simply curiosity. For adults, travel is one such exploratory
experience.

b. *Pattern recognition:* Exercise of existing patterns through perception would require periodic exposure to patterns already learned
and could be characterized as a desire to repeat familiar input
experiences.

c. *Sensations:* Since all perception and all formation of concepts about
external reality use sensory nerves, the exercise principle would call
for satisfaction from moderate stimulation of any of the sensory
modalities and dissatisfaction from lack of stimulation or from over-
stimulation. Sensations themselves unpleasant would then have a
positive and a negative component, a matter dealt with in 2.41.22b.

2.41.22   *Selector.* A similar distinction between learning new patterns
and repeating the old occurs here.

a. *Learning new motives:* This reasoning implies affirmative satisfaction from learning to like things that were previously unknown or
of neutral valence. Strictly applied it also means a positive valence
about learning to dislike something previously unknown or neutral.
That positive valence would be offset to some extent, perhaps more
than totally, by the negative valence attached to the thing itself,
and the model provides no prediction of the net balance. Hence we
will assume a net positive valence only for learning to like new
things.

b. *Exercising existing motives, primary or secondary:* We have already
discussed valences in connection with sensations. In addition, since
all emotions have valences, this aspect of the model would call for
affirmative satisfaction in simple arousal of an emotion. If the
emotion is itself positive this satisfaction would add a second
positive element. If it is negative any satisfaction from exercise
would be offset, perhaps overwhelmingly, by the unpleasantness
specific to that emotion. Nevertheless it is possible, at least in
humans, to arouse an emotion, as by story or picture, without
putting the individual in circumstances unpleasant to him. In this
way arousal of anger or fear could tap the pleasantness of exercise
without the negative real-life consequences. Even in real-life involvement the negative direct valence may nevertheless be accompanied by a sense that the emotion itself "felt good," as in a feeling

of catharsis, and modest activation might bring more pleasure of exercise than displeasure of the emotion itself. The same might be true for pain: slight pain may also have a pleasant component (see 2.41.21c). Frustration is discussed in 2.41.4.

2.41.23 *Effector.* The distinction between pattern-learning and pattern-using also appears in the effector.

a. *Learning new skills:* The same logic at the effector would mean intrinsic satisfaction in learning new motor skills or improving existing ones.

b. *Exercising existing skills:* This would mean intrinsic satisfaction in repeating things that one can already do well.

To perform the pattern of a well-learned skill means to execute overtly a pattern already developed in the brain—that is, to reproduce in one's environment a pattern that conforms to an internal pattern. We have posited that such overt reproduction of a motor pattern is positively motivated. But having stated it with respect to motor patterns the logic can be generalized. If you have a pattern of a vase or a landscape in your head, according to this logic you would receive intrinsic satisfaction from reproducing it externally by forming a vase of clay or a landscape of paint. Patterns not amenable to tangible reproduction might be externalized in words and produce intrinsic satisfaction in the mere talking. Similarly, internal patterns of likes and dislikes might be executed externally as actions that reflect those preferences, words that state them, or drawings that symbolize them.

These externalizations can also assist image-testing (2.11). To reproduce a pattern externally and then observe the externalization is one way to test that pattern. In fact, if we focus on neurological patterns there is an important way in which perception is to the formation of information concepts as overt behavior is to the formation of motor concepts. Each can also assist the other, as when your perception of your overt performance helps to improve your motor concepts and the overt reproduction of your information concepts helps to improve *them.* The execution of motor skills is image-testing of motor concepts since the actual performance mirrors the concept.

2.41.24 *Summary on exercise.* It does not seem to do serious violence to the customary meanings of self-expression, self-actualization, fulfillment, and related ideas to use a model in which we posit affirmative satisfaction to human beings, and probably to many other animals, in simply exercising one's behavioral faculties by learning new patterns and exercising old ones in the detector, selector, and effector. Self-expression in a narrower sense might refer solely to the effector process of externalizing inner patterns, but the model is not thus narrowed.

2.41.3 *Efficiency*

The third cross-system motivation involves efficiency. Other things equal, an organism that processes matter and information efficiently is more likely to adapt successfully than one that processes them inefficiently. The evolutionary process would therefore seem likely to selectively retain mechanisms for relatively efficient operations. The following paragraphs indicate briefly what some of these mechanisms may be.

2.41.31 *Detector.* It has long been known to artists and more recently to psychologists that the eye focuses first on high-information centers in a picture. (Saccadic movements may play an important part in this process.) These centers are intersections of lines or planes, junctures of contrasting colors, or textured surfaces as contrasted to smooth ones. The ear similarly attends to sharp changes of pitch or intensity. In addition, except for certain sensations of strong valence the system "adapts" so quickly to a sensation that one can almost say there is no sensation except a changing one. Since change of sensation contains more information than continuance, adaptation is a technique of ignoring low-information sensations. Other aspects of efficiency in the detector are discussed below in connection with motor-concept formation in the effector.

2.41.32 *Selector.* An important element of efficiency is found at the selector in the so-called expansion and contraction of aspirations. This means that we tend to want what we consider possible to achieve and not to want what we consider impossible. We are speaking, of course, of wants that select actual behavior, not of fantasies. This reaction can be construed simply as a special case of conditioning (Kuhn, 1963, p. 135). It makes efficient use of capacities on the one hand by preventing wasted effort on unachievable goals, since they cease to be wanted, and on the other hand by preventing capacities from going unutilized, as would occur if capacity outstripped desire.

2.41.33 *Effector.* Most complex actions consist initially of a sequence or combination of discrete subactions. Take, for example, the unskilled pianist who strikes a chord by three successive motions: the arm is moved, the fingers are spaced, the chord is struck. At an intermediate stage he positions his arm and spaces his fingers at the same time and then strikes. At an advanced stage all three motions blend into one smooth action. The fully learned motion may be thought of as a hierarchical superpattern consisting of three subpatterns, and efficiency arises in a saving of both time and energy.

2.41.34   *Summary on Efficiency.* With only moderately less confidence than I would assume that primary motives must be geared toward survival, so would I assume that for efficiency complex structures of either real or pattern systems are built hierarchically—again for reasons given by Simon (1965). Although there is sound reason why the experimental psychologist should refrain from reporting more than he observes, theoretical psychologists might pay more attention to the possibility that hierarchy is the basic organizing principle of complex learning, whether of motor or conceptual patterns. In this view, as the rat gradually converts its jerky run through the maze into a coordinated smooth performance the hierarchical description is that a sequence of separate small patterns of movement become coordinated into a single higher-level pattern. As a subintermediate piano player I have for years been watching my learning processes: I find the hierarchical structuring almost as tangible as the keys in front of me. In learning a composition finger movements become coordinated into short sequences that behave like a single movement. These become coordinated into still larger groups, and so on, until a whole composition becomes a single unit. In sight reading, individual notes gradually blend into patterns that become grouped into larger patterns. Facility increases as more and more patterns are learned and as new music is increasingly treated as new intersections of already stored patterns.

We have already mentioned the human capacity for making abstractions, which are higher-level patterns in a hierarchy. For motor performance this ability means that after you have learned a certain number of skills you may abstract from those experiences some common elements of learning, which we call learning skills or strategies. If you use these skills you will be able to learn considerably faster than before. A similar process also takes place in the detector, so that as you form more and more concepts you gradually develop concepts about the process of learning concepts. Important are methods for isolating similarities and differences, which when systematized are the essence of the scientific method. There also seems to be a preference for the simpler of two ways of grouping things, which is formalized as the rule of parsimony.

2.41.4   *Frustration*

The fourth cross-system motivation is that of frustration. Fundamental to the present model is the notion that human learning is the formation of patterns inside the behavioral system that correspond in some ways to patterns of inputs from the environment. Often, however, there is a mismatch of patterns. Such a mismatch is a way of describing (in this model) the responses that psychologists call frustration, conflict, or dissonance. Although a mismatch will be assumed to be of negative valence in itself, if reality turns out to be better than the expectation the ensuing positive feeling may overshadow the intrinsic negative one. We will call any such pattern mismatch *dissonance*. There is undoubtedly substantial coincidence between dissonance and uncertainty, though we need not debate whether the coincidence is complete.

We will first illustrate the problem and its evolutionary rationale in general terms and then turn to details. Some overt behavior might fail on the first attempt for fortuitous reasons. But it will fail repeatedly only if there is a lack of correspondence between the real pattern in the external environment and the learned pattern that represents it in the organism. In that case successful adaptation requires change. Modification of the internal pattern and the ensuing behavior is one such change, and it might work if the existing pattern is faulty in only minor detail. To abandon the effort entirely sometimes makes sense since there is no point in repeating a behavior indefinitely after it has been "proved" unworkable. If in frustration one abandons an effort a hair's breadth short of success Omniscience may deem it "irrational," but even sophisticated decision theory could make the same mistake (2.55.3). If patterned behavior fails repeatedly there is also sound logic in reverting to vigorous random behavior, such as banging the radio or kicking the door. Random behavior, after all, is the source from which all ordered behavior arose, and if an ordered pattern does not work there is sense in reverting to increased randomness as the possible source of a better pattern. Furthermore, because of the unpleasant valence that goes with it, frustration speeds the extinction of unsuccessful behavior by adding intensity to a conditioning situation otherwise characterized only by frequency. The valence, behavior, and emotion have already been shown (Table 1). With this background we will now trace dissonance through the three subsystems.

2.41.41   *Detector*. Regarding the relation of pattern formation to frustration, concept-learning is the process by which patterns are formed in the head in correspondence with patterns of information inputs. The result is a match of patterns—the formed one with the input. If the inner pattern is formed quickly no emotion is aroused. But continued failure brings the sense of restlessness or unease we call frustration. Subsequent satisfactory completion of the inner pattern relieves the frustration and brings the positive feeling we know as "Eureka!" Although these illustrations deal with simple sensory inputs the essential problem is the same for verbal or graphic inputs, as with the student struggling to grasp the meaning of diminishing marginal utility.

Regarding the relation of pattern-using or perception to frustration, dissonance can occur in perception when the cues are insufficient to identify their source or when the pattern to be identified is not yet fully learned. Dissonance is verbalized as: "I can't for the life of me figure out what that thing is." In the information model scanning processes have begun. But because no identification can be made, they are not terminated and search goes on indefinitely.

At the detector a negative valence to dissonance motivates the formation of concepts that show some correspondence with environment—in obviously adaptive behavior. It also motivates identification of incoming sensations, and this also is adaptive. When sufficient concepts have been formed to cope with the environment, however, dissonance occurs less often and the motive to form additional concepts declines.

I have indicated (2.11) that every perception is in some degree false and that by implication there could be no such thing as completely accurate perception unless a separate concept were formed for every exposure to the environment, and that almost instantaneously. No data could be discarded and no simplification of information-processing could be achieved by coding. Even the incredible capacity of the brain would shortly be overloaded, and language would not be possible since it depends on coded information (2.10b and 3.10.0g). Hence individuals (or societies) will at some point tend to stop adding new concepts and get along with what they have, that quantity constitutes an equilibrium as detailed in 3.18.61. It means in turn that in perceptually identifying a concept from cues some portion of the cues is inconsistent with that concept, which is another way of describing (cognitive) dissonance.

This dissonance can be removed or avoided by ignoring the

discrepancy or by modifying the code. The former is almost certainly easier and is presumably preferred unless the existing concepts are already unsatisfactory on other grounds—as when overt behavior is frustrated by a stock of concepts too small to provide the necessary discriminations. For example, a person with a low tolerance for ambiguity keeps his code list small and forces all information into it by ignoring discrepancies. A person at the opposite extreme would not accept any input as completely consonant with any existing pattern and hence would not give a stereotyped or "standard" response to anything. For every individual and for every problem there is presumably an optimal equilibrium level between generalizing and discriminating, depending on his history and his needs. A parallel problem will later be dealt with for a whole society in connection with language (3.18.6–.7).[1]

2.41.42 *Selector.* The selector is concerned with valences. The problem of frustration in the selector can be approached through a preference scale as in Figure 7. The right of the scale represents a strong positive valence toward something $(+)$, the left a strong negative valence $(-)$, and the middle indifference (0). The six lettered points along the scale represent various intensities of like or dislike for each of six different things. The problem is that of choosing between two items in the list.

If the choice is either between an A and a B or C or between a B and a C, no dissonance or frustration is involved. If a choice must be made between A_2 and C_1, for example, the internal patterns call for approach to the former and avoidance of the latter. By selecting alternative A_2 and rejecting C_1 the pattern of behavior is consonant with the valences attached to each. Similarly, in a choice between A_2 and B_1 the preference patterns call for approaching the former and ignoring the latter; thus to select A_2 is consistent with both patterns. The choice is also simple if it lies between B_1 and B_2 since both are indifferent. Because it does not matter whether either is approached or avoided, the selection of one and rejection of the other is not dissonant with the valence attached to either. But the choice tends to produce frustration

---

[1] As noted, this model assumes that overt behavior and perception are both based on concepts—motor and information, respectively. The psychologically oriented reader will recognize that certain experimental conclusions about generalization and discrimination of overt responses are being applied here to perceptual processes, on the assumption that the basic neural processes are the same for both.

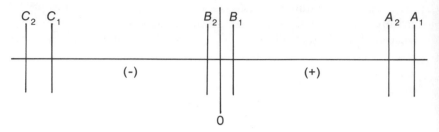

Figure 7. Preference scale.

when it lies between two strong approach responses (A_1 and A_2) or two strong avoidance responses (C_1 and C_2).

The problem is often viewed as one of conflict, and my purpose here is to show why it is also a matter of dissonance, pattern mismatch, and possible frustration. To illustrate, A_1 elicits a strong approach response. But so does A_2. Hence to choose A_1 means to avoid A_2, which is dissonant with the pattern that calls for approach to A_2. Similarly, if the situation requires a choice between C_1 and C_2, to reject the latter is consonant with the pattern of avoidance attached to it; but to accept C_1 is dissonant with the pattern of avoidance attached to *it*.

We have already examined several responses to frustration (2.41.4). One is to change a pattern to or toward consistency. In this case if A_2 is avoided in favor of A_1 the inner pattern is made more consistent with the overt choice if A_2 is pushed farther down the preference scale and A_1 is pushed up. Both kinds of behavior seem to be found empirically (Cofer and Appley, 1964, pp. 790f, from various works of Festinger). Another response to frustration is to abandon the effort. More explicitly, the difficult decision produces dissonance, dissonance is unpleasant, and unpleasant things are avoided. Avoidance may take the form of postponing or permanently avoiding the decision. Not to avoid an unpleasant decision means, in this framework, that it is a subset of a larget set within which it would be unsatisfactory to avoid the decision. In terms of habit, one can acquire a conditioned avoidance response to avoiding decisions.

2.41.43   *Effector.* Since the "will" to pursue certain behavior is a matter for detector and selector, dissonance in the effector can arise only in executing an action, as from physical blockage, inadequate strength or skill, absence of tools or materials, bad weather, and the like. How or why does such obstruction constitute dissonance? We are still assuming that the selection of behavior is based on information patterns that represent the environment, not on the environment itself. Similarly, the effectuation of behavior is the outward manifestation of a motor concept (as in the effector motions of throwing a ball), guided in an overall way by perceptions and motives (as in throwing to second base to get a player out). Once the motor pattern is released the whole sequence will follow, just as a perceptual cue in the detector will release a whole stored conceptual pattern.

At least in organisms as complex as humans the motor portion of the pattern is part of a larger information pattern that includes the feedback of its consequences—thus the throwing action and the ball's subsequent movement through space are both subconcepts of a larger one. To distinguish it from the action stage we call the feedback stage of the concept an *expectation*. Performance frustration is then seen as dissonance between the action (or attempted action) stage and the expectation stage of a pattern. Since dissonance is motivating, so is interruption of behavior already initiated. Later resumption of an interrupted task removes the dissonance, and actions that facilitate such resumption will, other things equal, be reinforced. Intrinsic motivation to "exercise" overt expression of patterns (2.41.23b) would add to this effect.

Frustration can also be usefully viewed as overload. Any mismatch of pattern means that the system is attempting to process patterns that are in some way beyond its capacity at that moment. Whether at detector, selector, or effector, once the system has learned to handle a pattern easily, only substantially different or more complex patterns will overload it and produce dissonance. There is no need to detail other forms of dissonance, since our purpose is merely to explain why conflict and dissonance can be viewed as pattern mismatch within DSE functions.

2.42   *Methodological Note*

In the language of J. G. Miller (1965) and elaborated by Berrien (1968, p. 25) there is a distinction between signal and maintenance inputs to a system. To express this model in that language, the hedonic aspect is parallel to the signal (channel-selection) aspect while the activation (energy) aspect is parallel to the maintenance. Two apparent discrepancies nevertheless occur here.

One is that in certain respects information is related to energy, as noted in Berrien (1968, p. 24). In those respects energy appears on the information side of the information-matter dichotomy whereas here it is joined with matter—as matter-energy. For the present purpose, I see no point in altering the information/matter-energy division. In any case, that division seems acceptable to Berrien when he later says (p. 26): "The primary source of action of a living unit arises from an internal energy system having as inputs only nutrients that contribute to, and organize, metabolism." And he explicitly identifies information with signals (p. 80).

The second discrepancy concerns the apparent multiple use of the idea of selection and the question of whether it really is multiple. As used here, *selection* is the function of the selector, operating on the basis of goals or preferences. Yet we have also referred to channel selection in the detector as a basic aspect of information-processing in the decoding stage of receipt of information. This problem may be resolvable if selection is viewed as pattern-matching. It can be handled unquestionably as a pure information problem in the computer recognition of patterns, as in the mechanized sorting of bank checks. Pattern-matching occurs in a simpler form in the McBee card-sort system. The screening of gravel by size of stone is a still simpler form.

But we have also dealt with intrasystem goal conflict as pattern mismatch. By implication, to select a preferred response without conflict constitutes a match of patterns. To cross the two halves of this meaning of *selection* let us drop to the level of primary motives and reflexes, as in the withdrawal of the hand from a hot stove. This reflex can be dealt with in theory as a two-neuron arc, even though it is more complex (Miller, Galanter, and Pribram, 1960, pp. 25, 29f). But if the two neurons have a fixed inborn connection they are logically the same as a single neuron connecting stimulus and response. If the inborn connection alone could achieve the desired result the organism would need no feeling or valence, as is the case with insects. The valence must therefore be construed as a device for facilitating learning, so that the hot stove can be avoided *before* it is touched, in an obviously superior adaptation. Thus in the simple reflex there is no distinction between informational and motivational "selection." To illustrate further, if heat-sensitive nerves are connected directly to muscle-contracting nerves that withdraw the affected part, the heat will be withdrawn from. But if they were connected instead to muscle-extending nerves, the reflexive response would bring closer contact. We can thus say that the mere fact that one channel is sensitive to one kind of input and a different channel to another is a "selection" in the same primitive sense that circular chips slide into circular depressions and triangular chips slide into triangular ones. At that level channel selection in the information sense coincides with behavior selection in the motivational sense.

In this model all primary motives are of this primitive, dual type, although they are supplemented by (a) connections of all inputs to information-processing centers; by (b) inputs, mainly from sight and hearing, that lead to information-processing centers only and not to valence centers or muscles; and by (c) connections of certain inputs to valence centers but not directly to muscles.

It is perhaps fruitless at this stage of knowledge even to speculate about how an information concept can become related to certain afferent nerves to produce an identification of the environment (as in perception), how an information concept can become related to a valence to produce a secondary motive, or how a motor concept can become related to certain efferent nerves to provide a complex muscular response. Nothing in the logic precludes motor concepts connected to valences, which would consist of overt behaviors that were intrinsically rewarding over and above satisfaction from their mere exercise. In its way each such relation can apparently be thought of as a channel selection. In the broad sense this means that channel selection occurs at detector, selector, and effector. It also seems to mean that there is only one basic kind of selection—channel selection via pattern-matching. (In this connection see Vickers, 1973.)

In this context it seems that the difference between informational selections and valence selections is the emergent one of level —the selection of a selection, or metaselection, instead of selection of an action. In any event, it makes a great deal of difference (in this model) whether the selected channel is connected to an environment identification, a valence, or a muscle. To keep the language straight, unless otherwise indicated *selection* refers solely to the function of the selector, with its basis in values and preferences.

2.43   *DSE: Mutual Modifications*

The opening of this chapter (2.4) notes how DSE systems may utilize one another. And in connection with cross-system motivations (Section 2.41) certain selector aspects of both pattern-learning and pattern-using were traced in all three systems. However, neither heading dealt with the way the content of one system may modify the content of another: the subject of this section. A quick scanning of each of three systems modifying each of the other two—six modifications—follows.

2.43.1   *Detector Modifies Selector.* Perhaps the most obvious effect here is that we tend to learn to like what we perceive to be possible and not like what we perceive to be unattainable—also called the expansion and contraction of aspirations, or the sour grapes principle. The effect is not universal, and it occurs more regularly for motives that effectually actuate behavior than for those that activate fantasy. The detector (or subdetector of the selector) also conceptualizes valenced sensations and other motives, after which they can be manipulated like concepts of externals: "Since I enjoy swimming I had better live where there is a pool." In the model of formal decisions (Section 2.5) goals are handled as data in a logical process.

Since secondary motives are conditioned connections between valences and concepts, every time a concept is altered a change in motive potentially follows. If you enjoy painting in thick, textured oils and then learn that acrylics can similarly be piled up on canvas, the scope of what you like has expanded with your concept of painting materials. The ability to make conditioned connections between thoughts (presumably related to higher-order conditioning) similarly allows us to attach valences to conceptualized instrumentalities. To the strict Puritan, pleasant sensations acquired a conditioned unpleasant component because they were instrumental to damnation. To endure pain can acquire a pleasurable component if it is seen as instrumental to receiving praise or relieving guilt. Even getting up early may acquire a positive valence when conceptualized as instrumental to going fishing.

2.43.2   *Detector Modifies Effector.* Until most motor skills are learned, the muscles must be consciously guided in some degree by detector processes: how to hold the tennis racquet or chisel, how to lean forward on the skis. Actual performance of most skills must also be guided by the detector: one must see where the ball is to hit it. The detector also provides the feedback necessary to identify an action as success or failure and thereby to guide the next steps in the effector.

2.43.3   *Selector Modifies Detector.* This is the obvious case of seeing what we want to see and believing what we want to believe. Carried to extremes it is the loss of contact with reality which we call psychosis. In part it may be viewed as simple application of operant conditioning to thoughts (we repeat pleasant thoughts and discontinue unpleasant ones) and to perceptions (we tend to decode sensations by selecting pleasant rather than unpleasant channels). Section 3.41 gives considerable attention to such modification of the detector in social interactions.

2.43.4   *Selector Modifies Effector.* Here we may be guided to learn those skills that produce pleasant sensations (to swim is cooling in

summer) or to avoid learning or performing those that produce unpleasant sensations (the ski boots hurt the ankles while swirling snow
bites the face). There are obviously many more complications, but
these illustrate the point.

2.43.5   *Effector Modifies Detector.* In this connection what we learn
often depends on where our muscles and skills can take us. We cannot
learn the feel of sitting in a swaying tree top if we cannot climb or of
riding the breakers if we cannot swim. Our knowledge of machines,
materials, and processes remains sketchy if we do not work with them.

2.43.6   *Effector Modifies Selector.* This modification overlaps that
of 2.41.23, that we often come to like that which we do well. Since
secondary motives are concepts with valences attached, to the extent
that the effector determines what concepts we may learn it potentially
affects the ones to which valences can be attached.

2.43.7   *Comments on Mutual Modifications.* In a system model it
must be assumed that if the states of the three subsystems are not
changed independently from the environment (the system is closed in
this respect), an equilibrium among them is possible and probable.
However, given the difficulty of creating even a simple mathematically
determinate system with three mutually interacting variables, we should
assume that this equilibrium position is in theory not predictable.

   Of the six mutual modifications, that of detector by selector
and the reverse are clearly the most interesting—perhaps for the same
reasons that the effector itself is relatively uninteresting for social
science. In this connection some comments on objectivity are in order.
In this language one way to describe a reporter's or scholar's non-
objectivity would be to say that his detector has been distorted by his
selector. It is at least theoretically possible to eliminate such bias, particularly in simple cases, and to care only whether an answer is correct,
not whether it is pleasing. I do not really care whether I must turn
left or right at the intersection, only whether the information is correct.
When it works well the community of scholars minimizes this bias by
rewarding conclusions based on evidence and punishing those that reflect the researcher's preferences. However rare in fact, objectivity in
the sense of a detector not modified by a selector within the boundaries
of a particular investigation is a conceivable condition.

   By contrast, one cannot even imagine complete objectivity
at the detector. Any mechanism that processes coded information can
handle it only through the codes already in the system. If I hold a
Ptolemaic concept of the cosmos, I cannot describe heavenly move-

ments in a Copernican framework. The Freudian may explain behavior by speaking of id, ego, and superego; the system analyst in terms of DSE; while the fundamentalist talks of God and the Devil striving for mastery. Thus while one may conceivably escape the influence of his motives on his perceptions, he cannot escape his conceptual set. To the large extent that his conceptual set is mentally manipulated in language, he also cannot escape from language. Although one's code may change with time, at any one moment he is stuck with it.

2.44   SUMMARY OF MODEL

The discussion of the model thus far has been fairly detailed, first to tie it to its evolutionary background and second to tie certain materials from psychology into the framework of systems and information. Although more complex than economic man, this model of social man is considerably simpler than psychological man and can be summarized fairly briefly as follows.

Man is a controlled system. His behavior can be understood (to the extent it can be understood at all) only by knowing the processes and content of his detector, selector, and effector and possibly *their* respective subsystems. His behavioral system is defined to include only the DSE processes that determine and control behavior, and his biological system is defined as part of its environment.

All three DSE subsystems build complex, hierarchical learned patterns on a base of relatively limited inborn patterns. The main process is conditioning, which the human is able to carry to very high orders. This high-order conditioning means that a learned pattern can be indirectly connected to an inborn pattern many steps removed and that the human can form patterns of patterns of patterns . . . to indefinitely high levels of abstraction. All three DSE subsystems show a fundamental similarity: they form patterns and they use them.

Except for reflexes and certain inborn connections of particular sensations to their environmental releasers, man's behavior is not genetically determined; man does not inherit behavior but a behavior-selecting system or metaselector. Geared to survival only in an average, probabilistic sense, the behavior it selects may be thoroughly antisurvival in certain cases and indifferent to it much of the time.

Man's behavior is initially random with respect to environment and becomes increasingly ordered to it through learning. Substantial randomness always remains. One can hardly imagine what "fully ordered" would mean, particularly since there is no sense in which the environment itself can be said to be fully ordered.

The unlearned base of the motivational system consists of

positive and negative valences attached by inheritance to relatively specific sensory inputs. Their content and relation to environment is not significantly different from the parallel inborn motives of the dog or ape. The vast difference between human motives and those of other animals lies in man's ability to form a complex conceptual structure and to make conditioned connections between those concepts and his inborn valences. The likelihood of achieving those complex levels is assisted by independent motivation within each DSE system with regard to exercise, efficiency, and frustration and with respect to both forming patterns and using them. Frustration can lead to an emotional reaction as well as to a simple valenced one. To it we add the emotions of love, fear, and anger, which produce approach, avoidance, and attack responses, respectively, toward other persons.

Despite its relative simplicity, this model seems sufficient for the purpose of building a deductive social science of the sort constructed in later chapters. Although not all details of this model are used explicitly, those that are included seemed necessary for an understanding of the model itself. Importantly, the few basic ingredients of the model, through combinations and interactions, can conceptually accommodate a varied and rich description of human behavior. For later purposes we note that both knowledge (2.10c) and motivation (Section 2.2, especially 2.22) are states of human nervous systems and can never reside in any higher-level collective social entity. And even though this model is naive as psychology, it might suggest how psychologists—purely for their own purposes—may nevertheless find it useful to have a succinct summary of the human behavioral system as a point of departure for understanding its details.

Chapter 6

2.5

Decisions

Why discuss decisions in a book on social science? The reasons are several, and to my mind compelling. Social sciences deal with interactions of people (among other things). But an interaction involves an action by each of the actors, each such action being a selection of one action rather than another by each of the actors. Often these selections are routine. But in the many cases where they are not they classify as decisions. Furthermore, this volume leans on psychology for its model of social man, and those aspects of psychology that deal with interactions are often considered social science. Psychology deals with the relation between stimulus inputs and behavioral outputs. At a different level that relation is also the central concern of decision theory, which is thus a bridge between simple stimulus-response analysis and the response selections made in complex social situations. In psychology, decisions arise in dissonant situations and in a "field" of multiple approach and avoidance elements.

More specifically, economics, sociology, and political science each incorporates important decisional elements, even if not always called that. In economics the theory of the firm is a special case of the general theory of decisions, even if historically the latter came by extension of the former. Viewed from a present perspective, decision theory is thus a direct basis for an important area within economics. Even outside economics, however, many propositions and terms from decision theory are needed in later chapters of this volume. Furthermore, although much behavior is determined by norms, they are not all-encompassing. Even people in highly traditional societies must often make decisions, sometimes because norms cannot be detailed enough to solve all problems and sometimes because they conflict. Sociologists often refer to role conflict, which in the present approach is an occasion for a decision. Every society also has deviance as well as conformity, which means that some individuals have selected nonconforming behavior. The term *adaptive change* often used by sociologists implies that at some point a new practice has been decided to be more advantageous than an old one. Buckley (1967, pp. 39, 79) explicitly argues the relevance of decision theory to sociologists when they study organizations, and the present approach closely coincides with that of Blau (1964, pp. 18ff, 220). And, finally, political science pays much attention to decisions in government and other organizations. For all these reasons decisions are relevant in some form to these social science disciplines. The relative roles of normative and decided behavior are discussed in more detail in Chapter Sixteen.

Preceding chapters examine how a person comes to acquire information, motives, and skills, but not *what* he acquires, a matter determined by culture and experience. Similarly, our study of decisions will assume that the person already has certain content of his detector, selector, and effector; we need not inquire here what it is or where he got it—a matter for later chapters.

This chapter deals with the relative strength of various motives as a basis for determining which alternative will be selected, a matter of *direction*. Near the end of the chapter we focus on the strength or magnitude of the effect produced when a selected alternative is executed, a question of *power*. Since decisions are concerned with behavioral selections of a complexity only man is capable of, all subsequent discussion deals with human behavior.

Definitions and propositions about decisions in this chapter apply whether the decision-maker is an individual, a group, a corporation, or other entity so long as it acts as a unit (see the definition of *party* in 3.10a). To simplify we will speak of the decision-making entity as a person. The special problems of decision-making within a group are examined in Chapter Twelve.

2.50   DEFINITIONS ABOUT DECISIONS

a. *Response selection* is the process of determining which of two or more alternative behavioral responses, including inaction, the system will perform.

b. *Simple response selection* is any response selection that does not qualify as a decision.

c. *A decision* is response selection under conditions of complexity. For this purpose *complexity* means uncertainty about detector, selector, and effector to such an extent that a response cannot be selected on the basis of a single scanning of the three subsystem states. The state of at least one subsystem must initially be uncertain about two or more alternatives, each of which must then be compared to the state(s) of at least one other subsystem. A decision may also be referred to as *complex response selection.* In the language of the preceding chapter uncertainty may be described internally as pattern mismatch (Section 2.41.4), which is a form of dissonance. If the mismatch is in the selector, a decision is an outgrowth of conflict.

d. *Decision theory* is the study of the decision process in humans with respect to the conscious content of all three subsystems, including consciousness of uncertainty. This definition does not imply that real decision-makers are conscious of the state of their subsystems. It means merely that decision theory ignores aspects that are not assumed to be consciously formulated.

e. A *problem* is any situation perceived as requiring a decision.

f. *Intuition (intuitive response selection)* is the use of information in the brain that is not consciously formulated or the use of unconsciously formulated connections between pieces of information, whether or not those pieces are themselves conscious. (The Freudian term would be *preconscious* rather than unconscious). Intuition involves main-level DSE rather than the subdetectors tentatively identified with consciousness (2.06). Intuition might be described as the tentative connection of existing concepts into a higher-level one (2.10f).

g. *Satisfaction* is the positive-valence state of the selector from achieving a positively valenced result or avoiding or decreasing a negative one.

h. *Utility* is the satisfaction that can be produced by an external when acquired or achieved; also, the ability of an external to create satisfaction.

i. A *good* is any external perceived as having utility. A good can also be referred to as a goal, goal object, valued object or event, or wanted thing. Although the term may be noxious to some behavioral scientists, we will use *good* consistently to avoid the possibility that a change in terms may incorporate an unintended change of meaning and analysis. A good is an external that generates an approach response; without a qualifying adjective, *good* is always positive. For some purposes, particularly regarding "coercion," we use the concept of a negative good (a *bad*), which has disutility and elicits an avoidance response.

j. A *preference ordering* is a list of two or more things arranged in rank order of the strength of the valences attached to them, with the strongest positive valence highest.

k. *To prefer* X over Y is the same as saying that X is higher than Y in a preference ordering.

l. *Advantageous:* The more advantageous of two things is the one that provides the greater satisfaction. Given rational behavior (see *u* below) the more advantageous coincides with the preferred.

m. *Value* is the position of anything in a preference ordering.

n. *Cost* is the satisfaction denied in the course of achieving other satisfaction; the satisfaction denied by a course of action in order to receive its benefits. It is sometimes feasible to define costs wholly externally as the goods sacrificed in order to have other goods. We then distinguish (1) *opportunity costs,* which are the dissatisfactions of having to avoid, give up, or do without what we would like to approach, keep, or acquire, from (2) *disutility costs,* which are the dissatisfactions of having to approach or accept what we want to avoid. Psychologically costs are a form of conflict or dissonance—see *conflict* below. Opportunity costs can be subdivided into those involving a relation of externals to externals and those involving a relation of externals to the decision-maker. For convenience these relations are diagramed and explained in Table 2.

Incompatibility is illustrated by the fact that one cannot be married and single at the same time, have a reputation for integrity if he is repeatedly caught lying and stealing, or get from New York to San Francisco in one day without leaving the ground. For the individual decision-maker scarcity means that both of two things are possible by nature, but that he lacks the time, money, energy, or other resources to do or acquire both. Economists apparently believe the only reason you cannot have everything you want is that things are scarce; psychologists seem to believe it is because they are incompatible. The difference probably reflects the economist's focus on the totality of an economic system and the psychologist's focus on the individual. Even at the level of the whole system, however, many economists recognize an incompatibility between zero unemployment and a stable or declining price level. For the present purpose we return to focus on the individual decision-maker.

TABLE 2

Costs

	Opportunity (avoid what we want to approach)	Disutility (approach what we want to avoid)
External-external (relation of one environmental element to another)	Incompatibility: X and Y are not both possible, at least at the existing stage of technology. To have one precludes the possibility of also having the other.	Note: Since disutility is an internal state, there is no external-external category under this heading.
External-internal (relation of environmental element to decision-maker)	Scarcity: X and Y are both possible, but the decision-maker lacks sufficient power to acquire both.	Unpleasantness: Y cannot be acquired without the acceptance of some unpleasantness: the toothache cannot be stopped without having the tooth pulled.

Decision costs and benefits are the costs and benefits entailed in the process of making the decision, without regard to the cost or benefit of any alternative. The costs may take such forms as time, effort, money, frustration, tension, and worry; the benefits may include the pleasure of choosing or of exercising decision skills, display of competence, pride in sound judgment, or satisfaction like that of solving a puzzle. Because they are sometimes highly personal, decision costs and benefits may be hard to identify. But if they can be identified they can also be classified. For example, to someone who takes pride in deciding impulsively—a deliberate pose of some people—a careful decision has the additional cost of undercutting that pose. For reasons to be seen in 2.51.31, decision costs can also be defined as the cost of ascertaining the final net benefit of the best alternative or at least of ascertaining that it is positive. Net decision costs are decision costs minus benefits that may be derived from the process. There is no need to use the term net decision benefits (see item o below), which would be the same as negative net decision costs.

o. Benefits are the goods or satisfactions available from one alternative in a response selection. Whereas goods and satisfactions are general terms, benefits apply to a specific set of alternatives. The gross benefit of Y is the total satisfaction to be derived from Y without regard to the cost of acquiring it. The net benefit of Y is, in a choice between X and Y, the gross benefits of Y minus the non-X costs of acquiring Y. The non-X costs of Y are those costs of acquiring Y other than that of foregoing X. The *final net benefit of Y* is the net benefits of Y minus the net benefits of X.

p. *Cooperation (cooperative relation):* X and Y stand in a cooperative relation when the same good produces both. To swim, for example, may both cleanse and cool.

q. *Conflict* is the presence of costs as viewed before a decision is made. If giving up X is the cost of receiving Y, then the desires to have both X and Y are in conflict. Once the decision is made and X is given up for Y the cost is incurred and the conflict resolved. In general the presence of conflict coincides with the presence of a problem, which is also the occasion for a decision. There does not seem to be any point, however, in debating whether this relation always or necessarily holds. Although the definitions are independent of party, conflict and costs can be intrapersonal or interpersonal. In intrapersonal conflict a party must give up X so he can have Y; in interpersonal conflict one party must give up X so the other can have Y. In these definitions cooperation and conflict are relations between desires or valued things, not between persons. Thus the *goals* of two persons can be in conflict, but not the persons. Similarly, their goals may stand in a cooperative relation, but given this definition we will never say that the persons cooperate. The reasons for this usage are discussed in Chapter Seventeen. Within the individual, conflict is an instance of dissonance.

r. *Tension* is a state of conflict, and hence of dissonance, that has given rise to negatively valenced feelings.

s. *Preference function* is the preference ordering of the decision-maker with respect to the alternatives he perceives to be available in a decision situation. It is the state of the selector regarding those alternatives. Although one might have preferences regarding actual but unperceived alternatives, these can have no part in the decision and are therefore excluded from the analysis. Preference orderings can be either gross or net, following the definitions in item **o** above. There cannot be an ordering of final net preference, for reasons that will be seen in the discussion of the opportunity function.

t. *Opportunity function* is the set of alternatives perceived to be available in a decision situation. Although not all alternatives may be perceived, unperceived ones are not really available and do not enter the decision. When necessary, as with *search,* we will distinguish the actual from the perceived opportunities. This and

⌒ the preceding term are from Boulding (1958). When discussing decisions we will assume that the detector takes into consideration both the objective state of affairs and what the effector is able to execute, and we will therefore join detector and effector processes under the heading of opportunity function. Thus the opportunity function states what is possible and the preference function states what is preferred. The decision results from the relationship of the two. Since this chapter is concerned primarily with decision theory, which was defined as involving the conscious content of DSE, we can simplify somewhat if we assume that we are concerned with the *images* of the opportunity and preference functions without questioning whether the images are in accord with reality.

u. *Rationality (rational behavior)* is the process of actually selecting the preferred alternative in an opportunity function. The precise meaning of rationality is amplified below with propositions about final net benefit and maximizing behavior (2.51.21 and 2.6). *Subjective rationality* is the selection of the alternative that is rational under the dominant motive or impulse at the moment of decision, in light of whatever detector and selector functions then prevail. Thus there can be no such thing as subjectively irrational behavior except as judged by other persons, as judged by the same person when his detector and selector states have changed, or when strong emotion prevents execution of behavior that is rationally selected (see 2.6). *Objective rationality* is a selection that is rational under explicit assumptions about opportunity and preference functions, consciously related. In light of these definitions, *decision theory* can be redefined as the study of objective rationality.

v. *Sunk costs and benefits* are costs and benefits that cannot be changed by the decision at hand. In practical terms they are costs and benefits resulting from past decisions that must be accepted as given and unalterable regardless of which alternative is now selected. We will assume arbitrarily that any subjective pleasure or displeasure in the past is forever unchangeable, and hence sunk, even if we later revise our estimates of how good or bad it was. The question of whether some past development can be affected by a decision is therefore relevant only to externals, as in 2.51.41.

w. *Variable cost* is any cost or portion of cost currently incurred to produce current benefit.

x. *Investment* is any cost or portion of cost currently incurred not for current benefit but to raise the ratio of benefits to costs at some future time. Making an investment in an act of delayed gratification.

y. *Fixed cost* is any cost or portion of cost currently incurred to prevent deterioration of the benefit/cost ratio, whether or not that ratio was achieved by past investment.

z. *Disinvestment* is the reduction of the future benefit/cost ratio, as by withdrawing investment, not incurring fixed costs, or accidental destruction.

aa. *Current:* Since costs and benefits are in most cases thought of as flows, not stocks, they are measured over time. This period could range from a few seconds, for the runner who conserves his strength during the first half of a dash to have more at the finish, to a lifetime, for the person who modifies this life to affect his next reincarnation.

bb. *Past, present, future:* For purposes of *making* a decision (as contrasted to measuring the magnitude of costs or benefits), the present is a dimensionless line separating past from future. That is, we assume that there is some moment before which the decision is still open and after which it is made and closed. Cases in which a decision is made at one moment to be effective at some future moment are dealt with later (2.52).

cc. *Time preference* is the relative strength of desires for present as contrasted to future satisfaction. *Positive time preference* is a time preference in which present is valued more than future satisfaction. *Time preference* without an adjective means positive time preference. *Negative time preference* is the opposite. Whereas all other definitions related to time spans (such as investment and fixed cost) are part of the opportunity function, time preference is part of the preference function.

dd. *Transformation* is any effector process that alters the environment or the system's relation to it.

ee. *Production* is any transformation that increases utility.

ff. *Destruction* is any transformation that decreases utility. Although nature often produces productive or destructive effects, by defining transformations as effector processes we confine attention to transformations effected by the system itself.

2.51   DECISION MODEL AND DEVELOPMENT

The following assumptions constitute the initial model for analyzing decisions:

a. A problem is already opened for decision and its scope is determined. It is irrelevant to the model whether circumstances forced the problem open, a periodic review was scheduled, or intuition suggested that a problem exists. The initial model does not deal with a prior decision about whether there should be a decision.

b. The decision is made rationally.

c. The available information is taken as given. This assumption has the same logical effect as to assume adequate and correct information about the opportunity function. It deals only with the relation between image and decision, not between image and reality. The meaning and importance of this assumption are nicely described by Tribus (1969, pp. 1f).

d. The decision process itself has no cost or benefit. The initial model thus deals only with the costs and benefits of the alternatives.

e. The preference function is given and known.

f. Variable costs and benefits only are considered; the time dimension of the decision is insignificant.

g. The choice is between only two mutually exclusive alternatives, X and Y.

h. The choices are unitary or "chunky": either alternative must be accepted or rejected as a whole, with no question of more or less.

i. There is no cost to the selection of X other than that Y must be foregone and vice versa: the only cost of X is Y and the only cost of Y is X.

j. The decision-maker is designated A.

k. However long the decision process may take, it is effective as of some given moment. All costs and benefits of the alternatives are divided accordingly into predecision and postdecision status, or past and future, as defined. It is irrelevant to the model whether costs and benefits occur instantaneously at some moment or are spread regularly or irregularly over an extended period. The model assumes only that the decision-maker can somehow assess their overall magnitude, allowing for time discount, uncertainty, time preference, and other factors.

l. All costs and benefits are those of the decision-maker, who makes the decision solely as a principal, not agent.

2.51.1   Under these conditions A will select the alternative that is highest in his gross preference ordering. This is the one with the largest gross benefit and the only one that can have a positive final net benefit. Even if no alternative is positive in an overall sense, the lesser of two evils is positive relative to the greater evil—a small loss is positive relative to a larger loss.

2.51.2   Relax assumption i and assume instead that both X and Y

have some costs in addition to the sacrifice of the alternative. In a choice between shore and mountains, for example, not only does each preclude the other but each also has other costs, as for surfboard or tent.   **2.51.21** In this event A will select the alternative with the larger net benefit; it will necessarily have a positive final net benefit and is the only one that can.

2.51.3   For assumption g substitute three mutually exclusive alternatives—X, Y, and Z in that order of preference. Then to select either X or Y necessarily sacrifices Z. If the net benefit of Z is charged against both X and Y in computing their final net benefits it cannot change their relative desirability, since it is the same for both. Hence as soon as Z is eliminated as inferior to Y the decision between X and Y can proceed as in 2.51.2.   **2.51.31** By extension of 2.51.21, among any number of alternatives one and only one will have a positive final net benefit and that one will be selected.   **2.51.32** Since the final net benefit is the only type of benefit that has been charged with all its costs, *benefit* will mean final net benefit unless qualifying adjectives are used. We can then simplify the language and generalize that the rational decision-maker will select that one and only alternative whose benefits exceed its costs. If the two are equal the model provides no prediction.

2.51.4   Any costs or benefits that are identical for two alternatives cannot be used as a basis for choosing between them, since their effects will cancel one another.   **2.51.41** By definition sunk costs and benefits cannot be altered by a decision and will be the same regardless of which alternative is selected. Hence they cannot be a basis for making the decision. To clarify, anything that has already happened cannot be undone, although it may be remedied or recovered in whole or part. If I have dropped my tie clip into my coffee I cannot undo the fact that it happened. But remedial action can recover it without harm to clip or coffee, and the act is wholly recoverable by a decision made now. If I drop in a piece of toast and decide to remove it promptly the coffee may not be hurt but the toast will: the coffee is recoverable but the toast is not. If I stir salt into the coffee instead of sugar, neither salt nor coffee is feasibly recoverable. Any part of the costs of a past act or condition that can be recovered by a current decision are not sunk for purposes of that decision. Similarly, any benefit created in the past that can be diminished by a decision is not sunk. The benefit of having made a flower garden in my side yard, for example, could be eliminated by paving it over for parking. (Nina, this is a hypothetical illustration.) However, since the destruction of any such benefit falls within the definition of cost we will speak only of sunk costs, not benefits.[1] For convenience we will say that any cost already incurred and not remediable or recoverable is past, sunk, and closed and that any cost still to be incurred or recovered is future, unsunk, and open.

2.52   *General Rule of Rational Decision*

If we accept assumption 2.51k that the moment of a decision is a point without time dimension we can join 2.51.41 with 2.51.3 to form the following rule of rational decisions: *A rational decision selects that alternative whose future benefit exceeds its future cost.* If costs and benefits are not incurred at the same moment in the future the time discount elements of 2.53 must be considered. If the decision itself is not to become effective until some time after it is made, then "future" is measured from that later moment. If perceived costs or benefits might change between the decision and its effective date the decision should be based on their presently estimated status at that later point. Uncertainty on this score is logically no different from uncertainty on any other score, and in any case this model is based on A's image, not on reality (2.50t). Even if the decision-maker is not rational it nevertheless remains true from the definitions that satisfaction would be greatest in any decision if that alternative is selected whose future benefit exceeds its future cost. We are speaking here of time as defining a boundary between relevant and irrelevant costs and benefits. Time as a factor in determining their magnitudes, as in time preference or time discount, is discussed later.

Paragraph 2.41.43 indicates that the human being tends to complete a cycle of activity once it is started and suffers dissonance if it is not completed. If any such cycle is reevaluated in midpassage and the remaining costs of finishing it are concluded to exceed the benefits, a rational decision to abandon the action will conflict with the psychological urge to finish it. No logical contradiction is necessary, however, if avoiding the dissonance of abandoning the task is included in the benefit of completing it.

To clarify the distinction between subjective and objective

---

[1] I have avoided my earlier wording (Kuhn, 1963) that "sunk costs are irrelevant," since that phrase may seem to mean that past events play no part in shaping the present problem. The opportunity function is a product of all its past causes, including the expenditure of costs that are now sunk, and the same may also be true for the preference function. *Irrelevant* should be used only if interpreted narrowly to mean that the magnitude of sunk costs does not enter a current computation.

rationality on the one hand and that between pastness and futureness on the other, let us suppose that a person gets part way through repairing an otherwise unusable table. Having put in much work and frustration he finds that the remaining work far exceeds his whole original estimate and that the benefit from the finished table will be less than the costs still ahead. Objectively the future cost exceeds the still-future benefit, and the objectively rational decision must be to abandon the task. But suppose he says, "I just can't stand having all that hard work go to waste" or "I want to finish the job just to prove that I can." So he finishes it, and thereby seems to violate the General Rule by counting sunk costs. But if we accept his expressed desires he has not necessarily violated the rule since the subjective benefits of finishing the task are still in the future. All costs and benefits are in the final analysis subjective, and the person who attends only to supposedly objective ones may be the least rational of the lot.

Economic analysis distinguishes the roles of fixed and variable costs in pricing in the long-run, short-run, and market periods. In the market period both fixed and variable costs have already been expended, and the seller will accept whatever price he can get in excess of zero. In the short run fixed costs are already expended or committed, and the seller will produce and sell for any price that covers variable costs. In the long run he will produce and sell only if total costs are covered. The present analysis agrees. But it also argues that the relevant characteristics are not fixedness versus variability but pastness versus futureness and that there are not merely three levels of time periods but infinite gradations. The same principle applies, for example, to the question of whether half-processed materials should be finished or whether a lagging sales campaign is worth an additional five million dollars in advertising. The doctrine of ignoring sunk costs is also familiar to the formal theory of managerial decision-making. The present analysis generalizes it to all kinds of decisions.

Previous chapters dealt first with DSE functions as pure cases and then observed possible relations among them. This chapter has dealt with costs and benefits as if similarly separable and independent. But these, too, may affect one another in several ways, notably in that benefit may be measured by cost or cost by benefit. We will deal first with the former.

We often want to demonstrate that we have power. To overcome large obstacles at large cost demonstrates this, and the larger the cost the greater the benefit in the form of demonstrated power. A related case is that of the pleasure of exercising our faculties (2.41.2), which may lead to exertion just for the fun of it. This situation creates no problem if we figure things right: if the activity is truly pleasurable it is not a cost.

By contrast some large measure of our appreciation of the delicate beauty of Chinese ivories, Byzantine mosaics, or Renaissance tapestries lie in sheer wonder at the years of patient skill that went into them. The response to the Pyramids is not so much appreciation of their form as sheer marvel at the labor of making them. Identical products made mechanically would have no such effect, and in a very real sense their benefit is high solely because their cost is high. The consumer who assumes that product X is better than its competing product Y just because it costs twice as much is behaving similarly. So is the man who assumes that his newly betrothed is a very special creature just because she was so abominably hard to pin down.

Viewed after the fact, any act whose benefit does not at least equal its cost must be judged a mistake. In such cases a decision-maker who does not wish to appear wrong, and whose costs cannot be hid, has an interest in overstating the benefits of his acts. If the cost includes death to thousands, he will feel an almost irresistable urge to glorify the benefit so that it seems greatly to exceed the cost. "An easy victory is not worth winning" and "Our sacrifice is the measure of our souls" are additional cases. If there is no independent measure of cost and benefit, and if cost is taken as the measure of the benefit, the reasoning is circular and rational decision becomes difficult if not impossible.[2] Extension of the same process may assign relevance to the magnitude of sunk costs when they should rationally be ignored. The result of these tendencies can be devotion to that for which one has sacrificed, whether or not it was worth any sacrifice at all. The opposite side of the coin is to assume that if something has high benefit it must also have high cost, and low cost if it has low benefit. Thus one might assume that effective medicine must be expensive or that a really worthwhile mate must have high upkeep. Since costs are often more objective than benefits, the errors of estimating costs are perhaps the less likely to go undiscovered.

2.53   We now relax assumption 2.51f—that variable costs and benefits only are considered—and add a significant time dimension to the period covered by the decision. Assuming the ability to make it, an investment cost will be incurred if and only if the utility of the

[2] Boulding (1968b) has observed that sacrifice tends to legitimate war. I agree but word it differently—first because it would dilute the meaning and analytic utility of *legitimacy* as defined later and also because I suspect that the central analytic problem is simply that of measuring benefit by cost.

expected benefit/cost ratio at future time t_1 exceeds that of the ratio of the present time t_0 by more than the amount of time preference between time t_0 and t_1. By corollary, the present value of anything whose benefit cannot be realized until some future time is equal to its benefit at that future time minus the amount of time preference between t_0 and t_1, whose difference is referred to as *time discount*. Again assuming the necessary ability, fixed costs will be incurred if and only if the utility of the amount of benefit/cost ratio thereby maintained until t_1 exceeds the time preference between t_0 and t_1. Disinvestment will be made or allowed to occur if the amount of the loss in the benefit/cost ratio at t_1 is less than the time preference between t_0 and t_1.

Because these terms are borrowed from economics some readers may assume they refer to money costs and benefits only. Yet the terms are used here broadly. To practice a guitar for skill, rather than merely for present enjoyment, is investment; practice aimed at maintaining the skill is a fixed cost, and loss of skill from inadequate practice is disinvestment. Hard fighting was the investment cost of acquiring freedom; eternal vigilance is the fixed cost of maintaining it. Strenuous dieting is the investment cost of losing weight rapidly; moderate eating is the fixed cost of keeping it down; and regaining weight is the disinvestment from renewed feasting—and so on across the spectrum of life.

2.54   We now relax assumption 2.51d and substitute that the decision process has costs or benefits.   **2.54.1** If the net cost of making a decision between X and Y is known in advance to be less than the final net benefit of the better alternative, A will incur that cost. If instead the decision cost is greater than the final net benefit he will choose between X and Y on an arbitrary or random basis. If the decision cost is less than the final net benefit of the best alternative, the decision process can be said to be "worth the cost."   **2.54.2** In a choice between X and Y, if nothing can be known in advance about the probability that the final net benefit of the better alternative will exceed the net decision cost, A will make either the main decision or the predecision whether to incur decision costs on a random or arbitrary basis. Partial information, even a 51/49 probability that the information costs are worth it, will move the predecision away from purely random choice.
2.54.3   Consider a choice among three or more alternatives whose net preference ordering is not known at all in advance. If the decision costs are expected to be smaller than the difference in net benefit between the best and worst alternatives, A will incur the decision costs.

If the decision costs are expected to be greater than that amount A will decide on a random or arbitrary basis.   **2.54.4** Consider the same choice. If it cannot be expected in advance that the decision costs will be less than the difference in net benefit between the best and worst alternatives, A will make the decision on a random or arbitrary basis.

2.54.5   A choice among three or more alternatives can be made in various ways. If it is made by paired comparisons and in various sequences, the total number of comparisons required to isolate the best alternative depends on the order of pairing. If X is better than Y and Y is better than Z, only two comparisons are necessary. But if both X and Z are better than Y, but by indeterminate amounts, a third comparison between X and Z is necessary. Potential complications rise exponentially with the number of alternatives, and having spelled out the general logic we need not go further. However, if it is possible to measure costs and benefits cardinally rather than ordinally, total costs of making the decision need rise no faster than the number of alternatives.

If one cannot know the expected decision costs in advance, he could incur preliminary costs to estimate the main decision costs, prepreliminary costs to estimate the cost of estimating the decision costs, and so on through an indefinite series of decisions about how or whether to make decisions. The content of such decision sets differs from that in substantive decisions and can probably be comprehended better if arranged hierarchically (see 2.10f). But these preliminary decisions need not be spelled out. They involve no new principles, and conclusions about them can be deduced from propositions already stated.

2.54.6   The benefit achieved from selecting the best alternative in a decision will be its final net benefit minus the decision costs.

2.55   *Information Costs*

2.55.0   Relax assumption 2.51c that the information is given and assume instead that significant items are not known. The acquisition of information, often referred to as *search,* can be treated as a decision cost, and in that role it is subject to all the propositions stated in Section 2.54. Information, however, has special characteristics that require special treatment as follows.

2.55.1   When a decision is made its execution is still a future event. Since every perception is in some degree false (2.11), one can never fully know the total situation in which he makes a decision, much less whether the decision can or will in fact be executed according to its terms. In describing what March and Simon (1958) call "bounded rationality" I have suggested (1963, p. 272) that since no one can know all the possible consequences of an act, "the beginning of wisdom for decision-makers is to recognize that they can deal with only a tiny fraction of what is possibly relevant." We will accept this proposition as an assumption and note several consequences for decision-making.

2.55.2   As a corollary of 2.54.4 and 2.55.1, the less sure A is of the consequences of his decision the more rational it is to decide by intuition or impulse. In cases 2.54.2 and .4 it is not at all irrational to make a purely impulsive decision. The model, of course, gives no hint about how many of life's decisions are of that sort. Nor can it state which consequences will be significant, though we can say that when one flicks a switch it is normally significant that the light goes on and insignificant that there is wear and tear on the switch. The simulation model (Chapters One and Sixteen) is the logical technique for dealing with such situations.

2.55.3   If we now relax assumption 2.51a that the decision is already open we may note that among the preliminary aspects of a decision are questions about when and whether to open the decision and about the nature of the opportunity function. And the latter includes questions about how many alternatives to consider and which ones, and how long and how far to search for additional alternatives. If all these questions had to be examined systematically most decisions would rationally be made by impulse, intuition, or arbitrary rule. As Boulding once suggested (in a lecture), perhaps the only sensible rule here is to search until tired. Yet even that prescription begs part of the question, for tiredness itself depends in part on the expectation of success, and one can never know at the time whether he stopped just short of finding the perfect alternative.

2.55.4   All in all there is probably no alternative in the normal case but to narrow and define the problem intuitively and then to attempt rational calculations of advantage only within the problem as thus narrowed. That is, one must use intuition to specify what is accepted as given before making comparisons among the alternatives viewed as open. In line with Section 2.56 below, for any decision there neverthe-less exists a theoretically optimum decision cost, and the marginal analysis there is just as applicable in theory to decision costs as to any other variable-quantity decision. The point is that an intuitive leap may often be the only feasible way to ascertain even approximately where that optimum lies. Once one has found its general neighborhood, he may then use more explicit calculations to locate it more precisely. As to *who* should make or participate in defining a decision, the obvious prescription is someone with sound intuitions, which normally means someone familiar with the area of the decision, perhaps checked by someone with different intuitions. Since the way a problem is defined may be the most important determinant of its outcome, our conclusion is that there are no specific rules for this essential aspect of decision-making. This conclusion should not be taken to deprecate the poten-tially high benefit from systematic decision-making within a defined decision set, since many real problems are reasonably well defined, particularly in firms.

2.55.5   To join 2.51.41 (on sunk costs) and 2.54.3 (on decision costs), if a decision-maker is employed under a contractually fixed salary, the cost of the information he can provide is sunk with respect to any decision he makes during the life of his contract. The future informa-tion costs of a decision are therefore substantially reduced, and the rationality of an informed, as contrasted to an intuitive or random, decision is substantially increased. However, the model provides no hint about whether the decision-maker is worth hiring in the first place, presumably another matter for intuition.

2.56   *Marginal Decisions*

2.56.0   Relax assumption 2.51h that the alternatives are unitary. Assume instead that A has decided in favor of some divisible thing and that his remaining decision is solely the quantity he will seek, which we will call the *number of units*. We also return to the assump-tion that there are no decision costs and need not relax it again.

2.56.1   If successive units of the same thing are contemplated, satis-faction will be greatest if the General Rule is applied separately to each successive unit whose costs and benefits are large enough to be dis-criminable, or to the smallest divisible unit if that is larger. This process is known as *marginal decision-making,* and its theory and terms are borrowed from economists.   **2.56.2** As successive units of the same good are acquired, the benefits from each additional unit decline after some point according to the law of diminishing utility—which is accepted because no one can imagine a situation that violates it. This law is apparently independent of whether one's theory of motivation is need-satiation, drive-reduction, hedonic selection, instinctive, biochemical, rational-instrumental, or status-seeking.   **2.56.3** As successive units of the same good are acquired, however, after some point the cost of each additional unit will rise. Under certain limited conditions the law of diminishing productivity will produce this result and, as with the law of diminishing utility, no one seems able to imagine a situation in which that law would not apply. If the decision-maker is operating within a range of quantities too small to involve diminishing pro-ductivity, his opportunity costs of additional units will necessarily rise after some point as more and more units of alternatives are sacrificed. (See Samuelson, 1970, p. 26, on the "law of increasing relative costs.")

2.56.4   Together 2.56.2 and .3 mean that satisfaction will be greatest when costs and benefits are equal. As diagramed in Figure 8 and ex-plained in 2.56.5, this point is where cost and benefit curves intersect.

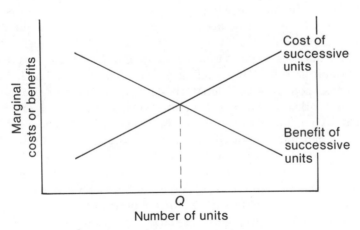

Figure 8. Cost/benefit curves.

2.56.5   For any unit to the left of the intersection in Figure 8, bene-fits exceed costs; hence satisfaction will be increased by doing or acquiring the unit in question. For any unit to the right of the intersection, costs exceed benefits; hence satisfaction will be decreased by doing or acquiring it. It therefore follows that the quantity of this good that will produce maximum satisfaction is at Q, the intersection of cost and benefit curves, at which costs and benefits are just equal. To diagram the problem with continuous lines implies that the quantities are infinitely divisible. If the good comes in discrete units, like chairs or trips to the movies, then except by accident the actually achievable optimum will fall above or below the theoretical one. In the case of the movie, incidentally, one normally goes or does not go, and an additional minute of watching is not identical with the preceding one and hence is not subject to diminishing utility—at least not in old-fashioned movies with plots.

2.56.6   Economics, of course, is replete with equilibria of this sort, as we shall see in Chapter Fourteen. The logic does not change, whether for economic or other analysis, if the benefit curve crosses the cost curve on a downward slope of the latter, so long as the slope of the benefit curve is the steeper.

2.56.7   Because of a common misconception that marginal decision-making requires cold and careful calculations, let it be clear that even the warm and impulsive individual inescapably makes marginal selections day in and day out. He puts two teaspoons of sugar in his coffee, not one or six. Even if he does not count the cost of sugar (he decides as if the cost were zero), up to two spoonfuls, the benefit exceeds the cost and he adds sugar; beyond two spoonfuls the cost exceeds the benefit and he stops. He spends, say, five hours on the beach because up to that point the additional satisfaction per unit of time exceeds the cost in other activities forgone; he leaves when the cost exceeds the benefit. When the alarm goes off the benefit of a few more minutes in bed is deliciously high and the cost low. But as the minutes slip by and he gradually wakes, the benefit of additional minutes diminishes while the cost in added rush, tension, bolted breakfast, and risk of being late goes precipitiously up. When the two curves cross he gets up; if they do not cross he stays in bed. If the poor soul is an instant waker they cross before he even gets the alarm turned off. We are, of course, describing the language of the analyst, not the mental states of the individual.

Sometimes we say a person has no choice, implying that marginal analysis is not applicable. What we really mean is that the slope of some cost or benefit curves is so steep that it leaves no reason-

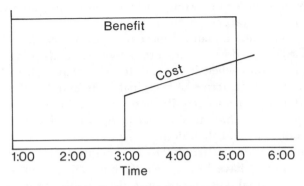

Figure 9. Cost/benefit versus time.

able doubt about the quantity at which the intersection will occur. Suppose you will be promptly fired if you leave your job before five without permission, but at three o'clock you suddenly feel an urge to get out. The curves then look something like those in Figure 9. There is nothing analytically wrong in saying that your choice is really unitary rather than marginal and that you must choose between working until five or not working at all. In that case we use the analysis of decisions between mutually exclusive alternatives rather than of marginal decisions about variable quantities. The former can be considered the limiting case of the latter.

The central point, however, is that one inescapably allocates time, effort, attention, affection, hate, strength, thought, money, materials, and a thousand other things. Such allocations are made constantly in every society from the most primitive to the most sophisticated. Habit is important, but it is only a stop-gap explanation since it does not explain why the habit was established in the first place, why it continues, or why variations take place around it. For a substantial fraction of the population the intrenched habit of going to work would be extinguished in a single day if its benefits dropped below costs. Furthermore, to say that a response is purely habitual is another way of saying that its decision cost is zero—though habit may have the opportunity decision cost of ignoring highly relevant information. In many response selections the difference in net benefit between two alternatives is too small to justify any decision cost at all—which shoelace to tie first. Since habit is one form of arbitrary selection, it is a rational technique according to 2.54.1. Economist Rothenberg (1966, p. 231) indicates that "habitual choice may be consistent with, indeed entailed by, utility maximization." Many such choices do not involve even the cost of conscious awareness of them.

Why does a man eat more or go to bed earlier when he has worked hard? Why does he do more indoor work when it is rainy and more outdoor work when it is clear? It seems that such questions can be answered only in terms of marginal costs and benefits. Although some marginal decisions require hours of careful calculation, many others require only that one do what he feels like doing so long as he feels like doing it, and then doing something else. That such satisfaction cannot be counted like dollars or doughnuts does not change its underlying logic. I have also argued (Kuhn, 1967, pp. 300–307) that to equate the marginal cost and product of labor in a firm—a task some theorists seem to think requires detailed knowledge of a sort not actually available—may often require nothing more than to hire the number of persons who can be kept reasonably busy making things that can be sold. The relative roles of culture and social controls in all these matters, as contrasted to decisions, are examined in Chapter Sixteen.

2.6   *Rationality*

To act rationally and to maximize are often closely associated ideas, as for example, "A unit of economic decision is said to act rationally when its objective is the maximization of a magnitude" (Lange, 1934). Yet they are not precisely the same ideas, and we should distinguish them.

I have defined rationality tentatively in 2.50u as the process of selecting the preferred alternative in an opportunity function. This definition accepts the formulation of opportunity and preference functions as given and implies that rationality—both subjective and objective—deals only with a selection consistent with the preference, made within the perceived opportunity function. We have already seen that there is no such thing as perfect information. Since all decisions are made in greater or lesser ignorance, a decision cannot be judged irrational on the ground of ignorance if it selects the best perceived alternative. Nor do the definitions of rationality question the goals themselves, as they are embodied in a preference ordering. A person may prefer to be rich or poor, powerful or meek. Rationality can question only whether his actions are instrumentally directed toward his goals and instrumental subgoals.

A difficulty is that the only sensible definitions of rationality come perilously close to including all behavior. The psychotic may very well be rationally pursuing *his* goals by techniques that seem instrumentally effective within *his* view of the world. And the "gut" feeling is certainly a selector state. Yet we all observe behavior we would unhesitatingly call irrational, and we will attempt to specify its nature within this conceptual set. All types nevertheless revert to the criteria of 2.50u—the judgment of other persons, the judgment of the same person at a different time or under different circumstances, or the emotional blockage of a performance once decided upon. Irrationality can presumably be said to exist only if the preference function or opportunity function is unstable, bizarre, or both.

Regarding unstable preferences, we all change our likes and dislikes. Furthermore in many matters we see some good and some bad, and even the "normal" person may show markedly different valences about something depending on what aspect of it currently dominates his attention. Hence *unstable* can only mean that changes occur more quickly or dramatically than usual. Something is hated one week, adored the next, and hated again the third. For slowly effectuated decisions the preference would often change and the decision be reversed in midpassage. If often repeated such behavior qualifies as irrational. As to *bizarre* preferences, some people dislike making telephone calls. But if the dislike leads to refusal to make a call that will save a life the preference would be considered bizarre and the decision irrational.

A view of opportunities can be considered *unstable* if it changes more often or more widely than usual—such as the person who alternately votes Republican and Communist as he switches views about who will save the country. (This vacillation is viewed as a change in opportunity function on the assumption that the goal of saving the country remains unchanged and only the view of how to do it alternates.) A *bizarre* view of opportunities would involve what others consider obviously contrary to fact—for example, a man who tries to drive to Boston by heading west from Denver or deliberately eats heavily to reduce.

2.7   *Maximizing*

Decision theory is largely an offshoot of economics and thus raises the question of whether the maximizing goal posited by economics might be better replaced with a "satisficing" goal (March and Simon, 1958). The distinction largely disappears if the terms are clearly defined. We can speak of maximizing something only within constraints. If I want to know how much corn I can husk I must specify such con-

straints as whether I mean within an hour or a year and whether I am
to use bare hands, a knife, or a husking machine.

Although all the constraints on a maximization could be
beyond the control of the decision-maker, usually some constraints involve his preferences. For example, how much corn can I husk in an
hour without getting blisters, cutting my finger, or buying a machine?
But as soon as a constraint is set by a preference we can no longer speak
literally of maximizing but only of reaching the highest level consistent
with conflicting goals, which means a satisfactory rather than a maximum level. As soon as we speak of multiple goals, their only common
denominator is satisfaction and the process involved is marginal
decision-making. That is, we pursue a goal until the next increment of
satisfaction would be less than that derived from some other goal. In
short, to maximize satisfaction, which is a conglomerate of particular
satisfactions, means to "satisfice" each subgoal, while to "satisfice" each
subgoal is automatically to maximize total satisfaction, assuming that
decision costs are taken into account. (Note that the question of suboptimizing *different* goals of the overall system is not the same question
as that of suboptimizing each subsystem in an organization, when all
subsystems are directed toward the *same* overall goal.)

In psychology the law of effect—which in basic form states
that the organism learns to repeat pleasant experiences and avoid unpleasant ones—is at bottom a maximizing mechanism. According to this
view behavior is determined by the motive or valence that is strongest
at the time behavior is selected, there being no measure of which is
strongest except to note which actually controls behavior. Subject to the
preceding reservations about the unstable and the bizarre, this psychological mechanism is inherently rational and maximizing since the
strongest motive is another way of describing the highest item in a
preference ordering, the selection of which fills the definition of
rationality. In this respect there is no difference in fundamental logic
between simple response selection and decision-making, but merely the
difference in detail, sophistication, and awareness that we have called
complex decisions.

For related reasons the concept of *free will* is not useful. One
might say, "I definitely prefer X but can demonstrate that I have free
will by deliberately choosing Y instead." The person is merely saying
that the gross benefit of X exceeds that of Y, but that the net benefit of
X, after subtracting the cost of losing an argument or a cherished
belief, is less than that of Y. His behavior is certainly subjectively
rational, and it is objectively so if we explicitly consider his desire to
win the argument or to believe in free will. All costs and benefits reside

in selector states and are subjective. Objective external things can be used only as temporary and arbitrary counters for them.

Compared to the subjective, objective rationality is the logical equivalent of a simplified coded version (2.10b) of the factors relevant to a decision. Like other codings it discards much data (2.10c) and thereby becomes more amenable to logical manipulation. Whether the objective version produces a better decision depends heavily on which data are discarded, a matter for which we apparently have no rule except trial and error. How much satisfaction one gets from something depends, of course, in important ways on his philosophy or attitude toward life. Some get great satisfaction from very little. Such a statement falls within this model, which merely uses different language in saying that people differ widely in the content of their selectors. It could also be a question of how the detector categorizes the feedback from an action in the learning stage of a behavioral sequence (2.07.2).

On the whole, decision theory is oriented more around organizations than individuals. The reason is simple—that some organizations, particularly business firms, hold a limited list of goals, often explicitly stated and relatively uncontradictory. Furthermore, because the organization's costs and benefits can often be isolated from those of its members (see Section 4.1) they may have greater external objectivity than for the individual. Hence the decisions of an organization may be much more amenable to decision theory than those of an individual. They reach their zenith of objectivity in the economist's model of the firm, where all costs and benefits are measurable in the single, quantifiable dimension of money—the model whose simplicity has been the takeoff point for most formal decision theory. But even money has no benefit unless we can do something with it, and it is ultimately the subjective utility of its uses, not the objective number of dollars, that is the basis of rational decisions.

2.8   DECISIONS AS EQUILIBRIUM

In 1.31.6 we noted that equilibrium of at least one variable seems to be the most probable state of a continuing open system, and we will examine certain aspects of decision-making in an equilibrium context. First we must note, however, that even when the discussion is carefully worded a common misconception arises about the notion of equilibrium. To avoid it, we must stress that (in this model) no system *seeks* equilibrium. Equilibrium is not a goal. It happens. The actor may have no notion about equilibrium at all. And if he did, he would probably have no idea where it is, when or whether he had achieved it, or whether he was better or worse off at that point. Equilibrium is a concept in the head of the analyst, not of the actor. The point of the analyst, however, is that certain actions do tend toward an equilibrium. His problem is to discover why this is so and where the equilibrium is.

Negative feedback, or oppositely paired mutual interactions, was identified as one of two main conditions for equilibrium (1.31), and that condition is also met if two variables are oppositely related to a third. The equilibrium quantity in divisible-unit decisions was seen to fall at the intersection of costs and benefits (Figure 8). Although costs and benefits are independent of one another, benefits vary inversely and costs directly with quantity. Hence equilibrium is possible, and an important aspect of human response selection is seen as an equilibrium. As diagramed in Figure 8 the benefits are those of the good whose quantity is being selected. The cost curve, however, is unspecific. It can represent one alternative good being sacrificed or a package of any or all other goods. It is thus another technique of representing the "actor within a field" (Lewin, 1951) and in theory can readily be broadened to a general equilibrium of all of a person's conflicting desires. This connection between decision and equilibrium is here couched in terms of system analysis but was not, of course, initially contributed by it. Samuelson (1965, p. 5), for example, noted it in connection with economic theory in stating that "the conditions of equilibrium are equivalent to the maximization (minimization) of some magnitude."

Earlier in this chapter a choice between two unitary alternatives was not treated as equilibrium. But conflict, which is the occasion for a decision, was viewed .s dissonance (2.41.42). In a choice between two positively valenced alternatives, for example, rejecting the higher preference item creates greater dissonance than rejecting the lower; hence selecting the higher minimizes dissonance. If the result is not equilibrium, it is the closest position possible under the circumstances. This conclusion is a logical extension of the statement that with successive but nondivisible units the achievable optimum may lie above or below the theoretical optimum (2.56.5). For obvious reasons the alternative with a positive final net benefit optimizes satisfaction and minimizes dissonance and hence expresses an equilibrium tendency.

With this background we now examine some aspects of DSE as questions of equilibrium.

2.81   *Equilibrium in Detector*

In all three DSE subsystems there can be an equilibrium with respect to the total content of the subsystem, the amount of a

certain kind of content, or the content relevant to a particular case.
2.81.1 Regarding *total* information in the detector, with no information about his environment one cannot cope at all. Initial information
about the nature and location of common objects and persons is
extremely valuable. Additional information is also useful but subject
to diminishing utility. Some information is acquired more or less
automatically through normal living, and its cost is low. But gaining
other information requires special effort, and its marginal cost rises as
more is acquired. At some equilibrium point the cost of further information exceeds its benefit and the individual ceases seeking it. Or, if he
wants to keep increasing his knowledge, he devotes neither all his time
to the task nor none of it. Instead he reaches an equilibrium at which
some average time is given to learning. The average, of course, varies
with person and circumstance.

2.81.2   The same principle applies to information for a *particular*
decision. The broad information that narrows it is valuable, but more
and more refinements are successively less valuable. We thus have a
curve of declining benefit. Meanwhile the cost of successive information rises, for two reasons. First the low-cost, readily accessible, best-
known, and intuitive information is utilized first. Second, as with any
other activity, more information means less of something else and
brings rising costs of alternatives sacrificed. Again there is an equilibrium of marginal cost and benefit. In this connection see Charnes
and Cooper (1958). Information for a decision may already be stored
in files or memory but may involve retrieval costs, in which case
equilibrium analysis may also be applied to determining how much
to retrieve.

2.82   *Equilibrium in Selector*

2.82.1   Regarding *total* motivation, as one goes through life he adds
to his store of likes and dislikes. Each new liked thing adds to his
sources of potential satisfaction and reduces the risk of having nothing
to enjoy. Except for changes of existing likes the movement is essentially one-directional, since a newly discovered thing that is disliked
does not itself decrease the stock of liked things. The selector thus faces
a problem of how many different things it is worth learning to like or
dislike. With the presumption of rising costs and declining benefits of
successive additions an equilibrium can be reached.   **2.82.2** And regarding motivation for a *particular* case, since the motivation is the
goal it is pointless to speak of acquiring a like or dislike for a thing as

part of deciding whether to select it. Moreover the question of whether to use an existing motive is not parallel to that of stored information. Since the motives are given for a particular decision there is no question of whether they will reach an equilibrium in it.

2.83   *Equilibrium in Effector*

2.83.1   Regarding *total* skills, since the effector is by nature instrumental its logic parallels that of the detector. As with information, there are both extensive and intensive margins for acquiring skills—how many and how high a level of competence. Both are subject to diminishing utility and productivity as well as to rising cost and hence are subject to equilibrium analysis.   **2.83.2** *Particular* skills and skill utilization for a decision present the same problem with acquiring skill as with acquiring information: How many skills and what degree of competence are "worth it" to perform a task? For the same reason as before, an equilibrium is possible. The same rationale applies in using skills already learned.

2.84   *Summary on Subsystem Equilibrium*

It is irrelevant to the discussion of equilibrium whether the decisions involved are conscious or unconscious, calculated or impulsive. Nor does it matter whether social forces impinge on the decision: these would affect the payoffs and the preferences but not the decisional logic. The effects of social forces are abundantly explored in later chapters. The point here is simply that every individual does reach a state or level for each subsystem variable and that the level varies greatly with the person and the society.

Little attention has been given to this question; yet I think it is significant. Sociologists and anthropologists tend to assume that the level is determined by culture. This can hardly be so if it differs widely, as it does, among persons in the same culture and from time to time in the same person. Across cultures there is the additional question of whether a culture prescribes the answers or simply sets parameters within which the individual finds his own equilibrium. This analysis has not even hinted at what traits or experiences produce a certain type or level of equilibrium—a matter for specialists. But it does suggest that a problem exists and that equilibrium is a tool for attacking it. Equilibrium analysis will also be applied later to certain aspects of communications, transactions, and other social phenomena.

2.9   INTRAPERSONAL POWER

Every decision is made within constraints, which in this model are reflected in the detector's information, the selector's goals, including the intensity with which they are held, and the effector's capacity to perform, including the energy available to it. The analysis of decisions takes these constraints as given and deals with quantitative and qualitative selections made within them.

But because different systems face different constraints, they also differ in the qualitative content and total magnitude of their effects. Given identical preference orderings the ninety-pound weakling may decide whether to carry ten bricks or one concrete block while the football hero decides between twenty bricks and two blocks. This difference in magnitude of effect is a problem in *power,* which we define as the ability to bring about or acquire wanted things. More simply, it is one's ability to satisfy his wants, but with reference to the ability to create desired external circumstances, as contrasted, for example, to satisfaction by fantasizing or reducing one's wants. The larger the magnitude of this effect, the greater the power.

2.90   *Definitions about Intrapersonal Power*

a. *Power* is the ability to bring about desired external states—that is, to bring about conditions that contribute to the satisfaction of wants.
b. *Intrapersonal power* is the ability to bring about desired external states by oneself—that is, to effectuate productive transformations.
c. *Interpersonal power* is the ability to induce others to bring about external states that one desires. "External states" include the internal states of others but not of oneself. As before, "internal states" mean only those of one's own behavioral system.
d. *Particular power* is the ability to bring about a particular desired state Y.
e. *Aggregate power* is the ability to bring about some total quantity or level of such states, $Y_1 + Y_2 + \cdots + Y_n$, considering both their number and individual magnitudes. We may note two things here. First, particular and aggregate power can be either intrapersonal or interpersonal. Second, according to the definitions a person would have low aggregate power if he produced an effect, however large, contrary to what he wanted. We will ignore Hitler's possible psychotic desire to destroy Germany and note an anomaly in definitions that would categorize Hitler as having low power because he failed

to produce an intended victory. There is perhaps no clear-cut way
of handling this situation. Let us simply say that a person who produces large effect with others has large power, but that his power
is less if he produces an undesired effect than if he produces a
desired one.

f. A is the person whose power we are examining.

g. Y is an external state desired by A—to have a wooden chair or clean
clothes, climb a mountain, or sleep till noon. The phrase "to get Y"
encompasses such varied actions as to do, acquire, produce, accumulate, or otherwise bring into existence the desired state Y. If one
wants to climb a mountain, then "to get Y" means to climb it.

h. X is the total cost of Y to A as defined in 2.50n. Whereas for decision-
making it is useful to distinguish the X costs and non-X costs of Y,
for this purpose all costs of Y are encompassed within X. "To accept
cost X" means to incur any and all costs of Y; "the capacity to accept
cost X" means that A has the time, skill, energy, money, endurance,
and so on to get Y. As with the opportunity function (in 2.50t), the
capacity to accept X includes the combined effect of detector and
effector. It contrasts with the willingness to accept X, which is a
matter of the selector.

i. AY is A's desire for Y, which is the same as the expected gross benefit
of Y.

j. AX is A's desire for X. In this context this is A's desire not to accept
the costs of Y and is the same as the perceived gross benefits of X. X
may be a single item or a conglomerate.

2.91   Given these definitions, the remainder of Section 2.9 deals
with intrapersonal power only.   **2.92** Assuming that A already has the
capacity to accept cost X, then his particular power to get Y depends
solely on whether AY exceeds AX—that is, whether Y is higher than X
in A's gross preference ordering.

2.93   The model includes the following main ingredients of A's
power. In the detector, the more accurate A's conception and perception of that segment of reality relevant to his purposes, the more
efficient will be his behavior with respect to it. In the selector, the
stronger A's desire for Ys the greater the costs he will incur to get them.
In the effector, the greater A's transformational skills in getting Ys (or
possibly the Xs that go into producing Ys), the more or larger the Ys
he can get. And as to the environment, since effectuation is defined as
a transformation of environment, the more amenable the environment

(say, in quality and quantity of resources), the greater the total Ys that A will get.

2.94 A's aggregate intrapersonal power is determined by his productivity, which in turn depends on his own DSE traits and the state of his environment.   **2.95** Viewed developmentally, personal power probably shows a positive-feedback relationship with motivation—the stronger A's motivation, the more Ys he will get; and the more Ys he gets, the stronger will be his motivation. Given chains of conditioning, his motivation to improve his detector and selector skills may also be raised. See also the tentative statement about positive feedback in 2.09.6, within which limits are imposed mainly by time, energy, and learning speed.   **2.96** Aggregate personal power shows a potential positive-feedback relation with investment—the greater A's productivity, the greater the external resources he has produced that can be invested to increase his future productivity.   **2.97** As a corollary of 2.95 and 2.96, *cet. par.*, an initial difference in aggregate power between two persons will tend to increase over time.

2.98   Figure 8 showed an equilibrium quantity selected where marginal costs and benefits are equal. That diagram and stage of analysis dealt only with a selection made within constraints that assumed but did not specify a certain level of power. By contrast, another type of diagram, also borrowed from economics, explicitly displays different levels of power. In the typical indifference diagram (Figure 10) the straight lines represent opportunity functions and the curved ones preference functions. The higher the opportunity function, the higher the preference function it will touch and the greater the resulting satisfaction. In present language: the higher the opportunity line, the greater the power it represents.

Indifference diagrams differ from benefit/cost diagrams (Figure 8) in that the former show multiple levels of power and the latter only one. But the logic of benefit/cost can also be seen in the relation between an opportunity line and the highest preference line it touches. If we read from the upper end of an opportunity line, as at point a, one has all Y and no X. As he moves down the line the benefit from each additional unit of X exceeds the cost of the units of Y sacrificed to get it. At point b the cost and benefit are just equal. For any additional unit of X beyond point b the cost exceeds the benefit. Hence the data represented by any such pair of lines could readily be transcribed to a simple benefit/cost diagram.

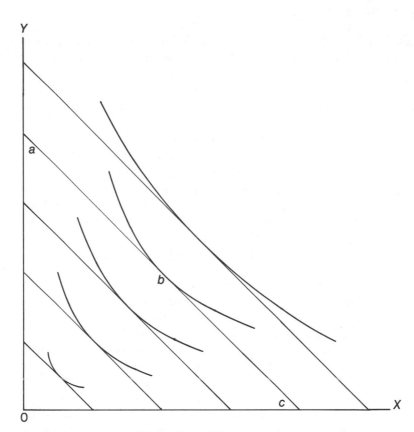

Figure 10. Indifference curves.

2.99   **SUMMARY ON DECISIONS**

The model of decision-making is tied to the model of social man, within which a decision is an interaction among detector, selector, and effector: detector and effector define the opportunity function and selector defines the preference function. The model cannot specify the content of either function, which depends on inheritance, culture, and experience; it can only indicate how the parts are related to produce a selection of behavior. Similarly the model can predict the consequences of future-oriented motives (with low or negative time preference) as contrasted to the urge for immediate satisfaction (high positive time preference); but it does not tell when either will occur.

The model avoids distinguishing rational man sharply from irrational, or "logical" from "psychological," though it does try to define *irrational*. The model thereby implies that in many of life's situations there is no real difference between behavior rationally calculated toward advantage (following the economic model) and impulsive or norm-oriented behavior (closer to the psychosociological models). A model that incorporates decision costs and insists on the ultimate subjectivity of all costs and benefits seems an adequate base for both types. There is no dichotomy here between "decided" and "self-actualizing" behavior, only some situations in which self-actualizing motives are part of the benefits and others where they are not.

Without changing the basic rules the model can accommodate decisions made under conditions ranging from no explicit formulation of opportunity and preference functions to complete, explicit formulation. In any case, the formally explicit, objectively rational decision in a complex situation, like the formal theory of any complex reality, is an abstraction of high degree. Like formal theory it has the important advantage that precise statements can be made about it with high confidence—and the important disadvantage that it may be wide of reality. Unfortunately for both theory-building and decision-making, we have no very good rules for making successful abstractions, except perhaps to use the intuition of informed persons and rigorously test their conclusions. If the specialist in decision theory finds some favorite topics missing here, it is because they *are* specialized and inappropriate for a general social science. As a social phenomenon, decision-making by a group is dealt with in Chapter Twelve.

Because equilibrium is a central feature of system analysis and because behavior selection is central to understanding the human system, this chapter looked at decisions as an equilibrating process. Regarding decisions about continuously divisible goods, the equilibrium approach is so obviously appropriate that one almost has to choose between using it and ignoring the problem. As goods become less divisible and more "chunky," the precise equilibrium position becomes less attainable but no less advantageous. Even when we reach the extreme choice between two indivisible alternatives, however, the selection of the one highest on a net preference ordering nevertheless maximizes satisfaction and minimizes dissonance—conditions that may properly be construed as an equilibrium under those constraints. Thus behavior selection, a central feature of psychology and of this model of man, can be diagnosed by a central tool of system theory: equilibrium analysis. To emphasize again a point that is sometimes abominably misrepresented, the individual does not *seek* equilibrium. He does what he wants and thereby stumbles into it—whether to his delight, consternation, or total indifference. Not only does he do this in his main system behavior but there may be some sense in using equilibrium analysis to explain the levels reached by his DSE subsystems as well. If one prefers the language of mechanics he might think of benefits as pressure and costs as resistance, in which case the likelihood that a certain alternative will be selected varies directly with pressure and inversely with resistance. A positive final net benefit then means that pressure exceeds resistance and determines the direction of action flow.

Finally the chapter introduced the concept of power—the magnitude of effect, not its direction—to provide the base for understanding interpersonal power. From this base we now move from the intrasystem stage to the intersystem level that constitutes social science proper.

Chapter 7

Communications

Chapter One indicated that the intersystem analysis uses a set of concepts directly paralleling those of the intrasystem. We have already examined the main features of the latter: the detector, selector, and effector functions; their interaction to produce a decision; and the decision viewed as equilibrium.

INTERSYSTEM ANALYSIS

We now move to intersystem analysis—analysis of the interactions of systems. An interaction presupposes that both systems are open and that in the simplest case an output of one system becomes an input of another. As to the initial models of the intersystem concepts, the *communication* transfers information between systems and reflects and affects their detector states. The *transaction* transfers something valued between systems and reflects and affects selector states. These two definitions are satisfactory for most purposes. When maximum rigor is required the terms may be defined as the transfer of anything between parties analyzed with respect to its information content or value content, respectively. For most cases information content can also be viewed as pattern. *Organization* carries out joint behavior by two or more systems. Whether it reflects and affects their effector functions in the same sense as the other two interactions is problematic but probably irrelevant.

Every real interaction is simultaneously communication, transaction, and organization. It does not seem possible to imagine a transaction's being completed without some communication, even if only that implicit in certain nonverbal acts. Nor does it seem possible to imagine a communication totally devoid of value components, since some value must attach to the time consumed or the alternative activities forgone. Furthermore a system is defined as two or more interacting components; organization is a system whose components consist of human beings (or suborganizations). Hence any two persons engaged in communication or transaction fill the definition of organization.

These matters will be dealt with in due time. For the moment we deal first with communication solely with respect to its information aspect and then with transaction solely with respect to its value aspect, in "pure" models. And while recognizing that either interaction also qualifies as organization, we will ignore its organizational aspect for the moment and concentrate solely on the interactions as such. If the nature and terms of certain interactions are functionally related to those of other interactions, those relationships constitute a pattern system. But we need not attempt to apply system analysis to them, for reasons that should become evident as we proceed.

When we examine organization as a unit we will focus on its DSE functions, and when we examine the interactions of its subsystems we will view them as communications and transactions. Organization is thus viewed not as an interaction but as a higher-level emergent unit, which means that communication and transaction are the only *inter*actions, strictly speaking, recognized in this social analysis.

Before moving to specifics we now further clarify the Chapter Two distinction between interactions of humans and their organizations and virtually all other system interactions. As to communication, humans typically transfer coded information by learned, arbitrary signs while most other systems transfer uncoded information only or learn about one another by perception—a type of information acquisition we will not consider communication (3.10.0j). It is of no moment here whether a sharp line divides humans from all other systems, which seems doubtful (and perhaps of more interest to theology and human pride than to science). In any event the aggregate differences are enormous. The general nature of the problem is therefore not affected by the presence of instinctive sign communications (as among the bees) or by apparently learned sign communications among complex mammals (such as dolphins) since this analysis of communication applies only to humans. The parallel problem about transactions is amplified in the next chapter. We first define some terms relevant to both communications and transactions and then some terms relevant to communications alone.

3.10   *Definitions About Communications and Transactions*

a. A *party* is a person, collectivity of persons, or organization engaging in an interaction. An organization can be a party to the extent that it acts as a unit. A collectivity not acting as a unit can be considered a party for certain purposes if all its individual members, A's, act so similarly that it makes no significant difference to B which A he interacts with. For convenience the following discussion speaks solely of persons. Any differences in analysis that arise when the parties are organizations or other collectivities are examined later. We will distinguish the following: A is the party to a social inter-action on whom primary attention is focused. B is a second party with whom A is interacting. C is a third party whose interaction with A, B, or both influences the A-B relation. C is not a party, for example, if he merely mechanically transmits something from A to B. The mere presence of C does not make him a party unless the behavior of A or B becomes contingent on C.

b. An *interaction* (repeated in part from 1.20i for convenience and amplification) is any change in A, including any action by him, that produces a change in B. For an output of A to become an input to B is necessarily an interaction, but the term here includes transfers that become inputs only in a figurative sense, as when ownership is transferred to B. Under these circumstances a "change" in B would at the minimum consist of information that he is the new owner. Among humans *interaction* means specifically communication, transaction, or both.

c. *Social interaction* is any interaction that involves contingency.

d. *Contingency:* An action by A is contingent on B if it is designed or modified to affect B or A's relation to B. Since that effect is still in the future, only expected future states or actions of B, not past ones, are relevant to contingency. As with a decision, the past is an im-portant determinant of the base from which an action is taken, but the past cannot be influenced by the decision. Suppose B dies and leaves indigent A a fortune. Certainly A's expenditures thereafter can be far different than without B's act. But, ruling out insanity and supernatural processes, A's expenditures can have no possible antici-pated effect on B and hence cannot be contingent. If A combs his hair out of habit or to please himself, the action is not contingent; but if he combs it in the hope B will see it, it is. Except for talking to oneself, all speech is presumably contingent. Among three parties an action of A is contingent on B if A alters his behavior toward C so as to affect the A-B or B-C relation. The future orientation in this definition of contingency is related to that in our model of decisions.

Contingency presupposes prior interaction but not necessarily prior contingency, and A must at least know of B's existence. A's reading about B's misfortune in the paper is an interaction according to item b above and may lead to A's contingent behavior of sending a note of condolence. But unless B generated his misfortune or the newspaper account of it for its expected effect on A, his action was not itself contingent on A. How explicit must contingency be to be worth considering? Our answer depends on the individual case, as in the vague case of throwing a bottle into the sea in the vague hope someone somewhere will get its message.

e. *Mutual contingency* is a situation in which B's response to A's contingent behavior is in turn contingent on A. This definition does not preclude simultaneous contingency, though any real interaction must presume some time lapse. In some usages *interaction* itself implies contingency. Although such usage might be feasible for social analysis alone, *interaction* has been defined broadly enough here to apply to any system whereas contingency is possible only in systems capable of building images of future states.

f. *Boundaries of an interaction:* An interaction is bounded analytically by specifying the systems (parties) involved, the type of interaction (such as transaction or movement of matter-energy), and the time duration. An interaction might range from less than a second (for two simultaneous nods of heads) to centuries (for a nation's struggle for independence). A long, extended interaction would almost inevitably involve analyzable subinteractions along the way. The relationships among them could in turn be analyzed in several ways, as with main and subsidiary transactions (3.28.5) and dynamic sequences in transactions (3.28.1–.45).

g. *Number of parties:* Most generalizations about interactions in this book are based on two parties, A and B—that is, on dyads. C's are involved only as they affect the A-B interaction. Cases involving additional parties (as in competition, coalition, or consensus) are clearly identified. In line with the distinction between acting systems and pattern systems, we will never treat interactions as themselves constituting systems.

BASICS OF COMMUNICATION

3.10.0   *Definitions About Communications*

a. *Communication,* for purposes of this volume, is any transfer of information (pattern) between systems or any movement of anything between systems analyzed with respect to its information content. However, to clarify the process we revert temporarily to general transfers of pattern (3.11), whether or not a system is involved at either end of the transfer.

b. A *sign* is a pattern, transmitted in a communication, that represents some other pattern which is its meaning. In successful operation a sign has the capacity to select a channel that contains or constitutes its referent. Signs are usually, but not necessarily, arbitrary.

c. A *referent* is the pattern represented by a sign or the meaning of the sign. Operationally the referent is the content of the channel selected by a sign (see also item h (1) below).

d. A *symbol* is any pattern produced by a person (system) in a medium external to himself that corresponds in some way to an internal pattern. A sign is one kind of symbol. Following the lead of White (1949, chap. 2) we may use the verb form *to symbol,* which we define as the process of externally reproducing an internal pattern (see 2.41.23b and 2.31). Although any nonrandom external behavior symbols the internal pattern that gave rise to it, our interest in symbols will center on relatively widespread and durable ones.

e. A *semantic sign* is a sign whose referent is an information concept. If the concept is itself a coded representation of some external reality, that reality might be considered a referent at second hand of the semantic sign. But for the analysis of communication we are directly interested only in the relation of the sign to the image that is its referent, which relation requires one kind of analysis, not in the relation of the image to reality, which requires an utterly different kind of analysis. Some nasty philosophical and analytic problems are avoided if we stick strictly to this formulation. The relation of concepts, and hence indirectly of signs, to reality is important to the pragmatics of communication but not to its basic theory.

f. A *syntactic sign* is a sign that indicates a relationship among semantic signs. Some words, primarily prepositions, indicate a relation among the referents (glass *on* table) rather than among the signs, strictly speaking, and will be considered semantic. Often the syntax indicates how the concept referents of semantic signs are to be intersected to produce the concept being transmitted by a message.

g. A *code* is a set of categories (and hence a pattern system) of coded information. For this volume we identify several types. (1) In sign communication a code is a list of signs and their referents. (2) In perception (receipts of uncoded information) a code is a set of concepts or images along with their perceptual cues. (3) Except for preverbal children and nonverbal animals, who can have only the second type of code, all codes of persons who already use language are presumably a mixture of the two types. One learns his set of images by some combination of sensory and linguistic inputs—some combination of personal experience and the communicated experiences of others.

h. *Code formation:* (1) For an individual, code formation is discussed in 2.10b and in item g(3) just above. We add that an individual cannot learn a language without attaining the set of concepts that are its referents—though the degree to which some students can successfully manipulate the specialized signs of an academic discipline with no notion of their referents has never ceased to impress me. In this model the referent of a sign to an individual is *his* concept, not the dictionary definition or another's statement. (2) For a society, general or specialized, the formation of a code occurs in the historical evolution of its language and in the experiences from which it grew. For some purposes a language and associated concepts can be deliberately created, as are specialized, technical languages. For communicational analysis the code must be viewed as detector states of the systems involved, even though external representations of the code can be assembled in a dictionary or glossary and may be referred to by sender or receiver. As with knowledge, the code is in the head.

i. A *message* is a piece of new or potentially new information transmitted by signs in a communicational act. The term "potentially new" evades the question of whether the receiver already knows the information; it asks solely whether the pattern transferred is capable of carrying new information.

j. *Perception, symbol communication, sign communication:* If I see a tree fall, my receipt of information is strictly perceptual (2.10m and 3.12c). If I see a boy fall, the information process does not change just because the thing perceived is another person. However, if I see him hit a nail I not only perceive an overt act; if I choose I can also infer something (right or wrong) about his internal DSE system states. I might also imitate him, in which case a pattern would be formed in my head similar to one in his head and constitute a communication in its broadest sense (3.17). Sign communications of messages, by contrast, require additional analysis and comprise the following two items. (1) Any clearly identifiable linguistic behavior, whether intended to communicate or not. Unintended communications include eavesdropping and reading someone's secret diary. (2) Any nonlinguistic behavior deliberately modified to communicate, including pantomime and the deliberate exaggeration of gestures. For such behavior to be construed as a sign it must be contingent on a hope or expectation that its referent will be understood by the observer. Except to the extent that such signs are involved, even fully mutual perception (A and B perceive each other and each perceives that the other is perceiving him) will be considered perceptual, not messages. Such mutual perception could occur even if the parties possessed totally different languages, or no language, and frequently occurs among animals devoid of language. Each party acquires his information by his own observation, not by transfer from the head of the other party. (See 3.16.61 below for amplification of this point.) As in any category, some cases are ambiguous. *Ouch!*, for example, is a word. Yet it is sometimes uttered like a groan or yelp, and to hear it can be as purely perceptual as to see someone beyond earshot bump his knee and rub it. Swearing is similar. The classification of such cases depends on whether perceptual or communicational analysis seems to describe it better.

3.11   *Levels of Information Transfer*

As a transfer of information between systems, communication can be understood if we examine it in the light of three lower levels of information transfer and one higher level. A is the initial location of the pattern and B is the location to which it is transferred.

Level 1: Direct transfer by contact—information uncoded at A and B. Examples of such pattern transfer are boot to mud, wind blast to window rattle, thumb to thumbprint. That the police may code thumbprints by type does not make the original print a coded transfer.

Level 2: Mediated transfer—information uncoded at A and B. Examples are the transfer of the pattern of a mountain via the medium of light waves and lens into its photographic image on film and the transfer of the pattern of a mammoth's foot via fossilized mud to a plaster cast.

Level 3: Mediated transfer—information uncoded at A, coded at B. This level is the typical perception of environment by humans and is treated in detail below. *Environment* for this purpose may also include other humans.

Level 4a: Mediated transfer—information coded at both A and B. This level describes "communication" and is treated in detail below, both with and without signs.

Level 4b: Social communication is the same as level 4a but with contingency. Aristotle wrote things I have read, and thus there has been a communication. But his writing was not contingent on my existence nor does my reading it affect his behavior. My reading Aristotle is obviously contingent on the fact that he wrote, but that is a different meaning of the word. As here defined, my behavior would be contingent only if I modified it to produce some effect on Aristotle.

3.12   *Steps of Information Transfer*

A transfer of information involves several steps, the number depending on its level.

a. A transfer of uncoded pattern by direct contact (level 1) is a one-step process and requires no further discussion.

b. A mediated transfer of uncoded information (level 2) is of no direct interest to social analysis. But the steps involved are a prototype for coded communications and will be examined. Here there is a transfer from A to M (medium) and then from M to B. Although the logic is the same if M remains still while A and B come into successive contact with it, for simplicity we assume that M moves from A to B. If we include the existence of the pattern at both ends of the process we have five steps: (1) A pattern exists at A. (2) The pattern is transferred to M. (3) M moves from A to B. (4) The pattern is transferred to B. (5) The pattern exists at B.

If we now call A the *source,* B the *receiver,* M the *medium,* the first transfer *encoding* (or modulation), and the second transfer *decoding* (or recoding or coding, depending on the circumstances), we have put into the language of communication theory the irreducible conditions for a transfer of pattern from A to B without direct contact. But receipts of information that are coded in the receiver require an additional step. To describe this step with respect to humans, the patterns in the medium must first become sensations, a step we will call *detection,* after which the pattern in those sensations is identified, or decoded. In marked contrast to decoding, detection is an essentially isomorphic transfer of pattern:

there is to a significant extent a one-to-one correspondence between the elements of the pattern in the medium and that detected by the receiver. No parallel problem arises in a coded source, so no additional step is specified there. We could therefore list six steps but will limit them to five, as before, by joining decoding with existence of the pattern in B. The detection step in a receipt of information is only one of many functions performed by the detector of a controlled system.

c. Finally we deal with a mediated transfer handled as coded information by B, whether uncoded (level 3) or coded (level 4) at A. We can illustrate these latter alternatives simultaneously, assuming that the receiver is a person and that the source is either a slammed door (uncoded source) or another person with an idea in his head (coded source). (1) A pattern exists in the source: the slammed door undergoes a pattern of vibrations or a person has an idea. (2) The pattern is encoded in the medium: the vibrating door sets up sound waves or the person does so by speaking. (3) The medium moves from source to receiver: sound waves travel. (4) The receiver detects the patterns in the medium: his auditory nerves are activated; he has auditory sensations. (5) The receiver decodes the pattern: he recognizes the sound as that of a door slamming or extracts the meaning of the spoken words. Upon decoding, the pattern exists in the receiver. For either perception or linguistic communication, if the person does not yet know the appropriate code this final step may be a coding instead of decoding (see 3.17.12).

   To deal strictly with humans and to clarify a previous point, perception is the receipt of uncoded and communication the receipt of coded information, but with both handled in coded form by the receiver. For sign communication the five-step sequence is the same whether the signs are words, nods, gestures, tone of voice, speed or loudness of speech, smoke signals, lanterns in Old North Church, or deliberately bent twigs. Human encoding and detecting are done at the system boundaries, and several intervening recodings occur. In human speech, for example, before an idea in A's head can take the form of modulated air waves it must first be recoded successively into words in the cortex, efferent nerve discharges, and muscular movements of diaphragm, larynx, tongue, and so forth. The neurophysiologist might insist on other stages as well. The pattern also goes through a succession of recodings in the hearer, starting with physical activation of parts of the ear by sound waves and ending with an identified pattern in the cortex. One could argue that one of these stages is also an encoding or a detection in some sense. We define both as occurring only at the biological boundaries since it is only between the boundaries of the two systems that the medium moves; all other stages are intra-system. Encoding is an output action, detection an input one.[1]

3.13   *Quantification of Communicational Content*

3.13.1   For purposes of discussing quantification of information, a communication by signs between humans must be divided into external and internal stages. (*a*) The stages *external* to the communicating systems begin when the message has been encoded into an external medium. They include transmission through the medium and conclude when the patterns in the medium cross the detector boundaries of the receiver. These stages are subject to quantitative measure of the number of *bits* of information. (A bit is the amount of information required to select unequivocally between two mutually exclusive and equally probable items.) The bit count may be made at any of several levels—such as the number of bits per word, per letter, per phoneme, or per second of modulated radio wave. Bit counts are of great importance in such areas as radio and telephonic communication, computer operation, and statistical linguistics, but they have little direct relevance to social analysis. Since a count of bits is done by a successive halving of doubt, in a certain sense it implies a hierarchical arrangement of information. And since a hierarchical structure is probably more efficient than a non-hierarchical one (2.10f) some things we have learned in counting bits may help us in efficiently ordering information internally. However, we need not deal with that problem here.

(*b*) The stages *internal* to the communicating systems in-clude the source's process of recoding his thoughts into words or other signs. This process in turn includes his choice of semantic terms and of syntax and his decisions about how much detail to include in a message. At the receiving end are questions, first of recognizing the signs and then of decoding their meaning. Recognition is presumably a straightforward perceptual process. Although a speaker who mumbles may present a practical problem, the socially interesting problem is not in recognizing the words but in extracting their meaning.

There is apparently no way of quantifying information con-

---

[1] Part of the problem here is that usage of *coding* and its prefixed variations is far from standard. I have arbitrarily defined *coding* and *decoding* to apply solely to "coded" information (2.10b), the former to forming a code and the latter to recognizing a code item from cues. By contrast I apply *encoding* and *recoding* to essentially isomorphic transformations of either coded or uncoded information, but without a shift between coded and uncoded levels. A better set of usages could probably be worked out.

tent by bit count at the initial coding and final decoding stages, since there is no way of knowing how much information is attached to a sign by the sender or receiver. The referent of a sign might be a simple notion, like *dot,* or an entire field of study, like *biology.* Furthermore the referent of *biology* is utterly different to the specialist than to an eighth-grader, to whom it is "that smelly class where they cut up worms." The referent of a word to a person is the product of *his* experience. It cannot be precisely the same for any two persons; it is private information to each and is not directly examinable by an analyst. Thus, although we can quantitatively measure its external stages and probably its internal neural transmission, we can never measure quantitatively the amount of information that constitutes the referent of the sign to either source or receiver. Another way of describing the difference is to say that all recodings are isomorphic, however much they may fail in precise point-to-point correspondence; whereas linguistic coding into words and decoding from words could be isomorphic only by accident.

3.14   *Accuracy of Communication*

3.14.1   Aside from the question of code identity between A and B, which is dealt with in 3.14.2, the degree of accuracy in a five-step model depends on the following major factors:

a. Source: the degree to which the pattern to be transmitted is clear in the mind of A.
b. Encoding: the degree to which the signs chosen by A represent that pattern. Accuracy may be enhanced by redundancy, as by repeating the message in different words or by giving illustrations.
c. Medium: the degree to which the signs are transmitted to B fully and without distortion. Again redundancy may compensate for such inaccuracies as noise or inadequate signal strength.
d. Detection: the degree to which the signs are correctly detected by B—that is, the degree to which he correctly perceives the words, gestures, and so forth. For example, a speaker's dialect or speech impediment could impede such detection.
e. Decoding: the degree to which the patterns selected in B's head correspond to the signs received.

3.14.2   The remaining question of accuracy depends on the degree of identity of codes between A and B. As indicated (3.13.1b), there is no such thing as complete identity. Hence there is no completely accurate communication but only a degree of accuracy proportional to the degree of similarity. If a message leads to an answering message or to another behavioral response, however, and if the response corresponds closely enough to the expectations of the source, we can say that the communication is accurate enough for its intended purpose. A response may, of course, correspond to expectations by accident. Or it might fail to correspond, even though understood, because of deliberate misrepresentation. But misrepresentation is a transactional question, and we need not propose a theory of accidents to accommodate it.

3.15   *Effects of Communication*

A communication from *A* may change *B*'s detector, selector, or effector states. In the parallel language of Ackoff (1960, p. 4), one can communicate information, motivation, or instruction. Strictly speaking, only information can be communicated. Once the message is decoded, however, it may motivate or instruct as well as inform. "Stay away from that chick or I'll slug you" is an instruction and a motivation. But unless it is first understood as information it cannot perform either function. Information about how to do something is also just information. Performance skills are acquired by practice, not communication, though communication may help direct the practice. A communication can, of course, describe the detector, selector, or effector functions of its source, and the classification of its content is not necessarily related to its intended effect.

Signs can also be conceptualized and be communicated about in metalanguage, as when we speak of "the word *blue*." These concepts can acquire conditioned attachments to valences or emotions, as with "dirty" words or those that arouse religious awe. If analyzed as concepts, they raise no new theoretical problems. The word itself must still be recognized from the signals that carry it, and the thing that moves from person to person remains information—a pattern. What happens to the receiver is a matter internal to him, not one of transferring the pattern.

3.15.0   Two widely used terms can be defined communicationally as follows: *Intellectual influence* is the ability, through communication, to alter the detector of others so that certain things are no longer conceived or perceived as before. Intellectual influence changes concepts or their associated cues. *Moral influence* is the ability, through communication, to alter the selector of others—that is, to change their motivational set about certain things. Since secondary motives, presumably the only motives that can be changed by communication, consist of conditioned associations of valences to concepts (2.23b), one might exert moral influence through the intellect by changing the concept to which a valence is attached. Both definitions are consistent with that of interpersonal power—the ability to bring about desired external states, one person's external states including the internal states of others.

3.16   COMMUNICATION MODEL

3.16.0   *Assumptions*

To distinguish communications from transactions, we deal initially with a "pure" model of a communication from A to B with the following characteristics:

a. The communication is based directly on the detector states alone of A and B, not on the selector or effector, and is an interaction between their detectors. Any selector or effector elements that A puts into his transmission must first go through his detector and be processed as information, and any effect on B's selector or effector must depend on their prior receipt by the detector. Any such effects are *intra*system; the *inter*system relation handles information only. Like the pure detector function, the pure communication is not distorted by feelings, though it may carry information about feelings.

b. It is irrelevant whether the message is true or false or even a deliberate lie. The only communicational question is whether, and how accurately, the pattern initiated by the source is reproduced in the receiver.

c. A does not lose information by virtue of having communicated it, nor does he gain any. A person often clarifies his own ideas in the process of communicating; but this activity is internal and not communicational. There is also a technical sense in which information cannot be increased in one place without being decreased in another (J. G. Miller, 1965, p. 196). That conclusion does not apply to the concept of pattern transfer used here; the pattern in a head, in a rubber stamp, or in a photographic slide is not much diminished by the amount of pattern transferred to another head, paper, or a projector screen.

d. The cost or benefit that accrues to A or B from the content or process of communication, if any, is ignored.

e. Contingency is not required, though it is not necessarily absent.

3.16.1   *Deductions from Model*

Before examining deductions we may note that concepts and generationalizations about pure communications are derived from such fields as information and communication theory, semiotics, linguistics, philosophy, and psychology, with no assistance needed from the purely social sciences.

3.16.2   Information is a pattern, communication is a transfer of pattern, and a pattern contains at least two elements. Thus to transfer by language any pattern except the referent of a single sign requires at the minimum (a) the transfer of two semantic signs, one for each element in the pattern, and (b) the transfer of a syntactic sign to indicate the relationship between them. For convenience *sign* and *word* are used interchangeably, even though a sign with a single referent may consist of more than one word (*blue grass*) or part of a word (*pre-*).

3.16.3   If *B* is to understand the communication—that is, to reproduce in himself the pattern transmitted to him—then in addition to detecting and identifying the signs he must (a) already have learned and stored the concept pattern that is the referent of each semantic sign, (b) have learned the relation between sign and referent so that each sign constitutes a sufficient cue for its referent, and (c) put those two patterns together in the relation indicated by the syntactic sign. To put the two patterns together means to form a concept or image, however temporary, that consists of the intersection of the two identified by the semantic signs. If you already possess the concepts of green and ball and already know the words for them, then *green ball* designates a concept represented by the intersection of the two, but which could not be represented by either alone (2.10b). If you know all the concepts, by intersection you can similarly create in your head the image represented by "I shoveled snow from my driveway last July." You may doubt my veracity, but you could not even do that until you had first got the image. We need not discuss syntactics except to observe that the order and forms of these words do not mean that the driveway shoveled my snow or snowed my shovel.

3.16.4   Since the pattern represented by a single semantic sign must already be in *B*'s head if he is to be able to decode the sign when he receives it, no new pattern (or change of pattern) can be created in *B* by receipt of a single semantic sign. None at all can be conveyed by syntactic signs alone; that their source issues meaningless syntatic words may provide information about him by perception, but the words themselves remain meaningless. Although two or more semantic signs are a necessary condition for conveying new information, they may not be sufficient to ensure that it *is* new.   **3.16.41** A single sign can, however, call to mind a pattern already there. If I say "bread" you may start thinking about it, but you cannot learn anything about it from the one word. To strengthen 3.16.2 by removing the possible exception it permits, a single semantic sign can evoke a pattern already in the receiver. But it cannot transfer that pattern to him and hence cannot meet the definition of communication.   **3.16.42** Similarly, by special arrangement a single sign can call to mind anything the parties have agreed it should. In the case of "One if by land and two if by sea," all concepts about the arrival of the British and their possible modes of transportation had already been communicated in advance. So had all the information about where the lights were to be put and what Mr. Revere was to do when he saw them. The only information remaining to be transmitted concerned which of the two already-formed concepts about the British was to be activated in Revere's head. Any simple prearranged dichotomy would do—yes-no, on-off, one-two, blue-green, sign-nonsign. If the British had tried something else, like retreat or attack by helicopter, neither sign could have conveyed that new idea. (Note that two lights were one sign as in Morse Code, where two dots represent *i*, not two *e*'s.

3.16.43   To combine and restate the two preceding points, a single sign can select or identify a channel. But it cannot put into it information that was not already there or provide information that would constitute a new channel. To speak of the intrasystem parallel of this interaction, a single sign can create the equivalent of perception, but two or more signs are needed to create the equivalent of concept-learning.   **3.16.44** Since the concepts conveyed by a message are made by intersecting existing concepts, it follows that the brain must be able to make intersections at the same rate it is able to comprehend the message. At least with messages about mundane affairs the process is rapid, and experience suggests that it is completed almost instantaneously with receipt of the words. By comparing this statement with 2.10p, it would seem that a message cannot bring its receiver to form a concept new to him by abstraction and move up the hierarchy of his conceptual structure, in the same direct sense that it can bring him to form a new concept by intersection and move down the hierarchy.

3.16.5   "To select" presumes more than one alternative, and to select one channel rather than another means that its sign was received rather than that of another. This means in turn that the sender had to transmit one sign rather than another and had to have available two or more signs from which *he* could select. This is the basis for the general statement that a sign can have no meaning unless at least one alternative sign was also possible under the circumstances (Ashby, 1958, pp. 123f).

3.16.6   The multiple signs in a message can be in different modalities. If I say "Ouch!" and deliberately rub my knee I can communicate that my knee hurts. As with pure linguistic communication, the message can be equally clear whether it is true or false. Another semantic element is speed or tone of speech, as when a slow, groaning pronunciation of *t-i-r-e-d* indicates that the word describes the speaker.

3.16.61   Beyond the cases where a nonlinguistic item is deliberately used as a communicative sign are numerous gradations of information acquisition. Although simple perception of another party is not communication—and neither is mutual awareness of mutual perception, taken alone—the latter can constitute one element of a pattern and can be combined with a linguistic element to produce the minimal two required for a message. If you shout "Stop!" as you see me preparing to ax your cherry tree the word is one semantic element and the mutually recognized perception of my action is the other. For such reasons it seems unnecessary to modify the conclusion that *any* communication of a pattern between humans requires no less than two semantic elements.

3.16.62   The intended meaning of a communication is sometimes different from its straightworward meaning, as with irony, sarcasm, humor, or ritual statements. Since we recognize as semantic signs such factors as context and tone of voice there is no additional theoretical problem in nonverbal signs that mean "not," "the opposite of what I say," or "you're off your rocker."

3.16.63   From the conclusion that a single sign cannot put new information into a channel (3.16.43), it follows that an idea communicated by language must have something in common with other ideas, for if it were strictly unique it would not be described by intersecting the concepts that are the referents of existing words. This means in turn that knowledge is necessarily structured hierarchically in its communicated form, whether or not it is learned or stored hierarchically in sender or receiver.   **3.16.631** In a verbal communication, concepts that have names are intersected to represent other concepts that do not have names. If we assume that names are more likely to be given to frequently communicated concepts than to infrequently communicated ones (see 3.18.62 below), we can also say that the concepts communicated by messages are relatively unique compared to the relatively repetitive concepts represented by particular semantic words.

3.16.64   Large communicated patterns, or message arrays, are constructed by similarly putting together smaller ones. To start at the bottom, two or more elements for which there are signs are combined by stating the signs and the syntactic relation between them. Such combinations constitute a sentence. The pattern represented by the whole sentence is combined with the patterns represented by other sentences. Each such pattern is combined with other such patterns to make a larger pattern, which is then combined with other larger patterns, and so on for the whole of a paragraph, article, or book. At the sentence and part-sentence level the syntactic signs are word endings, word order, prefixes, suffixes, and the like; at the next higher level they are conjunctions, articles, and prepositions. At still higher levels the syntactic units may be sentences (opening and closing sentences of a paragraph), paragraphs, or even whole chapters. These units are not syntactic in the conventional grammatical sense; but they are syntactic in the logical sense that they indicate the nature of the relationship among the parts. In a tightly structured monograph the whole article or book is a single hierarchy whereas in simple narrative it is a sequence of smaller hierarchical units.

3.16.7   Since all human information of a complexity relevant to social analysis is handled in the coded form of concept patterns and any language associated with them, to communicate by language requires the same two steps of learning and using patterns noted in 2.41.21, along with their improvement over time. The first, or learning the code, is to learn the signs, their referents, and the syntax along with the motor patterns that control the muscles of speech or writing. The second, or using the code, is to encode messages sent and decode messages received. As a result of using the code and detecting errors or difficulties in doing so, either the code itself or the mode of using it may be revised.

We can now recapitulate three levels of code-forming, code-using, and code-revising in Table 3 to show their parallel processes. The chart is confined to receipt of information, since perception has no sending process.

3.17   CULTURE

For all human beings living in societies, which effectually means the whole human race, much information is learned from others as part of the culture. For the analysis of this process several definitions are necessary.

3.17.0   *Definitions about Culture*

a. *Culture* is communicated, learned patterns. Like other system phenomena, these patterns can be classified into two levels. At the *matter-energy level* are external patterns produced in correspondence to internal ones. They are the symboled aspect of the culture —bowls, scythes, houses, radios, kinship behaviors, governments, rituals, roles, markets, pierced ears, baseball games. With respect to rituals and governments, the external, real aspect is the actual performance of people, not the concepts or categories of behavior. Strictly speaking the matter-energy level also includes the material form of the external stages of communication—books, films, drawings, gravestones. At the *information level* are information-processing patterns. These patterns include attained concepts and the linguistic code of signs, referents, and syntax. Attained concepts range from specifics like corn and eating to religion and worldview; they include moral and behavioral prescriptions. To the extent that values and preferences are regularly associated with a concept or with messages about it, they are construed as part of the concept. Here we are following 2.23b and .24 to the effect that one cannot have or express a valence toward any external until he has first formed a concept of it and that the valence is attached to the concept, not to the external. To the large extent that patterns produced externally by man are also given names, the information structure of the culture parallels the matter-energy structure. In addition to its man-made components, the information structure includes names and concepts of objects and events in the natural environment. Here

Table 3

Three Stages and Three Levels of Processing Patterns of Information Inputs

	Stage 1 Pattern Learning (Learning the Code)	Stage 2 Pattern Identification (Using the Code)	Stage 3 Pattern Modification (Revising the Code)
Level 1 Nonlinguistic (Sensory) Experience	Forming concepts	Perceiving	Modifying concepts in light of additional perceptions, especially faulty ones
Level 2 Linguistic Experience (individual as receiver of messages within a society)	Learning the language—signs and referents	Identifying referents of separate semantic words of message	Forming new concepts represented by the intersection of concepts in the message. Also learning more about the language and improving the ability to use it
Level 3 Science (specialized division of culture using formally created rather than natural language)	Forming hypotheses or theories and the associated classification system	Empirical testing of theories	Modifying theory in light of additional evidence

the signs and conceptual categories are cultural but the externals are not. Beyond these are abstractions and imaginary concepts that are part of the culture but whose relation to any external other than symbolic representations of them (such as totems) is problematic indeed. Conversely, though perhaps to a lesser extent, humans also produce external patterns to which they give no names—for example, many decorative patterns. Common patterns of behavior may not be given names, possibly because they are not thought of consciously or are not the occasion for comment. In English, for example, we apparently have no word to distinguish the full-length arm movement of walking from the bent-arm movement of running.

b. *The cultural content* (or *body of culture*) is the total set of cultural patterns common to a collectivity of people. For convenience, but without presuming that the terms cover all cases, we may designate the important types of culture content as *artifacts, sociofacts* (social structures and behavior patterns including rituals), *language* and its conceptual structure, *performance skills,* and the *values* attached to any of these.

c. *The cultural process* (or *cultural communication*) is the communication of cultural content to and among that collectivity, particularly to its new members. Under certain circumstances the process may cross societies, as in cultural fusion and diffusion below.

d. A *society* is a collectivity of people having a common body and process of culture.

e. A *perfectly transmitted culture* is a situation in which all members of a society possess an identical body of culture, accurately communicated to all new members by the cultural process. Such a situation is impossible in fact (like a perfect economic market or a perfect lever). The concept is analytically useful, however, and tight traditional societies come reasonably close to the concept in practice.

f. As to *levels of societies and cultures,* depending on the circumstances any body of people from two to the whole human race might be considered a society. Some generalizations about culture apply to all sizes, others to a limited range. The terms *subculture* or *subsociety,* and occasionally *superculture* or *supersociety,* indicate units smaller or larger than the one on which attention is focused. For the later analysis of government in clearly defined nations we will arbitrarily identify the whole society with the nation, but that usage need not concern us here.

g. *Cultural fusion* is the complete or partial merger of multiple cultures and their respective societies.

h. *Cultural diffusion,* by contrast, is a merging of cultural content of two or more societies by cultural communication, but without merger of the societies. This definition does not preclude total merger, though it is difficult to imagine in fact.

Thus a society is an acting system; its culture is a pattern system. The cultural process is a behavior of a society. As such it is not itself a system—certainly not an acting system—though certain relationships of the actions may constitute a pattern system.

The term *learned* in the definition of culture excludes from culture all inherited behavior or perception, as well as instinctive communications like those of bees. The term *communicated* excludes patterns learned by one individual but not passed on to others; thus it makes the phenomenon "social" by including more than one person. *Communicated* also excludes concepts or behaviors common to many people but learned independently by each; thus it inescapably implies that the people who constitute a society interact. Scratching an itch might illustrate independently learned behavior, though it is hard to know how much imitation is involved. Some animals other than humans apparently have the ability to transmit certain learned patterns from generation to generation but have little or no apparent ability to build new patterns on top of them. These transmissions are cultural, as here defined. The fact that humans can continue to accumulate a culture more or less indefinitely, while other animals do so minimally or not at all, represents a difference in learning ability, not in the nature or meaning of the cultural process.

Although social analysis is interested primarily in communicated, learned patterns common to relatively large groups, the preceding definitions do not specify or depend on size. However, to ignore size does not mean that it makes no difference. By analogy, the fact that the concept of the lever is the same for the tiniest watch and the largest engine does not imply that there are no other significant differences.

3.17.1 *Culture as Communication*

3.17.10 Given that culture is defined as a communicational phenomenon, some of its details also fit that definition. **3.17.11** The cultural process is communicational by definition. **3.17.12** The cultural content is wholly communicational, as can be seen by examining its various parts (3.17.0a). The linguistic code is obviously communicational. Attained concepts are, by definition learned by communication from others. To the extent that communication is nonlinguistic it is covered by the next point. Matter-energy level patterns also behave as communications, presumably uncoded, and show all the following five steps. (1) At the *source,* the pattern of (say) a bowl exists in the heads of present members of the society. (2) The pattern is *encoded* (symboled) in clay. (3) The clay bowl is itself the *medium.* (4) As to *detection,* new members of the society receive sensory impressions of the bowl by sight, touch, and so forth. (5) Those new members then *recode* the pattern of the bowl, which pattern then exists as an attained concept in their heads.

3.17.2   All external behavioral patterns are symbolings of internal patterns (2.41.23 and 3.10.0d). If we now assume a perfectly transmitted culture, all internal patterns of cultural content of one generation are learned from observing the symboled patterns of previous generations and this process goes on indefinitely. This is one basis for the idea that culture "has a life of its own" (White, 1949, p. 100)—which means simply that if the copying technique were perfect, a Xerox of a Xerox of a Xerox . . . would be exactly like the original and could be no other. A perfectly transmitted culture would transmit learned patterns as precisely as genes transmit inherited ones, and without the variations imposed by sexual reproduction. Having clarified the theory with the pure case we can defer the complications until later.

3.17.3   Humans have evolved biologically to the point where they inherit no inborn mechanisms (2.09.3) for providing food or shelter, procreating, or otherwise taking care of themselves. The likelihood that the individual would learn these things from his own experience soon enough to survive, even if brought through infancy by his parents, seems so small that we assume culturally acquired information about relatively durable patterns to be indispensable. Furthermore, unique information often cannot be made part of the culture. The culture can include information about lions in general, for example, but not about that particular lion now preparing to spring from yonder bush. Unique information may also be important to an individual and may be acquired faster and more reliably by exchanging with others than by relying solely on one's own perceptions. **3.17.4** Restated with different emphasis, the individual would remain primitive indeed unless he could build his own patterns on those accumulated by the whole society over many generations. Along these lines empirical evidence suggests that a child's mental development will be greatly and perhaps permanently retarded unless he starts receiving communications at an early age (Sherman and Key, 1932; Spitz, 1945; Pines, 1973).

3.17.5   To the large extent that the individual depends on both the general information embodied in the culture and the specific information exchanged by messages, he must both learn the language of the society, thereby automatically learning much of its culture when he learns the referents of its signs, and receive messages in that language.

3.17.6   *Assumptions and Propositions on Cultural Communication*

On the basis of 3.17.4 we relax the assumptions that truth or falsity is irrelevant (3.16.0b) and that there are no costs or benefits to the communicational act (3.16.0d). Instead we substitute the following assumptions:

a. The information embodied in the culture and in messages is consciously or unconsciously desired, and truthful and accurate messages are preferred to untruthful and inaccurate ones. By *truthful* we mean only that the message is in accord with the source's images, not that his images are in accord with reality.

b. The parties involved are essentially equals; thus we ignore possible differences in knowledge, competence, status, or other factors. It does not seem necessary to deal here with such situations as my deferring to a biologist in his own field; and questions of status, desire to conform, and related transactional effects are discussed in later chapters.

3.17.61   The accuracy of perceptions is then related to communication with others as follows.   **3.17.62** With respect to perceptions, observations of the same phenomenon by several persons and comparison of the results improves accuracy in the same sense that the average of ten measurements of the length of a table is probably more accurate than a single measurement, at least in the absence of systematic bias. Certainly confidence in the average of ten measurements can be higher than in a single measurement. More broadly, every perception is a test of one's hypothesis about his environment (2.11); another test is communication with others about one's perceptions.

3.17.63   Different perceptions vary widely in the degree of confidence felt by the observer and hence in the possible value to him of additional information from others. We will distinguish the following situations:

a. If the concept is clearly formed and the perceptual cues for it are highly redundant (much more than sufficient), *A* will feel high confidence in his perception, need no additional information from *B*'s, and reject alternative explanations if offered. If in full daylight with unimpeded view and opportunity for multiple observations, being sober and healthy, I see a car enter my driveway, I will confidently hold to my perception and consider others mistaken if they say it is a horse. Such situations create little difficulty, of course, since contradictory perceptions are unlikely.

b. If the concept is clearly formed but the cues are insufficient, *A*'s confidence in his perception will be low (by definition of insufficient) and will justify revising his perception in the direction of alternatives offered by *B*'s. Still assuming equality among all parties (3.17.6b), the average among these multiple perceptions is the most probable true value (3.17.62). Examples are perception of the amount of movement of a small light in an otherwise dark room when the only cues are the proprioceptive sensations of eye and neck movements (Sherif, 1936) or identification of a small grey object twenty feet away in the moonlight.

c. If the concept is not clearly formed or there are no clearly sufficient cues, *A*'s confidence in his own perception will be low. Here he is without significant information of his own and has little choice but to rely heavily on information from others. Obvious instances are those in which *A* simply lacks the capacity to observe the phenomenon in enough detail to form a clear concept for himself, as with earthquake, eclipse, lightning, magnetism, or the shape of the earth. Other instances would be those where externals appear in a seeming continuum of patterns (like trees) rather than in sharply separated ones (water versus ice). How, for example, could a person solely from his own experience classify the one-year oak seedling "the same" as its hundred-foot parent? In such cases the individual has no firm basis for questioning the perceptions and implied concepts provided by others, particularly if the others generally agree.

d. To generalize from items **a, b,** and **c,** the value of communication about perceived things varies inversely with the redundancy of cues and the clarity with which a concept is formed from one's own observations.

3.17.64   To perceive an event requires more information, *cet. par.,* than to perceive an object, since it involves at least the perception of two objects at different points in time, a comparison of their patterns, and a conclusion that the later pattern is a modified state of the earlier one rather than a wholly different pattern.   **3.17.65** The variety of events to which persons are motivated to give discriminable responses is much larger than the variety of objects. And the amount of information-processing devoted to events, including communication about them, varies in the same direction. The reasons are as follows. First, objects cannot interact. An interaction by definition involves a change, and the definition of an object specifies absence of change. Even if objects could interact, such interaction would be an event. Since an interaction involves a change (which is an event) in each interacting element, the number of events cannot be less than the number of entities interacting and can be indefinitely larger. Second, one usually responds to the behavior of others (events) rather than to their static states (objects), and the number of behaviors of persons is vastly larger than the number of persons. That is, it is behaviors of others rather than their static existence that give rise to satisfactions or dissatisfactions, and the same is generally true for nonhuman externals.

3.17.66   Regarding perceptions of events not involving behavior of others, since perceptions of events require considerably more information than perceptions of objects (3.17.64) it is less likely, *cet. par.*, that the cues will be sufficient. Hence under 3.17.63d communications from others about events are more valuable than about objects, and *A* is more likely to modify his own perceptions of events toward the perceptions reported by *B*'s.

3.17.67   Regarding perception of oneself and his own behavior, one's vantage point for observing his own behavior is poor. And when he speaks he knows only what he intends, not what meaning others extract. Hence he can perceive many aspects of himself and his own behavior only by inferring them from the responses of others. His hypotheses about himself are confirmed by expected responses and thrown into doubt by unexpected ones. What one can thus discover about himself is potentially complex and ambiguous, and sometimes deliberately distorted, for reasons to be seen in Chapter Eleven, especially Sections 3.46.4 and 3.47. Much information he must deduce from perceptions of transactional relations rather than having it transmitted to him by communication. Until we reach those sections, we can make only the vague but important point that one can form only a crude image of himself without communications from others.

3.17.68 However, a response of *B* to *A* depends on *B*'s traits as well as *A*'s. Hence *A* will require the responses of multiple *B*'s if he is to isolate aspects of the responses that reflect himself from those that reflect the traits of a particular *B*.   **3.17.69** As will be seen (Chapter Eleven), it is often advantageous to *A* to know how he is perceived by others. If he knows this, he will be motivated to observe others' responses to himself as a means of learning how they perceive him.

3.17.7   We now turn from perception of self to perception of behavior of others—a type of event. Since overt behavior is a symboling of internal states, for *A* to perceive *B*'s behavior well enough to give a satisfactory response requires information about *B*'s relevant DSE states. Except for relatively elemental motions (such as those dealt with by Harris, 1964) and obviously instrumental behavior like eating and dressing, many behavioral acts cannot even be identified without reference to the system states of the actor. Is Tommy really doing homework? Or is he just avoiding the snow shovel? Is that driver attempting suicide? Or doesn't he know the road is icy? We thus may note that in this model overt behavior is a function of system states, including that of the selector. In Harris's (1964) language, but with an opposite conclusion, this means that most units of behavior are *emic*, not *etic*—at least most of those relevant to social analysis.

3.17.71   If based on *A*'s perceptions alone, *A*'s information about *B*'s DSE states must be inferred from *B*'s actions taken in conjunction with their context. Projective tests are formalized sources of this kind of information.

3.17.72   If based instead on communications from *B*, such information must be based on *B*'s introspection and his ability to describe its results. Since system states and their interrelationships are complex, often badly understood by their owners, and difficult to communicate because the referents of the terms are so private, information received from *B* about *B*'s behavior can hardly be considered highly reliable. **3.17.73** Other things equal, communications from *B* are presumably more reliable concerning his explicit, overt acts and less so concerning the system states symbolized in that behavior. The main reason is that *B*'s overt behavior is public information. Hence under 3.17.62 (about multiple observations) *B*'s information about his own overt behavior can be improved in accuracy by comparing it with *C*'s observation of it, whether *C*'s observation is verbally communicated to *B* or is inferred by *B* from *C*'s responding behavior. And hence the referents of the terms used to describe the behavior are more likely to be similar for *A* and *B* because instances can be mutually perceived.   **3.17.74** There may often, of course, be reasons why B may misrepresent his system states to A or why he would communicate them more accurately to C than to A. These reasons are transactional and involve relations between transactions and communications (3.4). Hence they are not part of the model of pure communication, and discussion of them is deferred until later.

3.17.75   If based on communications from multiple C's, A's information about B's behavior may be improved in accuracy under the averaging effect of 3.17.62, the probable accuracy rising with the number of C's. The problem of getting reliable information about B merges with that of getting reliable information about anything from anyone, and the following observations (through 3.17.9) must be added. **3.17.76** Regarding matters other than B's own behavior, A can have confidence in information communicated from B and based on B's perceptions only if and to the extent that A has confidence in the accuracy of B's perceptions and of his communications about them. Since the ability to communicate is discussed elsewhere, we focus here on B's perceptions. **3.17.77** A will have confidence in B's perceptions if, *cet. par.*, A has confidence in his own perceptions and has observed that B generally perceives things as A himself does.   **3.17.78** Such coincidence in perceptions will be greatest, *cet. par.*, if A and B have similar codes of concepts and sufficient cues for identifying them.   **3.17.781** Conversely, a report from B that confirms a perception by A confirm's A's conceptual code as well as the perception.   **3.17.79** If we assume that A has observed the operation of 3.17.78, at least intuitively, he will then lack confidence in B's perceptions to the extent that he believes B's codes differ from his own, at least regarding the phenomenon involved. **3.17.791** Conversely, a report from B that confirms a perception by A supports a conclusion (at least intuitive) that B's code is similar to A's.

3.17.8   Little difficulty arises for perception of common objects with high redundancy (3.17.63a): I will trust even a dolt with almost any conceivable coding reliably to distinguish an automobile from a goldfish. Problems of reliability therefore lie in areas of greater complexity and particularly in behavior.   **3.17.81** Since the perceptual code is an important component of any culture, persons of different cultures presumably have different codes. If A knows or suspects this, and if he believes that B belongs to a different culture, he will lack confidence in B's perceptions.   **3.17.82** If A cannot know how closely B's code resembles his own but nevertheless wants to evaluate information from B, his best alternative, even if not necessarily reliable, is to generalize about B's code from any instances in which A and B have both perceived the same thing and communicated about it. Agreement on one perception will increase A's confidence in other perceptions by B, the degree of confidence rising as a positive function of sample size. Conversely, differences in one perception will decrease A's confidence in other perceptions by B.

3.17.83   If A also knows B's code, and if he can translate between the two, the degree of confidence is then presumably the same as if they shared the same code.   **3.17.84** However, the costs of information are higher with translation. By extension, information costs among any group are lower with similar than dissimilar conceptual sets.   **3.17.85** Hence if A prefers lower information costs he will prefer communications from B's who share his conceptual set. B's who share his culture will automatically fill this condition.

3.17.86   We now assume that A prefers information in which he can place high confidence. According to 3.17.81, A will then prefer communications from B's whose perceptual codes resemble his own. Again, B's who share his culture fulfill this condition.   **3.17.87** In parallel, since accuracy of communication is a function of the similarity of linguistic codes (3.14.2) A will get more accurate communications from B's who share his code. And since under 3.17.6a he prefers accurate messages, then *cet. par.* A will prefer messages from B's who share his communicational code.

3.17.88   Dissonance is created in A if he and B perceive the same thing through different perceptual codes and B then communicates his perception to A.   **3.17.89** Since dissonance motivates behavior that avoids or removes it (2.41.4), A will tend to avoid messages from B's who do not share his perceptual code.   **3.17.9** Statements 3.17.81, .85, .86, .87, and .89 are additive in creating a preference for communications from others of the same culture as oneself, who share the same perceptual and communicational codes.

3.18   **CONSENSUS BY COMMUNICATION**

Culture and the related process of symboling are subject to equilibrium analysis.   **3.18.1** If we assume learning in both concept-forming and performance skills, then as an individual acquires experience in both areas (a) his internal patterns will approach the external ones he observes and (b) the patterns he produces externally will approach the internal ones symboled. This positive-feedback interaction between internal and external patterns asymptotically approaches a stable equilibrium where internal and external patterns are identical. Precise identity is never reached—if, indeed, it is even definable—and the distance from it may be subject to an equilibrium of its own, as follows.

3.18.2   If we assume costs and benefits of modifying internal patterns and symboling them externally, the equilibrium falls some distance from identity. To illustrate, in the course of painting a picture, writing a book, or designing a house one starts with an incomplete image. His initial external reproduction is therefore an approximation, not a finished production, particularly for complicated patterns. Inspection of the external form helps clarify the image, from which a modified external pattern can be made and so on. In due time the cost of further modification exceeds its benefit, and the painting, book, or blueprint is construed as "finished." We have already noted that this process is one form of image-testing (2.41.23b).   **3.18.3** Even when some inner pattern is fully and sharply formed, if it is repeatedly reproduced externally (Stradivarius made many fiddles), then random variations—which are a partial mismatch between inner and outer pattern—provide a margin within which developmental change can occur. (See also Chapter Eleven up to but not including 3.44).

3.18.4   Also assuming learning, as two or more persons communicate successively (a) the more clearly they communicate, the more similar their internal patterns become; and (b) the more similar their internal patterns become, the more clearly they communicate. This, too, is a positive-feedback relation, with a theoretical equilibrium at identity. (For reasons to be seen in Chapter Sixteen, this situation constitutes the entropic process in a closed system.) As with the individual in 3.18.2, the equilibrium will fall short of identity since, possibly excepting science, virtually nothing is worth discussing until the parties see eye to eye on every detail. And as in 3.18.3 the impossibility of precise identity leaves room for developmental change.

3.18.5   *Definitions and Propositions about Communicational Consensus*

a. *Consensus by communication (cultural consensus)* is the achievement of similarity of pattern among persons through the positive-feedback interaction of 3.18.4, which may or may not be assisted by that of 3.18.2. Since the whole of culture is here defined as communicational, this consensus process can presumably be generalized to the entire content of culture, including values, artifacts, and sociofacts as well as the conceptual and linguistic structures. This definition, of course, does not specify when, whether, or to what extent such consensus will in fact occur. Identical patterns thus achieved would meet the definition of a perfectly transmitted culture if they were also passed on without change to successive generations.

b. *Collective concept formation* is the same as consensus by communication, but with emphasis on the developmental change as new concepts are formed and old ones modified, in contrast to the cross-sectional emphasis on equilibrium. To be collective, however, new concepts must be subject to the consensus effect as or after they are developed. Another form of such equilibrium, producing consensus by quantity rather than quality of communication but requiring the assistance of other factors, is found in 3.41.79. For the same reason as in item **a**, this process can also be generalized to the entire content of culture.

3.18.51   We will see in Chapter Eight that multiple transactions may be linked through involving the same people or the same things exchanged. Multiple communications are similarly linked by involving the same people or the same ideas. And just as the values of many different things are also related, as in the whole of a market economy, different ideas are also related in the totality of a conceptual set—or, more broadly, all patterns in the totality of a culture are related. (See the discussion of anthropology in Chapter Seventeen.) Different ideas are also related within a message by the nature of the communicational process, as in 3.16.63.

        The definition of a perfectly transmitted culture does not require consistency among its parts, only identity among all persons. The degree of consistency would depend on such factors as the logical capacity of the people and the compatibility of their goals. Consistency would presumably be greatest within consciously created aspects of the culture, such as science or engineering. In addition to logical or motivational inconsistencies, an "imperfect" culture would also show inconsistencies among individuals and subcultures, the consequences of which are dealt with later. Because a culture is a pattern rather than an acting system, system analysis, as such, can presumably throw no light on the relations among its parts (1.20g, i, j, k, l, and 3.17.0h). And although the form and content of a particular message are predicated on the conceptual structure embodied in the code being used, this chapter deals only with generalizations that apply to all conceptual structures.

3.18.6   We have noted (3.18.3) that the practical inability to achieve a perfectly transmitted culture leaves a gap for developmental change. Changes in nonhuman environment can also generate such change. We can illustrate with a change discussed in messages and simultaneously clarify the relation between concept structure and message. If corn in some society customarily grows five feet tall, that height is part of the concept referent of corn. If one year it grows to six feet people say, "The corn is taller this year." And if it grows to six feet indefinitely thereafter, for some years people will say, "Corn is taller than it used to be." But eventually these messages stop because corn itself now means a six-foot stalk. Economics already has a well-developed model for similar changes that seems worth exploring. Starting from equilibrium, a sudden but permanent shift in supply or demand sends the market through a temporary short-term equilibrium on the way to a new long-term one. Messages about the height of corn are similarly a transitional step between initial and terminal positions. In the final position of the market the marginal cost and benefit of an additional unit are consistent (in fact, identical) with average cost and benefit. If we view the initial changed height of corn as an increased demand for communication, the marginal benefit of additional messages rises during the transition and their marginal cost is incurred. Eventually the marginal benefit of each message goes down and its marginal cost is no longer incurred. There is also a certain way in which the stock of concepts incorporated in the code are "existing capacity" developed by past investment, and messages are current output produced by incurring variable costs. In both the market and communications an increased demand is met at first by straining existing capacity but in due time by increasing the capacity.

3.18.61   Since it is not possible for all information to be incorporated in the code (3.17.3), the long-run equilibrium operates by the same logic as the short run—that is, an equilibrium of costs and benefits— but with different content to the costs and benefits. At any given moment average total cost of information might be less if the capacity were larger, but since it is not there is no alternative but to incur the higher marginal costs of more and longer messages. To illustrate, before there was a word for osmosis every time the idea was communicated a paragraph or so of message was required. However, once the investment was made in giving it a name and agreeing on its meaning the same idea could be made part of a message by using only a single word.

3.18.62   If we assume some satisfaction-maximizing motive of the sort that seems difficult to escape (2.7), people would seek the lowest average total cost of communicating rather than the lowest variable or investment cost, taken separately. (For this problem we assume no significant fixed costs.) This problem resolves into a question of how many ideas will be given names and how many will be transmitted by messages. Since the general theory is merely that of equating the costs and benefits of additional words, additional detail would consist of identifying those costs and benefits.

The benefit, of course, is a saving in communication activity. "Install new brake linings, check and if necessary repair or replace wheel cylinders, master cylinder, and brake lines, and bleed and refill brake fluid. Oh, and don't forget to check the brake drum and have it turned if it's scored" thus becomes "brake job." One cost is increased effort to learn a new concept ("brake job" means all those tasks only after considerable experience and explanation). Another cost is increased communication until the new term and its meaning are agreed on and standardized. A third cost is that an increased vocabulary may increase the likelihood of semantic error. A fourth is that as more words are added the most discriminable ones are presumably used first (*lock* versus *key*) whereas less discriminable alternatives are left for later additions (*locks* versus *lox*)—though natural languages grow in ways that seem to make this aspect minor and unreliable. This list of costs is suggestive only.

3.18.63   The total benefit from introducing one new term is apparently a function of two variables. The first is the saving per usage, which would depend on two subvariables: (a) the reduction in message size (compared to the longer phrasing, "brake job" is a 95 percent saving in number of words) and (b) the value of the saving, which would depend on the value of communication time, paper, and the like. The second variable is the number of usages. If "brake job" must be referred to five times a day there is more point to the shortening than if it is used five times a year.

3.18.64   This equilibrium theory, like that of a perfect economic market, is to be construed only as a first approximation; deviations from the equilibrium occur in both directions and for numerous reasons. A distinction I have made elsewhere (Kuhn, 1967, p. 322)—

between *frictional* imperfections that slow or block movement toward equilibrium and *causative* imperfections that actively push away from it—might be useful for understanding this and other equilibria. Because of the absence of quantifiable measures of cost and benefit in most linguistic communications, deviations from equilibrium are presumably greater than in economic markets. Nevertheless many new words adopted by contemporary society, at least, seem to fall in the pattern of a cost/benefit theory.

3.18.7   Besides the question of assigning names to extant concepts is that of developing new concepts or modifying existing ones. To modify an existing concept means to change the meaning of its name; to add a new concept means (a) to describe it by several existing words, (b) to name it by an existing word, which will then have an additional meaning, or (c) to coin a new word. (The Germans combine (a) and (c) to make a new word that consists of several existing ones, thus illustrating the hierarchical structure of ideas.)

Without specifying how new concepts would be named, we can identify the possibility of an optimal rate of change that would balance the conflicting short-run and long-run values. Viewed cross-sectionally (and in the short run), messages would be most efficient and accurate if all persons held identical codes, as in a perfectly transmitted culture at equilibrium. Viewed developmentally (and in the long run), messages would be most efficient and accurate if the code were progressively adapted to the changes in concepts that arise from new learning or changed reality. Assuming that the changes are not adopted simultaneously and with identical new meanings across the whole society, such change means nonidentical codes at any one moment. As contrasted to 3.18.61, which deals solely with the number of signs used to communicate a set of concepts, this paragraph deals with a qualitative change in the concepts themselves, along with additional concepts—whether or not the number of signs changes.

3.19   **CONCLUSION**

All interactions between persons (or aggregates of persons acting as units) are classified here as communication, transaction, organization, or some combination. A major reason for this scheme of classification is that in basic cross-sectional form the analysis of the three is distinct and nonoverlapping.

With respect to *real* interactions I have no quarrel with those who insist that "every communication is a transaction and every transaction a communication"—as will be abundantly clear in Chapter Eleven. I would equally agree that every *real* political behavior has its social and economic aspects, every social behavior its political and economic aspects, and so on. But with respect to *analysis* I insist that good social science can be born, or even conceived, only to the extent that its main concepts and propositions can be rigorously isolated. If they cannot be rigorously isolated I would doubt their analytic value. In line with that approach this chapter has spelled out certain traits of communication, first as a purely informational process and then as a social one. As an informational process communication is related to human detectors. Some motives were introduced later in the chapter. But since these incorporate only desires for information they fall analytically in the selector subsystem of the detector and do not take us outside the main-level detector. Their effects are communicational, and certainly no transactional analysis was used or implied. The next chapter similarly explores pure transactions, with no mention of communications. The interrelations come later and in abundance.

Aside from delineating the basics of communication among humans, which need not be recapitulated, this chapter makes several points important to the overall framework. For one, the intersystem communication of humans is tied to the intrasystem detector functions, and relevant propositions suggest the parallel in psychological processes between concept formation and perception on the one hand and code formation and particular messages on the other. Moreover the chapter sharply distinguishes information about another person acquired by communication from that acquired by perception. In the first case, patterns go from the head of the source to the head of the receiver; in the second, the head of the source is not involved directly if at all. Furthermore, in sign communication the information is coded at its source and the transmission process can be understood only through linguistic analysis of sign and referent; in perception the information is uncoded at its source and no signs are issued or received. To insist, as some do, that to perceive someone else is a "communication" from him is to guarantee sloppy science. The distinction between sign communication and perception is also basic to the present categories for classifying systems, in which semantic communication is a major basis for distinguishing human systems from all others.

A second point is that since culture is crucial to social analysis, this framework must incorporate it. And unless the framework is to be massively breached, culture must be analyzable as communication, transaction, organization, or some combination. Once we have described all external behavior as symboling inner patterns and have stated the steps of communication, culture seems obviously communicational. In that guise it also seems more scientific and less mystical than in some other contexts. Whether sociologists and anthropologists will agree is another matter.

Third, an equilibrium approach to the tendency toward identity of internal and external patterns, both within an individual and among communicating persons, provides at least a first-approximation theory in an area presently characterized more by description, empirical findings, and assertions about norms than by theory. This approach also suggests at least a beginning theoretical structure for dealing with the relations between static equilibrium, long-run equilibrium, movement from one to the other, and an optimal rate of change. All such approaches seem essential for dealing with nonstatic systems or relations of systems. That they here seem borrowed from economics means merely that economics is the most obvious of the social sciences to have stumbled into a system approach. At the same time the impossibility of reaching a perfect equilibrium or perfect cultural transmission provides one basis for developmental change—a matter for Chapter Seventeen.

Chapter 8

3.2 Transactions

Basics

The previous chapter indicated that the communication is the intersystem parallel of the intrasystem detector function in that both deal with information. We now move to the transaction, which is the intersystem parallel of the selector in that both are concerned with valuations. In other words, a transaction reflects and affects—molds and is molded by—the selector states of the parties. Such terms as *desire* and *satisfaction* are frequently used below in preference to the clumsier *selector states* and constitute major variables in the analysis.

In exchanges among lower levels of life an interaction between two systems typically takes the form in which an output of system A becomes an input to system B. This statement is also valid in a reasonably literal sense for humans with respect to communications. Most things transacted, however, do not literally get inside the human who receives them, nor were they initially inside the human who gave them up. They can therefore be construed as outputs or inputs only in the figurative sense that they move from one orbit or sphere of influence to another.

Viewed subjectively, a decision is an internal partial analog of an exchange, in which the costs are the satisfactions given up and the benefits are the satisfactions received. Viewed objectively, the external things forgone in a decision (given up if already possessed, or not received if not yet possessed) are the costs while the things acquired are the benefits.

The same decisional process occurs in an interpersonal context in the transaction. Here (in the initial model) A gives something to B contingent on B's giving something to A. As seen by A, having to give up one thing is the cost to him of the benefit he receives from B. As in any other decision, A's satisfaction will be enhanced if the benefit to him of the thing he receives exceeds the cost to him in the thing he gives up. The same can be said for B.

It is important to the overall framework (and parallel to the later distinction between formal and informal organization) to note that the two decisions involved in the transaction are mutually contingent, but that they are not the same decision. They mesh, or compliment one another, but do not coincide. It is true that the parties agree to consummate the transaction. But the things they decide on are quite different. As will shortly be expressed in symbols, A decides to give up X in return for Y while B decides to give up Y in return for X —which are not the same decisions at all. In fact, if both parties did make the same decision—if, say, both decided to give up X in return for Y—the transaction would not be possible. Since the transaction involves two decisions rather than one, it hinges on four valuations: the costs and the benefits of the exchange to each of two parties. The distinction between joint decisions and mutually contingent decisions will be clearer after we deal with decision-making in formal organizations.

Every decision involves an interaction between detector and selector functions. Here, as elsewhere, the detector deals with information about the opportunity function and the selector handles the preference function. In a transaction the opportunity function includes two distinct ingredients. The first concerns one party's ability to do or provide something wanted by the other; the second concerns the other party's response. What one party wants constitutes his preference function relative to the transaction; what the other party is willing to give him in return is his opportunity function.

For the main analysis of transactions only this second ingredient of the opportunity function is involved. We accept as given that each party can do or provide the item; the only question is whether he *will* do so on terms that will induce sufficient response from the other party. However, the sense in which this ability can be taken as given differs as between transactions in positive goods (goods) and negative goods (bads), as follows.

In a given transaction my power and bargaining power in a lump of gold are independent of whether it took me ten minutes or

ten years to find it. And my power and bargaining power in being able
to release a man from a bear trap, if I want to use it, is independent of
whether he stumbled into the trap or I personally sprang it on him.
Whether for goods or bads there is thus the transactional relation itself,
which is an interpersonal relationship. Antecedent to it is a pretransactional stage that establishes certain levels of power factors from which
the transaction starts. In the case of threats there is also a potential
posttransactional stage. My deliberate acquisition of the nugget can be,
and often is, independent of any party with whom I might eventually
exchange it. Hence for goods it is feasible to analyze the transactional
stage separately from the pretransactional and to take the presence of
the goods as given. But if I deliberately spring bear traps on people to
extract concessions for releasing them, the springing is intimately related to the releasing—though I might simply set traps and bargain
with whomever steps into them. With bads, and particularly with
threats, it is therefore not feasible to separate the transactional stage
from the pretransactional—the releasing from the springing—and the
two will therefore be joined. However, we will still take as given the
ability to perform the pretransactional stage—to spring the trap or to
make a credible threat of it.

   Although some of the following points are expressed in the
language of game theory, the transaction seems a broader and more
fruitful analytic tool. However useful the game analogy in certain
multiparty relationships, particularly under conditions of uncertainty,
it seems sterile for general social analysis. As defined, in game theory,
a game is more likely to be a tactic or a strategy designed to influence
a transaction than to be a transaction itself—as those terms are defined here.

   Despite its broad contours and basic conclusions, this chapter is heavy with minutiae. I do not like it that way and would prefer
merely to illustrate the method. But given the present wide skepticism
about unified concepts it seems important to demonstrate the fertility
of the model. The spawning process may have gotten out of hand and,
once started, was not responsive to hints about planned parenthood.
The teeming progeny have by now oozed out in many directions. But
since a central question is their ability to do precisely that, I have kept
most of the squirming brood.

INITIAL TRANSACTIONAL ANALYSIS

3.20   As noted, a transaction is any interaction between parties
analyzed with reference to its value content to the parties. Although

that definition should be retained for basic analytic purposes, it is often satisfactory to define the transaction as a transfer of anything of value between parties. It may be useful to think in passing of an elemental prototype of transaction that consists of an exchange of matter-energy as contrasted to the transfer of information that constitutes communication. Such a prototype might be useful for integrating human and nonhuman systems into a single basic analysis. But as soon as we recognize that valued things given in transactions include information, affection, praise, and simple permission, it seems pointless to narrow the model even initially to matter-energy items. As noted, when information is transferred as part of the content of a transaction, only its *value* is relevant to transactional analysis. *Party* has been defined, and for simplicity we speak of parties as persons, though the analysis of transactions is the same for any collectivity that acts as a unit.

3.21   *Initial Transaction Model*

The initial model of the transaction is based on the following assumptions:

a. Two persons only are involved, A and B.

b. The model is selfish-indifferent (hereinafter shortened to "selfish"), by which we mean (1) that each party tries to maximize his own benefit from the transaction—to give as little and to receive as much as possible—but will conclude the exchange if its benefit equals or exceeds its cost; and (2) that each party is indifferent to the position of the other, wishing neither to help nor to hurt the other. That is, neither party feels hostile or generous toward the other. This condition partly coincides with the legal concept of an "arm's length" relationship. Wherever it might otherwise make a difference, *selfish* should be read as *selfish-indifferent*. The indifferent component distinguishes it from generous and hostile attitudes dealt with later (3.26.1–.2).

c. Only two goods are involved: X is the good initially held by A; Y is the good initially held by B.

d. Both A and B know their own preferences for X and Y and make no error about them in the terms they accept. The model makes no assumptions about the nature or source of these preferences. A desire may be objective and instrumental, as for a nail, or subjective and complex. To "know one's own preference" does not imply precise formulation but simply enough sense of the preference to say yes or no to proffered terms.

e. Neither party initially knows the other's preferences. But for purposes of exposition we assume that the observer-analyst knows all four preferences.

f. Both X and Y are goods. They have positive valences for both A and B, and they reinforce approach, not avoidance, responses.

g. Both A and B are principals acting in their own interest, not agents of others.

h. The transaction process itself has no cost, as of time, information, frustration, or loss of face. It also has no benefit, as in the fun of bargaining or pleasure in one another's company.

i. Contact between A and B has already been established and the possible exchange has already been narrowed to X and Y. The only questions are whether, and on what terms, the transaction will take place. This means that the analysis is focused on the "distributive" aspects of bargaining (Schelling, 1956; Walton and McKersie, 1965). By contrast, "efficiency" aspects of bargaining (Schelling, 1956) deal with attempts to find more satisfactory solutions to exchanges between parties who are, and under the circumstances of the case will continue to be, in a transactional relation. It is obviously to the advantage of both parties to discover some X that is more valued by B than one they initially focused on and/or less costly to A to give up—and with the appropriate substitution of terms the same can be said of Y. But the ability to find this X depends on the parties and circumstances and does not seem amenable to a broad model. Having identified the problem and indicated why it is not discussed here, I do not deal with it further.

j. The transaction is unique, by which we mean that it stands solely on its own merits. Neither party is concerned about its possible effect on any other transaction with either the same or different parties in either the same or different goods. Neither party considers the alternative of dealing with some other party.

k. Neither party has desires relevant to the transaction other than those for X and Y.

l. No question of delivery is involved. Either the transfers are simultaneous in both directions or there is perfect confidence that delivery will be made as agreed.

m. Each party can withhold the thing desired by the other. Among possible other things this assumption excludes theft or the removal of one's privacy without his consent. (Coercion is already ruled out by assumption f).

n. Each party is able to give the good in question. To illustrate the opposite, one might not be able to give love or respect if he feels fear or contempt.

o. The necessary communications are effected. To illustrate the analytic separation of transaction from communication, the model does not assume that communications are absent. Instead it takes information for granted and attends solely to the valuations.

p. The preferences of the parties are given and do not change during negotiations. The initial model is thus strictly cross-sectional, not developmental.

q. Neither party is stubborn enough to abandon negotiations without first offering the most that he will in fact give. Thus the transaction will not fail through either party's miscalculating the other's intent.

r. A and B are equally good tacticians. Operationally this means that the total potential gain from a transaction will be split equally between them or (in later language) that the transaction will be settled at the midpoint of the overlap of effective preferences (see 3.21.2f).

3.21.0   *Some Initial Propositions about Transactions*

3.21.01   If neither X nor Y is divisible (say, a horse and a cow) the exchange will be made or it will not be made. But there can be no question of "terms." It is, of course, possible, to exchange the horse for the cow plus a bale of hay. Such an exchange would violate 3.21c by introducing a third good, however, and is unnecessary since any such problem is automatically analyzed in connection with divisible goods (3.21.3).

3.21.02   If neither X nor Y is divisible the transaction will be completed only if A and B have opposite preference orderings such that A prefers Y to X and B prefers X to Y—that is, if $AY - AX \geqq 0$ and $BX - BY \geqq 0$. (See symbols in 3.21.2.)

3.21.03   Measured in satisfactions achieved, a model transaction is thus necessarily a positive-sum game except in the limiting case (which we will ignore), where it is zero sum. If only material goods are transferred, the transaction is overtly zero sum. If X or Y is a service, no statement can be made about the overt result since the thing given up by one party is not the same thing received by the other.

3.21.04   Because at least one party gains from a transaction and normally both do, the *fact* of the transaction is cooperative. In the words of Blau (1964, p. 255), each party advances his own interest by advancing the interests of the other party. It is similarly Adam Smith's invisible hand at the level of the individual transaction. As seen below, in exchanges of divisible goods the *terms* of the transaction are in conflict, however, since the better the terms from A's point of view (the less he gives for Y) the worse they are from B's view (the less he gets). **3.21.1** The terms of a transaction involving only two goods can vary only if one or both are divisible in some degree. With one or both goods divisible, no gain can occur unless A and B have a difference in their relative valuations of X and Y. (If both consider one orange worth two apples neither could gain from an exchange at that ratio, and one or the other would refuse any other ratio.) If A and B do differ in their relative preferences for X and Y, with A preferring Y to X and B preferring X to Y, the terms of an exchange will fall within the limits of the two preferences.

3.21.2   *Symbols and Relevant Definitions*

The following symbols represent the preferences—that is, the system states of A and B—which determine the transaction:

a. AX is A's desire for X. If X is material, AX is A's desire to keep it or its expected subjective utility during the period relevant to the transaction. If X is a service to be rendered for B, AX is A's desire *not* to perform it, which is the same as his reluctance to do so or the cost to him of doing so. If X is some state of affairs under A's control, AX is the perceived cost to A of changing from the status quo to the state desired by or requested by B—for example, by granting B some permission. Depending on the circumstances, AX may also be thought of as the cost to A of acquiring Y or of concluding the transaction on certain terms. In the economic model of the firm, AX is its cost of producing X.

b. AY is A's desire for Y or its expected subjective utility. If Y is a material good, AY is A's desire to acquire it. If Y is a service or state of affairs to be effected by B, AY is the desire that B perform, grant, permit, or accomplish Y. Depending on the circumstances, AY can also be thought of as the value to A of acquiring Y or of concluding the transaction. It can also be viewed as the cost to A of not having Y.

c. BX is B's desire for X. With appropriate substitution of symbols it is identical with AY.

d. BY is B's desire for Y. With appropriate substitution of symbols it is identical with AX.

e. *Give* and *receive* are construed broadly here. *To give* includes such acts as to do, permit, perform, or even cease doing something that annoys another. More explicitly, whenever A's behavior (including any change in behavior) affects B's state of satisfaction and is contingent on behavior by B that affects the state of A's satisfaction, the potential for a transaction exists. Giving one's presence, company, conversation, trust, friendship, moral support, or affection is encompassed within the definition of *giving,* as is giving up one's privacy. With respect to less tangible items, however, this stage of the model is not applicable unless the items fulfill assumptions 3.21m, n, and q. Hereinafter "A gives X to B" means any of these behaviors. *To receive* encompasses the recipient's role with respect to anything given.

f. *EP (effective preference)* is the selector and effector states of a party toward a transaction.

g. *A's EP for Y* is A's net desire (or net preference) for Y in the context of the transaction at hand, or $AY - AX$. A transaction involves two mutually contingent decisions. A's EP represents the ingredients of his decision in that AY is the benefit he expects to receive in acquiring Y while AX is his cost of acquiring it. More explicitly, A's EP is the maximum amount of X he would be willing to give for Y— his "reservation price." *Effective* means that A can make his wishes effective by being able to give X for Y and that the pretransactional stage is successfully accomplished. Any decision reflects both preference and opportunity functions, and the "preference" half of effective preference similarly represents the preference function while the "effective" half represents that part of the opportunity function which is within A's control. The transaction requires that A appeal to some motive of B; the effective half of effective preference is his ability to provide something that will do so. This formulation is in accord with Thibaut and Kelley (1959, p. 34): a person must be willing and able to make an exchange. It is also a one-person version of effective demand in economics: people not only want some good but also have the purchasing power to buy it. B's EP for X is symmetrical in all respects, and is a one-person version of supply.

h. *An overlap of EP's* is any situation in which at least one set of terms of a transaction is acceptable to both A and B. It includes the limiting case where only one set of terms is possible, which would be represented by the EP's overlapping only at their limits or just touching.

i. *Power factors* is a convenient term for the EP's of both parties to a transaction, along with the strategic modifications of those EP's.

      With these definitions in mind we now see that, in the context of a transaction, A's EP is his own preference function and B's opportunity function while B's EP is *his* preference function and A's opportunity function.[1]

3.21.3   *Transaction Diagrams*

The interaction of the two contingent decisions represented by the effective preferences of A and B are shown in Figure 11. A's EP

[1] The reader familiar with other bargaining power models will note that the nearest equivalents to the value components of this model (AX, AY, and so on) are expressed as a ratio—as in Chamberlain (1955), Pen (1952), and Cartter (1959). The following background may be useful for showing why this model is preferred.

The concept of an effective preference, reflecting the costs and benefits in one of a pair of mutually contingent decisions, is meaningful only if both its components, AX and AY, are significant magnitudes of about the same order. If they are of widely discrepant magnitudes the notion of making a decision between them becomes meaningless. For example, as AY approaches zero it becomes meaningless to ask how much X A will give in return for Y, since he will give virtually none. And when AY reaches zero A will accept Y only as a gift. Similarly, as AX approaches zero it becomes irrelevant to ask how much Y A will insist on getting in return for X, since he will insist on none and will be prepared to give X as a gift.

When the factors in the transaction vary over extremely wide ranges it seems useful to view the EP as a ratio instead of as a difference—that is, AY/AX instead of $AY - AX$. In the ratio formula, as AX approaches zero the EP approaches infinity whereas in the difference formula it approaces AY. Here the ratio makes more sense: it means that as AX gets very small A will be willing either to give much X in return for Y or to accept little Y in return for X—depending on which is the discrete and which the continuous variable. But to say that the EP equals AY has no meaning since an EP for Y is, by definition, measured in units of X. This dilemma can be resolved arbitrarily by saying that in the difference version, which is the one used here, we also consider the EP to be infinite, rather than equal to AY, when AX becomes zero. When AY approaches zero the ratio also seems more sensible. In the ratio formula, A's EP would then become zero; in the difference formula, it would equal minus AX. But AX is expressed in units of Y, which has now disappeared. We can again say arbitrarily that A's EP approaches zero, rather than approaching minus AX, as AY approaches zero.

Although the ratio is more felicitous for the extremes, the difference is better for the range in which settlements would probably fall. Within that limited range the two analyses are identical. Whether we say that the buyer's desire for a car minus his desire to keep his money is such that he will give \$300 for it, or that he values the car and money in a ratio of 300 to 1, his EP is \$300 in either case. All things considered, two main points favor the difference formula. First, it retains absolute magnitude so that a choice between match and cigarette lighter does not look the same as a choice between cruiser and battleship. Second, the worst position possible in the ratio formula is zero, if costs are huge and gains small, whereas in the difference formula it is a large minus quantity. The inability of the ratio to show a loss, rather than mere zero benefits, seems compellingly against it.

is at the left; B's is at the right. The easiest case to visualize is that in which X is a divisible unit, like money, and Y is an indivisible one, like an automobile, with A the prospective purchaser and B the seller. Figure 11 illustrates that A is willing to pay as much as \$600 for a used car while the seller is willing to accept as little as \$400. The EP's thus overlap between \$400 and \$600. The midpoint of the overlap is shown by the dotted line at \$500. There is no need to change the basic diagram

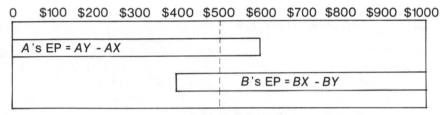

Figure 11. Transaction diagram: case of two goods—one divisible.
Diagrams of this sort were introduced by
Boulding (1955), and similar ones are
used by Walton and McKersie (1965).

if X and Y are both divisible. The scale then can indicate the number of units of X to be given for some stated or assumed quantity of Y. Although the scaling becomes more complicated, the basic diagram is still applicable when several types of goods are involved—as in the contemplated exchange of six fighter-bombers plus economic aid plus tariff relaxation for a naval base plus a treaty of friendship plus exclusive rights to explore for oil. In such cases a general form of the diagram might be used with top and bottom scales as in Figure 12. Here we can think of X and Y each as a good potentially exchangeable between the parties. The numbers can represent any units relevant to an exchange —successive units of the same goods X and Y, fractions of X and Y, or

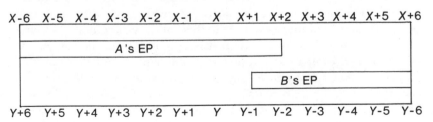

Figure 12. Transaction diagram: general case.

different goods. In the latter case, for example, X might be a horse, $X + 1$ a bale of hay, and $X + 2$ a saddle. If they are different goods their order as well as their number will be significant; if they are of differing values they will not be spaced uniformly on the diagram —complications we need not go into in this volume. To return to the model of two goods only, in Figure 11 A's EP is identified as $AY - AX$. Strictly speaking, the EP extends to the point where $AY - AX = 0$ (paragraph 3.23.6).

3.22   POWER AND BARGAINING POWER

3.22.0   *Definitions about Interpersonal Power*

a. *Interpersonal power* is the ability to acquire wanted things from others, whether on good or bad terms. Just as individual power is the generalized concept of instrumentality at the personal level, interpersonal power is the generalized concept of instrumentality between persons. As indicated earlier (2.5 and 2.9), power is concerned with the magnitude of the effect one creates in his dealings with others.

b. *Particular interpersonal power* is the ability to acquire a particular wanted thing Y from others. As used below, "an increase in A's power to get Y" means the increased power to get Y if Y is a discrete item, the power to get more Y if Y is a divisible item, or both.

c. *Aggregate interpersonal power* is the ability to acquire a series of wanted things $Y, Y_1, \ldots, Y_n$ from others. The larger the number and the greater the value of wanted things that A can get from others, the greater is his aggregate interpersonal power.

d. *Bargaining power* is the ability to get good Y on good terms—that is, by giving relatively little in exchange. The base from which to measure whether terms are good or bad is discussed immediately below.

Regarding usage, *bargaining power* is always referred to by those words. *Power* alone refers to the general term (item **a** above), though in certain contexts the adjective *plain* is used to distinguish power from bargaining power. Whether plain power is particular or aggregate is stated explicitly when it is not clear from the context. We never directly compare the bargaining power of A and B. That is, we do not say that "A has more bargaining power than B." We say, however that "A has more bargaining power with B than with B_1," that "A has more bargaining power than does A_1 in dealing with B," or that

"after event k, A's bargaining power with B was greater than before k."
To compare A's bargaining power directly with B's makes sense only
if some set of terms is assumed in advance to be proper or standard, in
which case one might want to say that "A's bargaining power is greater
than B's" or that "A got very good terms" if A is able to get terms
clearly more favorable than usual. But by definition of the unique
transaction, no prior standard is presumed to exist. The terms of transaction fall where they will, and any terms falling within the overlap
represent a net gain to both parties.

We need not deal explicitly with potential, or unexercised,
power of any sort because when we identify the ingredients and conditions of power, if exercised, we simultaneously identify the ingredients
of potential but unexercised power. In any case the discussion of
internal and external power factors and related items in Chapter Ten
identifies certain power factors; and the identification does not hinge
on whether they are put to actual use.

The model proposes no measure of power or bargaining
power independent of goods actually acquired and terms actually
reached. Our focus is on the *ingredients* of power and bargaining power.
These seem independently measurable in terms of goods available to
give in exchange and in intensity of desires. Although I doubt whether
we could ever compute numerical values of effects from numerical
values of the causes, I have no doubt that we can often measure *changes*
in EP's, or in their components, well enough to predict the direction,
and perhaps the approximate magnitude, of *changes* in terms or quantities acquired.

e. *Bargaining advantage* is a useful phrase when a set of terms becomes
   a point of reference in negotiations, as in an explicit demand or
   proposal or some sense of a "fair exchange." To accommodate such
   situations we define *bargaining advantage* as bargaining power
   forces that enable one party to get terms more favorable than some
   reference terms to which they are compared. For certain cases, particularly when strategic bads are employed, as in the strike in
   union-management relations, it is useful to shift attention from the
   EP's for what the other party can provide to the EP's for a particular
   set of terms of possible settlement. Those terms might be a proposal
   by either party or by a mediator.

f. *Negotiations* (or *bargaining*) are the communications, tactics, and
   strategies by which A and B determine whether and on what terms
   they will conclude a transaction. Negotiations can also include
   attempts to establish contact, especially if the method provides

apparent clues to EP's. Most of the following propositions indicate
the effect of *A* on both *A* and *B*. Since all such relations are symmetrical, their converse need not be spelled out.

Propositions about Interpersonal Power

3.22.1   A transaction will take place only if the EP's overlap. To
state that the transaction will take place if the EP's just touch is an
arbitrary assumption from 3.21.2h to simplify discussion. At that point
it is a matter of indifference to both parties whether the transaction
goes through or not.

3.22.2   *A*'s particular power to get *Y* is a direct function of his own
EP. That is, if *B*'s EP is given, a long EP from *A* will overlap it and *A*
will get *Y*. If *A*'s EP is not long enough to overlap, he will not get *Y*.
We say arbitrarily that *A* has more power with a large EP than with a
small one, even if there is no overlap. He is then "less unable" to complete the transaction and get *Y* and has "less distance to go." We
similarly say that a large overlap represents more power than a small
one since *A* is then surer of getting *Y*.

3.22.3   *A*'s particular power is a direct function of *B*'s EP.

3.22.4   Assuming overlap, *A*'s bargaining power is an inverse function of his own EP.

3.22.5   *A*'s bargaining power is a direct function of *B*'s EP.

3.22.6   The greater the overlap of EP's (the greater the difference
in relative preferences of *A* and *B* for *X* and *Y*), the greater the subjective gain of the parties from completing the transaction. For the single
transaction this is the logical equivalent of the economic concept of
consumer surplus.

3.22.7   The preceding relationships are diagramed in Figure 13,
which shows the initial position of the parties (at *a*), an extension of
A's EP with *B*'s remaining unchanged (at *b*), and an extension of *B*'s
EP with *A*'s in its initial position (at *c*). At *a* the midpoint terms are
$500. At *b* the extension of *A*'s EP to $800 brings the midpoint terms
to $600 and displays the rise in *B*'s bargaining power and the decline
in *A*'s. The reverse condition appears at *c*, where *A*'s bargaining power
has risen and *B*'s has declined. At the same time both *b* and *c* show a

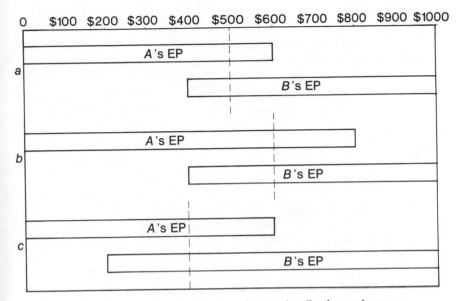

Figure 13. Transaction diagram: changes in effective preferences.

larger overlap than does *a*. This means that the transaction is surer of being consummated at either *b* or *c* than at *a*, which means in turn that *A* is surer of getting the car and *B* of getting money. By the same token the plain power of both parties to get what they want, without regard to relative goodness of terms, is greater at both *b* and *c* than at *a*.

For some purposes it is convenient to express power and bargaining power as a formula. Any factor that extends either EP enlarges the plain power of both; any factor that shortens either EP contracts it (3.22.2–.3). Hence if P represents power and subscripts A and B represent the parties: P_A or $P_B = (AY + BX) - (AX + BY) = (AY - AX) + (BX - BY)$. By contrast the bargaining power of a party is enhanced by any factor that extends the other party's EP (3.22.5) and shortens his own (3.22.4). Hence if BP represents bargaining power: $BP_A = (AX + BX) - (AY + BY)$. The bargaining power of B reverses the two halves of the formula.

3.22.8   The preceding propositions starting with 3.22.2 are methodologically important. *A*'s particular and bargaining power are both a direct function of *B*'s EP. Thus in the model *A*'s position is always helped by *B*'s greater desire for X. By contrast, *A*'s particular power is a direct function of his own EP and his bargaining power is an inverse function. Hence the greater his desire for Y the more likely he is to get it—but the worse the probable terms. It is my strong suspicion that this two-sided effect of one's own desires, taken with the general failure of discussions of power to distinguish sharply between power and bargaining power, has contributed much to the failure of *power* to constitute the central concept in social analysis that many feel intuitively it must. In this connection see such sources as Chamberlain (1955), Dahl and Lindblom (1953), Lasswell and Kaplan (1950), Russell (1938), Merriam (1934), Tawney (1931), Weber (1947), and March (1966).

In some formulations A's power is viewed as consisting of B's EP (Emerson, 1962) while B's power consists of A's EP. The effect of the four evaluations (AX, BX, AY, BY) is the same either way. However, I have tied A's power to the strength of his own urge and ability as well as to B's "dependence" (Emerson's term) on what A can do for him, since only this formulation is consistent with the definition of intrapersonal power in Section 2.9. Moreover B's EP alone avails A nothing; his own EP must reach far enough to meet it.

A's power has also been viewed as his ability to get B to do something against his will. According to this view, however, A's ability to get thousands freely and enthusiastically to follow his merest suggestion would not be power—a view I feel makes little sense. The situations in which B behaves "against his will" are dealt with in connection with bads (3.25).

Others define A's power as his ability to get B to do as A wants. The distinction between that and the present formulation is purely terminological. In present language, to say that A receives Y is the same as saying that B has done what A wants. It is also the same as saying that A exerts power (or influence) over B; that B does A's bidding or follows A's instructions. For reasons clarified later, A cannot exert power over B unless he can do something B wants, which means that A must in some way also do B's bidding. The reasons why coercion does not violate this approach—as at first glance it seems to do—are clarified later (3.25.0e). The present formulation is clearly at odds with those (see Lasswell and Kaplan, 1950) that view all power as coercive. The present concept of power coincides closely with that of Gold (1968) except that Gold treats power in bads as a distinct category.

Weber's (1947, p. 152) concept of power includes the ability to carry out one's will despite resistance. In the present model this notion is explicitly incorporated as a question of what A must do or give to overcome B's resistance. Tawney, 1931, p. 229) speaks of power as the ability to modify the conduct of others in ways one wants while preventing modification of his own conduct in ways he does not want. In the present model this notion is phrased as the problem of getting B

to give as much as possible while A gives as little as possible in return. Given these examples, there is no need to translate other variations of the concept.

For reasons such as these, *influence, control,* and related terms are not used here as a name for this concept: *power,* with or without modifying adjectives, will suffice. Since power is a trait of an acting system alone, in this terminology we speak of an idea or pattern as having influence but not power (3.15.0). And although the line may be difficult to draw in real cases we similarly distinguish the power of a person from the influence of his ideas. One sharp line is, however, possible: his power ends at death, although his influence may continue for centuries. (See also 3.33.)

In the present framework we do not use the phrase "A has power over B." The notions customarily described by that phrase nevertheless appear at various points, particularly as strategy and at the pretransactional stages of transactional relationships. "Power over" the nonhuman environment would consist of the ability to transform the environment, as discussed in Section 2.9.

We defined *intellectual influence* as the ability to modify the detector states of others and *moral influence* as the ability to modify their selector states. Although this line, too, may sometimes be difficult to draw in practice, whenever such influence operates directly or indirectly to improve A's terms in a transaction it will be said to constitute intellectual or moral *power* rather than influence. The exercise of such power with explicit reference to a particular transaction is encompassed within the discussion of tactics and strategy (3.23 and 3.24 below).

A transaction represents a *flow,* the analysis of which is crosssectional and in which AY and BX may be thought of as force or pressure and AX and BY as resistance. By contrast, aggregate power represents a *stock.* To the extent that such a stock is transactionally acquired (and the extent is considerable), it rises or falls depending on whether the inflows from transactions are greater or less than the outflows. Because the size of the stock is the net consequence of a series of transactions, each of which depends on the outcome of the previous ones, the process of accumulation is developmental—as we shall see in more detail in Chapter Ten.

3.23   *Tactics*

3.23.0   *Tactics* (by A) are defined as a communicational adjunct to transaction in which A seeks to modify perception of EP's—that is, to change the detector states of either party regarding them. Although the definition is not thus restricted, typically A seeks to learn B's EP while concealing or misrepresenting his own.

3.23.1   Let us momentarily relax assumption 3.21r (that A and B are equally good tacticians and that terms will fall at midpoint of overlap) and substitute that A fully understands the bargaining process, that B knows only his own EP and is naive about bargaining, and that A knows B is naive. If so, A will seek to learn B's EP, which B (being naive) will consciously or unconsciously reveal. A will then represent his own EP as just touching B's, and the transaction will be concluded on terms near A's end of the overlap. (As soon as A is sure that he can get certain terms from B, we can say that his own EP shrinks to that amount and that thenceforth he will, in fact, be unwilling to give more. See 3.27.33 for the effect of alternatives on EP's.)

3.23.2   If we now return to the assumption that both parties are good tacticians, it follows that each will try to learn the other's EP without revealing his own.   **3.23.21** If A makes an offer it automatically gives B information about A's EP. From 3.23.2, each party will (a) avoid making a first offer or (b) understate his EP.   **3.23.22** For A to take an adamant stand on an offer of less than he is willing to give has a bivalent effect on A's position, one aspect working through B's detector and the other through his selector. Regarding B's detector, an adamant stance raises the credibility of A's statement about his EP, *cet. par.*, and therefore raises the probability that B will accept it if it falls within the overlap. And regarding B's selector as reflected in his EP, A's adamant stance raises the risk that he will not get Y at all since the foreshortened version of A's EP may not overlap B's. If A does not relax his adamant stance the transaction will fall through, even though it would be advantageous to both parties to complete it. On the whole, adamant stances are probably more important in transactions with bads than with goods.

3.23.23   For A to relax an adamant stance and improve his offer also has a bivalent effect, similarly working through B's detector and selector. (See Cartter, 1959, p. 122, regarding the double-edge of compromise.) Regarding B's detector, by revealing that A's initial adamant position was a misrepresentation, the relaxation decreases A's credibility (brings a loss of face). Unless A can show convincing reasons why it is his "really final" offer, the relaxation suggests to B that even more will be forthcoming from A if B holds out. This is the appeasement effect. And regarding B's selector, the relaxation increases the likelihood that B will accept A's offer since it is more favorable than A's prior offer. A's new offer may also overlap B's EP when the previous one did not and thus increase the probability of being accepted. **3.23.24** Assuming that A understands and expects the appeasement effect, he will not relax an adamant stance unless he can represent it as a changed situation, new information, or the like, and not as an admission of prior misrepresentation.   **3.23.25** If A can make an original or revised offer wholly credible as representing the limit of his EP, and if it overlaps B's EP, the transaction will be concluded on the terms offered by A.   **3.23.26** Since A's EP is private, his allegation about it can be wholly credible if he can present B with overt evidence that A's alternatives are at least as good as those of B's offer (see 3.27.33). **3.23.27** If neither side can make its position wholly credible the terms are indeterminate within the overlap.

3.23.3   It may be speculated that a large fraction of all lying consists of tactical misrepresentations with respect to anticipated or possible future transactions. We may note in this connection that since AX, AY, BX, and BY are the only desires relevant to the transaction (3.21k), the parties have no preference between truth and untruth as such. Uniqueness similarly implies no concern for the effects of reputation on other negotiations.

3.23.4   For this paragraph only, relax assumption 3.21i that contact has already been established between A and B and negotiations opened. Then if A's EP for Y is given, but he perceives that B's EP is too short to overlap it and hence anticipates that B would be uninterested in entering negotiations, his tactic could be to exaggerate his own EP sufficiently to persuade B that a transaction is possible. Having brought B into negotiations he would then use appropriate strategies to lengthen B's EP.

3.23.5   *Methodological Note*

When economic analysis deals with perfect competition, all prices and quantities are theoretically determinate. Thus economic theory has gradually moved through imperfect competition to bilateral monopoly, which is thoroughly indeterminate. The present analysis reverses the order: bilateral monopoly is the most general transactional model and the perfect market is a limiting case. This sequence reflects the purpose of this volume—to start with analysis common to all social sciences and then move toward special applications.

By *determinate* in this context we mean that if no forces

are at work other than those specified in the model, and if they work according to the model, then only one solution is possible—as at the point of intersection of supply and demand in the perfect market. The present model is determinate to the extent of saying that a transaction will not be completed if there is no overlap of EP's and that it will be completed on some terms within the overlap if there is one. But it is explicitly indeterminate about the location of the settlement point *within* the overlap. This means that the model has not incorporated enough assumptions to provide the logical conditions for determinacy. Nor can I at this point imagine what those assumptions might be. Since the vast majority of human transactions are indeterminate in the sense of this model, it would seem inappropriate to use a determinate model even if we could find one.

This indeterminacy is related in an important way to the definitions used here. Whereas bargaining power as often defined is concerned with the location of a settlement within the range of overlap, it is here defined as determining the range itself and nothing within it. It should by now be apparent that to define bargaining power in the former and more customary way, which is indeterminate, is not only to forfeit the hundreds of propositions herein that deal directly or indirectly with bargaining power but also to eliminate the base for the deductive approach to these matters.

3.23.6   Nevertheless, as seen in 3.22.1, if the EP's just meet only one set of terms is possible—at the point of contact—and under this limiting case the solution *is* determinate. There is also the broader sense in which the terms of a transaction depend on the EP's of both parties. Among A's EP, B's EP, and the terms, if we know any two we can deduce something about the third even if the content of the deduction is not precise.

To return now to the case in which the two EP's just touch, if we add information about the utility functions of the two parties, this point of contact will also be seen as an equilibrium. Assume that X is a variable unit (money) and Y an indivisible one (an automobile). Let a stand for the number of units of X that A will give to B in a possible exchange. In Figure 14 the line mm represents the utility to A of the package $Y - aX$ as a increases. This line crosses the zero axis at P, where A is indifferent between Y and a units of X. The line nn, read from right to left, represents the utility to B of the package $aX - Y$ as a decreases. This line crosses the zero axis at the point, P, where B is indifferent between Y and a units of X. When the points of indifference coincide, the solution is determinate. If the two curves cross above the zero line the EP's overlap. Except under the assumption of equally

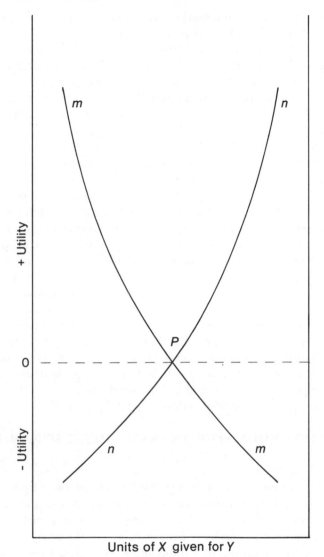

Figure 14. Determinate transactions as equilibrium.

good tacticians, the solution is indeterminate within that overlap. If they cross below the zero line, there is no overlap and no transaction is possible.

We may consider A's preference for Y as his "demand" for it and regard B's preference to retain it (his cost of giving it up) as the "supply," even if it is definitionally inappropriate to apply these terms to only two persons. But whereas a determinate outcome of supply and demand for an economic market will occur *wherever* the two curves intersect, here the solution is determinate only at the zero utility line of each. (This application of utility curves to the derivation of EP's was suggested by R. T. Riley of the University of Cincinnati.)

The Edgeworth contract curve, another way of representing the two-good, two-person transactional relation, has particular usefulness for helping to conceptualize the problem of total welfare for multiple parties. However, I am skeptical about its wider usefulness as an alternative model. Its analysis of the allocation of a universe of two commodities between two persons seems satisfactory for certain cases involving tangibles, where the thing given up by *A* is identical with that received by *B*. But for the services that make up most human exchanges, the thing received by *B* is not what is given up by *A* and vice versa. I therefore do not use a box whose boundaries imply a fixed total quantity of the two goods since I cannot conceive of a fixed total of encouragement, threat, relief of stress, hostility, union security, or helping mother with chores. Moreover the Edgeworth box clearly distinguishes the more from the less preferred distributions of the two goods, for each party separately and for the two jointly. But it does not indicate the relative acceptability of various distributions to either party—a serious shortcoming. Furthermore it gives rise to only a short list of deductions about a special transaction, in marked contrast to the plethora of deductions derivable from the present model. Hence the mere passing mention of the Edgeworth box, an excellent analysis of which appears in Boulding (1952).

3.24 BACKGROUND FOR TRANSACTIONAL DYNAMICS

3.24.0 *Strategy*

We now define a *strategy* by *A* as any attempt by *A* to modify the *actual* EP's in a transaction— that is, to change the selector or effector states of either party. This change is accomplished by manipulating the preferences (as by communication) or the actual or perceived alternative opportunities that mold the preferences. Most, but not all, strategies operate on the other party.

3.24.1   Relax assumption 3.21p (that both EP's are given) and substitute that they may change during negotiations. Thus, *A* can raise his power and bargaining power if he can expand *B*'s EP—which means to increase *BX* or decrease *BY*. And *A* can raise his bargaining power if he can contract his own EP while maintaining overlap—which means to increase *AX* or decrease *AY*. Either can be changed by a change in his detector (perception of alternative opportunities) or his selector (subjective preferences).   3.24.2 This relaxation changes the analysis from cross-sectional to developmental and the model as thus far developed provides no conclusions about the direction it will take. (See the discussion of developmental analysis in Chapter Sixteen.) The cross-sectional analysis, however, does provide the following advantages. First, it permits analysis of the strengths and weaknesses of each party's position at any given moment and the terms within which a settlement is possible—if any. Second, it permits prediction of the effect on power, bargaining power, and midpoint terms of any change in power factors. Third, by the same token it indicates to each party what he must do to improve his position. And fourth, it provides an analytic language in which to describe changes.

As we shall see (3.25), developmental change within transactions in bads is more likely to follow predictable patterns than with transactions in goods.

3.24.3   *Promises and Credibility*

We now return to the initial model of the transaction but relax only 3.211 (that delivery in both directions is simultaneous or that there is perfect confidence that delivery will be made). Instead we examine the conditions under which A will now deliver X in return for a promise that B will deliver Y later. For this purpose we need three additional definitions:

a. A *stake* is any valued thing S, other than X or Y, access to which is contingent on performance of some act relative to a transaction. As a valued thing, a stake includes (among other things) money, position, prestige, credibility, an easy conscience, or avoidance of legal sanctions.

b. *Delivery* is the act of *giving* in a transaction, as defined broadly in 3.21.2e.

c. A *contract* is a promise to make delivery as agreed. A stake may or may not be specified. But if the term is construed as broadly as defined in item a it is difficult to imagine a contract without some stake or that either party to a selfish transaction would accept the other's promise except under the conditions indicated in the propositions immediately below.

3.24.31   For the assumption of trust (3.211) substitute that neither party has confidence the other will fulfill his promise of delivery unless it is in his interest to do so. Hence B will deliver Y in fulfillment of his promise if his stake in delivery, BS, is greater than BY; and A will accept B's promise if he perceives BS as exceeding BY. Thus if B wants his promises to be accepted by A, he must be able to demonstrate a stake exceeding BY. Although this proposition is stated briefly, its social ramifications are wide. It means that for all potential transactions in which X is delivered after Y, A cannot get Y unless his promises are credible. This means in turn that in many circumstances A's power and bargaining power hinge heavily on his credibility—that is, on trust in him.

3.24.4   Relax assumption 3.21i (that the parties are already in contact and have narrowed the possible exchange to items X and Y). This situation should be noted, though little can be said about it. If A and B do not know of one another, presumably nothing happens. If A knows of B he may seek him out, depending on the expected costs and benefits of doing so. To say more than this would require many additional assumptions that do not seem worth making here.

3.25   **BADS AS STRATEGIES**

        We now return to the initial model of the transaction but relax assumption 3.21f (that only positive goods are involved). Before we can examine how bads may be used in transactions, however, the following definitions are essential.

3.25.0   *Definitions about Bads*

a. *A bad (negative good)* is any object or event that reinforces an avoidance response—that is, has a negative valence. Just as we did not identify what constitutes a good, neither can we specify bads, which depend on the preferences of the persons involved. For the transactional analysis, something is a bad if a party is willing to accept a cost to get rid of it. We can nevertheless note that a bad is usually, but not necessarily, a service rather than a commodity, as is its removal. Bads include force, violence, bodily injury, pain, and destruction of property at any level up to and including war. An insult is often a bad, particularly if it diminishes self-esteem or signals the end of a friendship, and so is the withdrawal of a good already possessed. Cessation or diminution of a flow of goods is often a bad—cutting a child's allowance, a worker's wages, or an employer's work force during a strike. Even reducing a good's rate of increase can be a bad. In any event, a bad is determined not by examining the thing, but people's behavior toward it. In the pure case (which we are assuming here, and in contrast to physical blockage defined just below) the effect of the bad applied by A is solely to change B's detector or selector states (mainly the latter), not his effector. That is, the bad affects his will, not his ability, to act.

b. *Stress* is an act or situation that constitutes a bad when applied to someone in a stress transaction (item f below) and generates a desire that it be removed.

c. *Physical blockage* is an action by A that prevents B from effectuating overt behavior—death, injury, physical constraint. Blockage is transactional only to the degree that B's consent is involved. If blockage is a stress that B will pay A to remove, it behaves like other stress in a transaction, as detailed below. In other cases, such as death or jailing, blockage totally removes B from the transactional scene and leaves A free to do as he likes. Transactional analysis is then irrelevant because consent is irrelevant, the only question being the technical one of whether A has the physical power thus to remove B.

d. *Force* is a subcategory of bads consisting of physical blockage, destruction, pain, or injury.

e. *Coercion of B by A* is a situation in a stress or threat transaction in which A is able to make BX so large that B's likelihood of refusing to give Y to A is small. Although the two categories overlap substantially in fact, force is not always successfully coercive (B may die rather than concede), and coercion (such as blackmail) does not always involve force. Positive inducements such as bribes or seduction can be large enough to have the same effect. But only bads are included in this definition. Successful coercion is coercion that does induce B to give up Y.

   To clarify, we say that coercion is a means of getting consent, not of bypassing or overcoming it. A person's actions normally originate inside his head. Only crude actions induced by direct electrical stimulation of motor nerves, or more complex ones induced by hypnotism, can apparently be induced from outside, and we ignore both here. Any conscious actions originating from inside are said to be performed by consent, under one's own control, and constitute directed behavior. Thus we say that a person's consent had been coerced, but not that he had been coerced into acting without his consent. Any other usage seems fuzzy.

f. A *stress transaction* is one in which a bad is applied unilaterally by A to B prior to negotiations (or to a certain stage thereof), along with the express or implied promise that it will be relieved if and only if B does as A requests. Here the bad is applied pretransactionally.

g. In a *threat transaction* an express or implied promise is made by A that a bad will be applied to B if and only if B does *not* do as requested. Here the bad is applied posttransactionally, if at all, as will be seen below. To the extent that a threat is also stressful (itself a bad) the threat transaction behaves like the stress transaction, and removal of the threat follows the same logic as removal of stress. Certain distinctive traits of the threat nevertheless justify separate attention (3.25.2–.3).

We may note that stress and threat involve the two steps noted in the introduction to this chapter. The pretransactional, or application, stage reflects a unilateral decision by one party to apply or not to apply the stress or threat. The transactional stage involves the mutually contingent decisions of both parties as to what will be given in return for the relief of the stress or threat. The model does not formally use the terms *voluntary* or *coerced,* though the latter term is defined above. The distinction between negatively and positively based inducements seems adequate and is often less ambiguous than these two terms. If a father says, "You can have a new tricycle if you clean up your room, and I'll spank you if you don't," is the ensuing clean-up voluntary or coerced?

Strategy is mainly the attempt to alter the other party's EP. Although that term includes side transactions that have that effect, as in Section 3.27 below, a strategic move is not itself a transaction unless it independently qualifies as such. For present purposes, A's application of stress to B does not constitute a transaction. It would seem to stretch the meaning of *exchange* to include the case in which B gives a good to A solely because A has a knife at B's throat. The present transactional analysis successfully handles the relation, however, and is thus a more useful and comprehensive concept.

By definition, a bad creates an avoidance response. Hence B will not give a positively valued Y to A for a negatively valued X. B would instead be willing to give some positive value to *avoid* X, to which problem we now turn.

3.25.1   *Stress Transactions*

While holding to all assumptions of the initial model except 3.21f (that only goods are involved), we now add the following:

a. That A applies some stress to B as a form of strategy and promises to relieve it if B will give him some good Y.

b. That the promise is believed by B. (This assumption reinforces 3.211.)

c. That the stress generates in B an EP to have it removed.

d. That continuance of the stress has no cost to A.

e. That relief of the stress has no cost to A.

f. That initial application of the stress may or may not have costs to A.

g. That B applies no counterstress to A.

    To illustrate, imagine the following case. MD: "How much will you give me to set your broken leg?" *Joe*: "Nothing. I haven't got a broken leg." MD: Breaks Joe's leg. "I'll set your leg for \$100." *Joe*: "Okay."

3.25.11   The components of such a transaction are as follows. X is relief of the stress previously applied by A. Y is the good possessed by B and wanted by A. AX is A's desire to continue (or not to relieve) the stress. BX: is B's desire that A remove the stress. AY and BY are the same as in the initial model, and both are positive.

3.25.12   A transaction using bads as a strategy can be a negative-, zero-, or positive-sum game, depending on the circumstances. Objectively, for example, the result is zero sum if B gives money to A in response to a threat that is then not executed. The same amount of goods exists as before, though in different hands, and nothing negative has actually happened. If A injures B before B gives up the money, the result is negative sum because a bad now accompanies the good. It is difficult, however, to imagine a situation with a positive-sum overt result. Subjectively a net gain or loss can be definitely established only if both parties gain or lose satisfaction. If one gains and the other loses we could know the net result only if we assumed that interpersonal comparisons of utility are possible—which we do not. It is unlikely that a mutual gain would result, although it could if B valued the excitement of being robbed more than the money he lost. A mutual loss is clearly possible, as when A spends \$100 preparing for the robbery and gets only \$10 from it while B also undergoes a loss. If both positive- and negative- sum results are possible, so is a zero sum.

3.25.13   *Application Stage of Strategic Stress.*   **3.25.131** To exert power or bargaining power on B, A must be able to apply the stress, which ability is assumed for purposes of the stress transaction (3.25.1a). **3.25.132** Even if A is unable to relieve the stress he has applied to B (he is too weak to open the bear trap) he may nevertheless get Y if B can provide it, since we assume that the promise of relief is believed by B (3.25.1b).

3.25.133 A will not apply stress to B unless (a) he believes it will be sufficient to induce B to give up Y and (b) the cost of applying the stress is less to A than AY since otherwise he would be engaging in an action whose cost exceeded its benefit. **3.25.134** If either Y or the stress applied by A is a variable quantity, A will not apply stress unless it costs him less than the value of Y he expects B to give him in return.

3.25.135 The greater the stress induced by A, the greater will be BX and the greater B's EP for X—hence the greater A's power and bargaining power. **3.25.136** Hence, by joining .134 and .135, we see that the greater the amount of stress A can achieve and maintain on B at a given cost to himself, or the less the cost of a given amount, the more is A's power and bargaining power enhanced by applying stress. This statement encompasses the well-known formulation: The less the cost to A of imposing a cost on B, the greater is A's power. **3.25.137** Once the stress is applied, however, its cost is sunk and its magnitude is logically irrelevant to the outcome.

3.25.14 *Transaction Stage of Strategic Stress.* **3.25.141** In the transaction over relieving stress all factors behave as in the initial model (3.22.1–.6). That is, A's power to get Y varies directly with AY and BX and inversely with AX and $BY;$ his bargaining power varies directly with AX and BX and inversely with AY and BY. At this stage the question is not whether A can or will *create* the stress; he has already done so. The transaction stage concerns his ability and willingness to *relieve* it.

3.25.142 If (relaxing assumption 3.25.1e) relieving the stress involves costs to A (the MD's cost in time, effort, and materials to set Joe's leg), this cost raises the value of AX. It thereby increases A's bargaining power since he is less willing to perform X. But it decreases his power to get Y since he is less willing or likely to perform X.

3.25.143 Relax assumption 3.25.1d and substitute that the stress has continuing cost as in a strike or war. The effect of this continuing cost is ambiguous. (a) At the transactional stage it is the equivalent of a negative cost of relieving the stress—that is, a positive value of doing so, as if the MD enjoyed setting the bone. As such it decreases the magnitude of AX, increases A's EP, and increases A's power to get Y but reduces his bargaining power. (b) But it also reopens the question of whether A is able to maintain the stress or finds it worth its cost—a return to the logic of the prestress situation. Here A's power varies inversely with the cost of imposing and maintaining stress, and a continuing cost decreases his power (3.25.136). Which of the two effects will prevail depends on the situation. If A is clearly able to continue the stress despite its cost, then the poststress power factors at (a) prevail and the question is solely whether and on what terms A will relieve the stress. But if the question is whether he is able to meet its continuing costs, the prestress power factors at (b) prevail. In the latter case, if A decides to relieve the stress he does not do so in exchange for Y but because to continue the stress is beyond his capacity, beyond its value to him, or both.

3.25.15   By way of clarification, we have seen that the more A wants Y, the greater is his power to get it—but the less his bargaining power (3.22.2 and 3.22.3). In the stress transaction the more A wants Y, the more cost he will be willing to accept to impose the stress on B. Hence the greater will be his power to get Y, as before. This stress will also increase his bargaining power. But it will do so because of its effect on B's EP, not the effect on his own. Hence this conclusion does not contradict the earlier statement about the opposite effects on A's power and bargaining power of his *own* EP. The maximum power position for A obviously consists of extreme stress that can be imposed at low cost and whose continuance and relief also have little or no cost— for example, managing to string up B by his thumbs without a struggle. However, when continuance of the stress has fixed costs to A, his pretransactional decision to impose it requires an estimate of when, not merely whether, B will capitulate. For the same reason, B's decision when or whether to capitulate involves not only the value to him of having the stress relieved but also his estimate of when or whether A will relieve it without any concessions from B or in return for concessions smaller than those asked for. Confirming 3.25.15, we thus see that, whether viewed as a factor affecting either the application or the relief of stress by A, a fixed cost to A of continuing the stress raises B's bargaining power. In bargaining over relief it reduces AX and hence increases A's EP. But as seen by B it also increases the likelihood that A will relieve the stress unilaterally and thereby shrink B's own EP.

We note that the application of stress has a parallel in transactions of goods. In advertising, for example, the seller A can incur a cost designed to extend the EP's of his potential customers B's. He must first decide whether the initial extension is worth its cost. Once done, its cost is sunk and is replaced by the question of whether maintaining that extension is worth its fixed cost. The parallel is closest when the advertising is directed toward the sale of a large single item, like London Bridge.

If we remove assumption 3.25.1g (that B uses no counterstress) and assume instead that he does, we find that no new analysis is needed. Like other propositions about transactions, those about stress transactions are symmetrical. The stress applied by B can be handled by introducing a second Y to the formula, in which Y_1 is the good wanted by A and Y_2 is relief of the stress imposed by B. The magnitude of AY would then equal $AY_1 + AY_2$ and BY would equal $BY_1 + BY_2$. The transaction could also be completely symmetrical with A holding a good desired by B. This added factor would then be symbolized as above using X_1 and X_2. Situations in which there is no good on either side seem more likely to arise in threat than in stress transactions. Strictly speaking this modification involves multiple goods, to be dealt with in 3.26 in Chapter Nine. This item seems easier to follow, however, if dealt with in this context.

In real cases, as in a fight or war, it may be difficult to tell whether an action by B is an attempt to apply counterstress, to prevent or remove A's application of stress, to impose total physical blockage on A, or to prevent or remove A's blockage of B. These are questions of identifying pieces of reality with pieces of the analytic structure. When identifications cannot be made with reasonable confidence we cannot diagnose the situation or state propositions above it—as in the "mushy ice" problem of the Preface. Whether a broader simulation model will help depends on certain imponderables (Chapter Sixteen).

3.25.151   We now remove assumption 3.25.1b (that the promised relief of stress is believed by B) and substitute that B does not believe A will relieve the stress even if B gives him Y. B will then expect to receive nothing in return for Y since X now seems nonexistent. BX will then be zero and B will have no EP for X. B's power to get X will drop to zero, and so will A's power and bargaining power in getting Y.

3.25.16   As in transactions in goods, tactics and strategies may be employed seriatim by one or both sides. The analysis then shifts to the developmental, where the cross-sectional model provides no overall guide. However, the cross-sectional analysis of transactions in bads can be used for the same purposes as in 3.24.2. The content of tactics here is the same as in goods: A's bargaining power is greater if AX is overstated and AY understated, and his EP for Y appears shorter than it is.

For bads as well as for goods, at any point in a dynamic negotiating situation, as in union-management bargaining, the static analysis indicates whether the parties would be willing to settle on a specific set of terms—say, those proposed by a neutral party—if we use the following interpretations of the symbols. AX is the cost to A of settling on the proposed terms, as measured by the loss of better terms that might be achieved by holding out, appropriately weighted by some probability factor. AY is the benefit to A of settling on the proposed terms, as measured by the saving of costs that would otherwise accumulate if a settlement were postponed—or by the worse terms than those proposed that might have to be accepted if B cannot be budged from his present position. BX and BY correspond to AY and AX, respectively, with appropriate substitution of symbols. If the EP's overlap on the basis of these factors a settlement will be reached; otherwise not. (For four details of this situation, see Kuhn, 1967, pp. 120–123).

3.25.17 In a transaction using bads with continuing costs it may be difficult, especially in compromise cases, to distinguish between A's motivation to relieve stress because it has worked (B has conceded part of Y) or because it has not worked (B has not conceded much, and seems unlikely to concede enough to justify the cost of continuing the stress). To the extent that they are distinguishable, however, the former is amenable to transactional analysis while the latter is amenable only to unilateral cost/benefit analysis by A, the benefit being a change in B's EP.

3.25.2 *Threat Transactions*

3.25.20 For an initial model of the threat transaction, we modify 3.25.1a to read that A threatens to apply some bad to B unless B gives A some good Y. **3.25.21** Under these assumptions, if the threat is successful and A receives Y, the threat has no cost to A other than that of communicating it to B. **3.25.22** If the threat is unsuccessful, however, subsequent analysis depends on the content of the threat, as follows. If A has threatened to apply some bad that could have been used in a stress transaction, and if it seems worth its cost to A, A can then transform the threat to a stress, in which case the preceding analysis is applicable. Assuming that A is prepared to use it, the benefit to A of threatening stress before applying it is that the cost of the stress may be saved. On the other hand, the threatened act, if carried out, may not be a relievable stress. To clarify, if A holds a flame under B's hand at a distance that is painful but not injurious, the stress is removed by removing the flame. But if A burns down B's house this stress cannot be relieved by A in return for a concession. The burned house is a sunk cost to B and therefore irrelevant to subsequent dealings with A. The burning of B's house in itself gives no power or bargaining power to A since both depend on the ability to give B something he wants—namely, an unburned house.

3.25.23 Suppose the content of A's threat cannot be converted into a stress. Then if the threat fails, the attempted transaction to which it was applied has also failed. A's decision to execute the threat or not must therefore depend solely on the anticipated effect on subsequent transactions. That is, the question of executing the threat is post-transactional. If the transaction is unique (our current assumption), A has nothing to gain from executing the threat, even if it is costless to do so.

3.25.24   Let us now drop the assumption that the threat transaction is unique (3.21j) and assume instead that it can affect subsequent transactions with B or with others. Since a threat is wholly ineffective if it is not credible, then not to execute a threat when necessary may affect the credibility of A's threats in future relations with B or with others who are aware of the present relation. Failure to execute the threat will, *cet. par.*, reduce the credibility and hence the effectiveness of future threats. That is, to fail communicates at the minimum that "A does not execute all his threats." The cost of making a threat therefore includes the risk that A may have to execute it or else lose face or credibility.

      A's threat will not be credible unless he appears to B to possess sufficient EP to execute it—that is, both the will and the ability. (Like any other tactic, it involves the creation of beliefs or images, and B's response is based on his image, not on reality as such. However, B's knowledge about the realities of A's position influences B's beliefs.) Hence if A wishes his threat to be credible he must (a) build or maintain the capability of executing the threat, (b) make the threat known to B, and (c) appear to have sufficient stake in executing the threat to justify his doing so. That is, a threat is a promise and a promise must have enough apparent stake behind it to be credible (3.24.3). There are innumerable ways to communicate willingness to execute a threat —among them executing smaller threats on others, partial execution on the other party, and displays of strength. (Although the main discussion of observed transactions as communication appears in Chapter Eleven, advance mention seems in order here.) In any event, within the assumptions A has nothing to gain and much to lose by making a threat that is not credible.

3.25.25   Remove assumption 3.25.1g (that B uses no counterthreats) and assume instead that he does. The preceding generalizations about threats by A are symmetrical. But a counterthreat raises additional problems. To make his own threat credible A must convince B that AY is larger than the cost to A of blocking execution of a credible threat by B—or, if he is unable to block it, greater than the cost to A

of having B's threat executed against him. If B is able he can make A's threat noncredible by convincing A that he, B, is able to block it or to execute a counterthreat whose cost to A would exceed AY—thereby preventing A from making the threat in the first place. This is the deterrent effect.

3.25.26   Within limits set by the value of the goods to be received or lost as a result of threat, the mutual threat relation is subject to positive feedback. That is, A can overcome the deterrent effect of B's potential counterthreat by increasing his own potential threat, which new position of A requires B to increase *his* potential and so on. A positive-feedback relation is by definition unstable unless there is a limit to one of the variables involved. If the stakes are great enough (especially if they are viewed at a life-and-death level), the maintenance of deterrent capacity may take precedence over positive elements in the potential exchange relation. In that case attention to means may completely overshadow attention to any ends toward which they might be applied.

The mutual threat relation is unstable, *cet. par.*, for two additional reasons. First, if A feels sure he is in the dominant position he may use a preemptive strike to weaken B before he can strengthen *his* position. Second, long disuse of the threat may lead B to believe that A is no longer able or willing to use it. Since the execution of a threat not converted into a stress has no value except for credibility, and if credibility is seen to be slipping from disuse, A may not find it feasible to wait until he wants some positively valued Y; he may instead demand some Y he doesn't really want as a pretext to exercise the threat.

3.25.27   If we remove the assumption of uniqueness and assume instead that some third party will use pressure against A and/or B for using or threatening to use bads, the cost of this pressure must be included in the decision whether or not to use or respond to threat. The effects are covered in Chapters Nine and Ten, mainly in connection with multiple goals including generosity and hostility (3.26), multiple parties (3.27), coalitions (3.27.5), interrelated transactions (3.28), and aggregate power (3.3).

3.25.3   Although the logical bases for these conclusions are presented later, particularly in 3.26 on multiple goals, it may be indicated here that if fear or anger is generated in B by A's stress or threat, the addition of those emotions to the "straight" desires concerning them will have two main effects. First, fear will intensify the cost of the stress or threat to B and lengthen his EP to eliminate or avoid the bad. Second, hostility toward A will shorten B's EP.

3.25.4   The social disadvantages of power based on bads are perhaps obvious; yet it may be useful to spell them out. First, and in contrast to transactions in goods, those in bads are normally negative-sum games. They reduce total utility rather than increase it. Second, whereas aggregate power based on goods is limited by the ability to produce satisfactions for others, as well as by competition and coalition, a stronger coalition is the only apparent limitation on negatively based power (4.12.42). Third, the power relations based on threats are highly unstable (3.25.26), and the power in stress transactions can shift suddenly and dramatically—as when the prisoner snatches the gun from his careless guard. Fourth, the effect is highly unpredictable. One may give all he owns when his life is threatened or he may refuse to give a nickle. Fifth, bads can destroy health, comfort, security, and property. Hence they can also destroy motivation if persons have no assurance they can keep and enjoy the fruits of their labors. Sixth, since stress or threat can be imposed unilaterally by A, the *fact* of the transaction is not voluntary for B even though the eventual terms may be agreed to by him—and he may, in fact, concede nothing. Seventh and finally, because bads that dwarf goods in value can often be imposed at small cost they make for highly one-sided transactions when used by only one party.

3.25.5   *Concluding Note on Strategic Bads*

We have treated the use of bads as pre- or posttransactional, unilaterally imposed strategies designed to alter the other party's EP. The transaction is not over imposition of the bad but over its removal or nonperformance. This means that the things bargained over in all transactions are positive. Their degree of positiveness is measured from whatever position a party has been put into by his own prior actions, by the other party, or by other forces. This emphasis can be clarified by noting its wording. In stress the significant statement is not "If you don't give me what I want I will keep hurting you," but "If you give me what I want I'll stop hurting you." In the threat it is not "If you don't give me what I want I will hurt you," but "If you give me what I want I won't hurt you." When we use the positively oriented formulation virtually the entire remaining body of transactional analysis can be applied to these transactions in bads. Under the negatively oriented wording such applicability is clumsy at best.

Chapter 9

Transactions

Details

The preceding chapter laid out the basic assumptions of transactions and deduced many propositions about transactions in goods and bads. This chapter deals with such obvious complications as multiple goals, more than two parties, and interrelated transactions. And the following chapter concerns aggregate power—an accumulated stock of ability to get wanted things, normally accrued by a series of successful transactions.

3.26   MULTIPLE GOALS

In this section we relax assumptions 3.21b(2) (that each party is indifferent to the position of the other) and 3.21k (that neither party has desires in the transaction other than those for X and Y). Instead we assume that either party may wish to help or hurt the other. All other assumptions of 3.21 apply.   **3.26.0** For this purpose we need two additional definitions. A *hostile transaction* is one in which, besides the selfish desires of both parties concerning X and Y, one or both wish to reduce the satisfactions of the other party in the transaction—that is, to "hurt" him. The underlying motives for this wish are irrelevant and may reflect dislike, retaliation, anger, fear, frustration, psychopathic personality, or a desire to improve the other party's soul or subsequent behavior. The hostile transaction includes disciplinary action by a loving parent ("this hurts me more than it does you"). A *generous transaction* is one in which, besides the selfish desires of both parties concerning X and Y, one or both wish to increase the satisfactions of the other—or at least the amount of goods received by him—in the transaction. Here, too, the underlying motives are irrelevant. They may reflect love, enthusiasm, empathy, sense of obligation, gratitude, custom or ritual, or a desire to relieve one's own distress over the other's condition. The overtly generous stance could also reflect inner tensions, such as a desire to punish oneself or relieve guilt. In the context of interrelated transactions (3.28) it may also reflect an instrumental desire to change the other party's attitude—to "soften him up" for some later relationship or raise his sense of obligation. The consequences of such motivations are detailed later in this chapter.

Since a transaction includes two parties, the attitudes of both must be included for a complete classification system. In "games against nature," for example, nature is indifferent. The selfish (selfish-indifferent) player, as here defined, is also indifferent to the opposing party, and the distinctive characteristic of nature is its indifference to *both* parties. Whatever her status as a player in a game, nature cannot, of course, be a party to a transaction, precisely because she has no wants to be satisfied from another party. Incidentally, this approach differs from that of Riker (1962, p. 35), who labels B's attempt to maximize what he can get from A as "malevolent."

It should be clear from the definitions that there is no analytic relationship between (a) transactions in goods or bads and (b) the selfish, generous, or hostile attitude. Goods and bads can both be used with any of the three attitudes. There is almost certainly an empirical correlation, however, particularly between hostile attitudes and bads.

3.26.1   *Hostility*

3.26.10   In an otherwise selfish transaction in goods we now introduce hostility by A toward B, such that A to some degree wants B not to have X. Since the way to accomplish this is to keep X, this amount of A's desire to keep X is added to the initial AX, thereby increasing AX and decreasing A's EP.   **3.26.11**   In consequence, hostility by A toward B increases A's bargaining power and decreases B's while **3.26.12** it also decreases A's power to get Y and B's power to get X.   **3.26.13** In this connection we define *boycott* as a hostile shrinkage of an otherwise overlapping EP to the point where there is no overlap. (There is no point, of course, in the notion of a boycott against someone with whom there would be no dealings in any event.)   **3.26.131** A boycott thus effectively eliminates A's power to get Y and B's power to get X.

3.26.14   Change the assumption of 3.26.10 to that of a stress transaction.   **3.26.141** If A wants to create dissatisfaction in B, which in this case will occur when he applies the stress, he will get some satisfaction from this act itself. This means that the net cost of doing so is reduced and that A's power to get Y is increased (see 3.25.136). **3.26.142** In the transaction stage of the stress transaction, A's desire that B undergo stress increases A's cost of relieving the stress.   **3.26.143** This increases AX and hence increases A's bargaining power in the transaction and reduces his particular power.   **3.26.144** In a threat transaction A's hostility reduces his anticipated cost of executing the threat and therefore increases his power to get Y (see 3.25.2).

3.26.15   If A is sufficiently hostile toward B, he may give the bad to B as a "gift" without receiving any Y in return. A does this when his desire that B have the stress equals or exceeds the cost to A of applying it (the initial AX), so that the net value of AX is zero or negative. A might then harm B for the sheer pleasure of it.

3.26.2   *Generosity*

We now return to the positive-goods model as in 3.26.10 but substitute generosity for hostility by A toward B, such that A to some degree wants B to receive X from A. The amount of this desire is subtracted from the initial AX, thereby decreasing AX and increasing A's EP. Generosity by A thus decreases his bargaining power and increases B's, while it also increases A's power to get Y and B's power to get X.

3.26.21   Let us now change the assumption of 3.26.2 to that of a stress transaction. If A dislikes creating dissatisfaction in B, which applying the stress will do, A will get dissatisfaction from applying it. This means that the net cost of applying the stress is increased and that A's generosity in a stress transaction decreases his power to get Y. At the transaction stage, A's displeasure at B's stress decreases the cost to A of relieving the stress, decreases AX, decreases A's bargaining power, and increases his power to get Y. In a threat transaction A's generosity increases his anticipated cost of executing the threat and thereby decreases his power to get Y.

3.26.22   We now return to the positive-goods assumptions of 3.26. If
A is sufficiently generous toward B, he may give the good to B as a gift
without receiving any Y in return. A would do this when his desire that
B have X equals or exceeds the intrinsic AX, so that the net value of AX
is zero or negative. If A generously wants B to have X, any tactics he
might use would have to reverse those of 3.23.1. That is, he would have
to represent his EP as being greater than it is—by exaggerating AY,
understating AX, or both. In the case of a pure gift, which involves no
Y, the appropriate tactic is to represent AX as very small or negative
("I was about to throw this out. Could you use it?").

3.26.23   Whereas subjectively the selfish transaction is necessarily a
positive-sum game (3.21.03), the generous transaction can be negative
sum if tactical misrepresentations are used and thus can bring a loss
to both parties. Suppose your penurious aunt A offers you B a bowl
of her favorite, expensive mussel soup. Though you detest the stuff
you eat it with apparent gusto—whereupon she destroys additional
satisfaction for both of you by ladling out a second bowl. (We are still
dealing with the desires of the parties concerning the things exchanged.
In 3.26.4–.5 we add the effect of satisfactions or dissatisfactions in the
transaction itself.)

3.26.24   In a selfish-indifferent or hostile transaction, credible tactical
misrepresentations by A work to B's detriment. Except for the condition
like that above in which B dislikes X, tactical misrepresentations by A
in generous transactions work to B's benefit. Hence A's lies will be
more acceptable to B in the latter case and are typically thought of as
small and white rather than big and red. Later discussion, mainly in
Chapters Ten and Eleven, indicates that the immediate cost of a gift
produces longer-run benefits in addition to any pleasure in the giving
itself. To that extent the gift constitutes investment, as in building or
maintaining a friendship.

3.26.25   *Methodological Note on Generosity.* It is, of course, possible
to formulate the analysis so that power in generous transactions resides
in A's ability to get B to accept his generous offering. Such a concept is
consistent with the idea of power as the ability to achieve what one
wants, which in this case would be that B receive X. We do not use
that formulation, however, as it would considerably confuse the analysis
(Kuhn, 1963, note pp. 356f). Nevertheless, in connection with aggregate
power A's power resides in his ability to create or provide an X to offer

B. That emphasis is consistent with many possible motivations of *A*— as, for example, that *X* can extract *Y* from *B*, that *B* shall have *X*, or that receiving *X* will put *B* under obligation to *A*.

Generous and hostile transactions are probably more primitive than selfish-indifferent: they reflect emotional, impulsive responses. A mother's generous treatment of offspring and her hostile response to enemies appear at relatively low levels of animal life, and both responses appear more reliably in primitive societies than do selfish transactions. By contrast the selfish transaction, indifferent toward *B*, cannot be based on emotions alone. It requires a relatively detached analysis of costs and benefits and typically appears only in "advanced" levels of social development. Once all three types (selfish-indifferent, hostile, generous) are known, however, it seems analytically easier to start with the indifferent stance toward the other party, as in the initial model of 3.21. Precisely because it involves some calculation of advantage, even if not quantitatively measurable, and permits clear propositions to be deduced, the neutral stance is the simplest point from which to begin *logical* analysis. Simple modifications of those propositions easily accommodate generosity and hostility.

There is a growing suspicion among some anthropologists (see Sahlens, 1965) that in societies which make much use of gifts the long-run content and terms of trade may not differ much from what they would be if the same transactions had been selfish. It would be easy to modify the model to accommodate this result if we construed each EP in a generous transaction to include a component representing its anticipated long-run payoff. Moreover transactional analysis can be applied directly from the selfish model if we assume that each gift is "paid for" immediately in the form of a socially enforceable "account receivable." We will not, however, use that formulation here, although the discussion of normative versus decided behavior in Chapter Sixteen is useful in this connection.

In the selfish transaction we can predict *A*'s action on the basis of his own preference function (EP) alone, without knowing who *B* is or how *A* feels about him. That is, *A*'s EP for *Y* is the same for all possible *B*'s. An initial model of transactions would be hopelessly complex if it assumed that *A*'s EP might differ for every possible *B*. It therefore seems that a structure which moves from the rationally selfish toward the hostile or generous can more easily be made rigorous than one which moves in the opposite direction. Separate theoretical structures for generous and hostile transactions are, of course, possible and do exist, presumably to be retained only if the present type of coordinated explanation is unsatisfactory.

3.26.3   *Transaction Costs*

Let us now return to the initial model but relax assumption 3.21h (that the transaction process itself involves no costs). Instead we assume that the process does have costs—as in time, frustration, money, inconvenience. For the moment we exclude satisfaction from "losing" the negotiations. This means that to acquire Y involves, in addition to the cost of giving up X, the cost of the transaction itself. This cost enlarges the net value of AX and therefore reduces A's EP. That is, AX now consists of $AX_1 + AX_2$, where AX_1 is the cost of giving up X itself and AX_2 is the cost of conducting the transaction. In this case transaction costs to A increase his bargaining power and decrease B's while they also decrease A's power to get Y and B's power to get X.

3.26.4   *Transaction Benefits*

Substitute the assumption that the transaction process itself provides benefits—in conversation, friendship, release from other duties, self-expression, an excuse to travel, pleasure in spending money, and the like—but continue to exclude pleasure from "winning" the negotiations. By reasoning parallel to 3.26.3 transaction satisfactions to A decrease his bargaining power and increase B's while they also increase A's power to get Y and B's power to get X. In addition to any shift it brings in power or bargaining power, such intrinsic satisfaction from the transaction process increases the total benefit to the parties while transaction costs similarly reduce it.   **3.26.5** These effects of cost or benefit in the transaction process (3.26.3–.4) can presumably be superimposed without change on negative-goods, hostile, and generous transactions.

3.26.6   We now return to the initial model, but to assumption 3.21k (that neither party has desires other than those for X and Y) we add A's desire to "win," to enhance his self-image as a tough bargainer, or to wring concessions from B. For any terms that A considers a bad bargain, to give X for Y incurs not only the initial cost AX but the additional cost of dissatisfaction with the terms. For any such terms the net cost of the transaction rises for A, which means an increase in AX and a decrease in EP. Hence with respect to any terms that A considers a bad bargain, his bargaining power rises and that of B decreases. If A also develops hostility toward B because B seems to be driving too hard a bargain, then (according to 3.26.11 and .12 on hostility) that feeling adds to the above effect. Hence, too, for any set of terms that A considers a bad bargain his power to get Y decreases, as does B's power to get X. If for some reason A wants to "lose" the negotiations or drive a bad bargain, these conclusions are reversed. All aspects of 3.26.6 can presumably be superimposed without change on any other type of transaction.

3.27   MORE THAN TWO PARTIES

In this section we relax assumption 3.21a (that only two persons are involved) and substitute additional persons in various relations.

3.27.1   *Pressure Transactions*

Let us now assume a third person C through whom A might have a tactical or strategic effect on B that would affect A's position in a transaction between A and B. A tactical effect would occur if C modified B's belief about A's EP. A strategic effect would occur if C modified B's preference function inside or outside the A-B transaction or modified some of his transactional alternatives that are outside the A-B transaction but affect B's EP in it. Numerous relations of this sort are possible, a sampling of which illustrate the technique of analysis. For this purpose we define a *pressure transaction* as a three-party relationship in which C employs tactics or strategies on B to improve A's power or bargaining power in an A-B transaction. As an example of pure tactics, if A knows that B and C communicate from time to time, A might "inadvertently" communicate to C something about his desires or tastes relative to B, X, or Y in the hope that C might pass the information to B and thereby create in B some belief about AX or AY. Or through a transaction A might get C to communicate deliberately to B information, accurate or inaccurate, desired by A. This transaction between A and C could be selfish, generous, or hostile, using either goods or bads, but no new principles are involved. If this communication from C to B seeks to give B some impression about A's EP, it is tactical with respect to the A-B transaction. If it seeks to change B's EP, it is strategic. As another alternative, through a transaction A may get C to engage in a transaction with B that will change B's EP.

For the analysis of three-party transactions we use the following symbols. Z is any interaction (communication or transaction) between C and B engaged in by C on behalf of A, designed to change the state of B's detector or effector with respect to A, X, or Y. CZ is C's desire not to engage in interaction Z—that is, not to perform it for A. x is any good offered by A to C in return for Z. y is any good given by B to C in transaction Z—when Z is a transaction that affects B's EP for X. And z is any good offered by C to B in Z if Z is a transaction. In short, x, y, and z are goods that may be exchanged as part of tactics or strategy of the negotiating process. Ax, Cy, and so forth have the same meaning as in the base transactions.

Any reticence C might have to perform Z, such as fear of spoiling the C-B relationship or of hurting his own credibility, would be a component of CZ and would reduce A's power and bargaining power to get Z and hence Y. Any reticence A may have to ask C to perform Z would be included within Ax and hence, like any other component of Ax, would reduce his power to get Z and hence Y.

According to the model (3.21) there can be no base transaction to which C is party since he would then automatically be redesignated A or B. The special complications that arise when communicated information becomes the subject of transaction are dealt with later. As an example of a pair of strategy transactions the United States, A, might offer aid to Egypt, B, if Egypt in turn will negotiate an arrangement to let Israeli shipping through the Suez Canal so as to improve American chances of selling merchant vessels to Israel, C.

The analysis places no limitation on the identity of the third party. For example, nothing in the analysis precludes the use of a deity as the third party, and parties ranging from individuals to nations have often invoked such help to visit curses, disease, or loss of strength on the other party. In fact, much religious doctrine deals with transactional relations between God and believer, and some faiths are quite explicit about what behavior of each is contingent on what behavior of the other. A tentative survey (namely a class report by a student) of denominations in the United States suggests that members of rich churches have relatively good transactional but bad communicational relations with God whereas the churches of the poor show the reverse.

Strictly speaking, this framework incorporates no three-party transactions as such because all their significant aspects are covered in some other fashion. For example, Section 3.27.2 deals with a third party as agent. Competition (3.27.3) involves A dealing with B while B_1 is an available alternative. Coalition and collective bargaining (3.27.5–.6) cover cases where B and B_1 join in dealing with A. The analysis of multiple transactions (3.28) delineates the tactical and strategic effects on the A-B transaction if B observes A's negotiations with B_1. The discussion of organization in later chapters deals with the problem of achieving a single decision binding on multiple parties, whether or not that decision is unanimous. But the decision is not a transaction, and other sections are relevant to the same problem. The classic triangular trade is a simple case of three separate two-party transactions; they are interrelated because each party's EP in one transaction is dependent on the outcome of a prior (or anticipated) transaction with the other. In short, although this framework deals directly with many situations involving more than two parties, it does not specifically acknowledge a transaction between more than two—for the simple reason that I can think of no transactional relation that is not covered by this book. Furthermore, no matter how numerous or complicated the relations of A or B with other parties (of Israel or the Arabs with the United States, France, Japan, China, Russia), before an agreement is reached all benefits or costs of those other relations, whether they are conceived as pressure strategies or tactics, can be translated into ingredients of $AX, AY, BX,$ or BY. Only when they are put in that form is it clear why the two principal parties behave as they do.

3.27.2 *Bargaining Through Agents*

An *agent* is a third party C who acts on behalf of A in a transaction between A and B. In contrast to the general situation, where C attempts to influence B in ways favorable to A, here C does all the negotiating with B. Symbols remain as in the general third-party relation except that the interaction Z is now broadened to include the whole process of negotiating for Y. Any such relation involves two transactions: the A-C (agency) transaction, which determines the terms on which C will act as agent for A, and the main C-B transaction over X and Y.

3.27.21 The following assumptions constitute the model of agency:
a. C is also selfish, with no generous or hostile motives toward A or B. He has no other motives outside the transaction.
b. The A-C and C-B transactions are both unique.
c. A communicates his EP for Y accurately to C.
d. The reward x, which C receives for performing negotiation Y, is flat and not contingent, and clear criteria are available to determine whether C has performed satisfactorily.
e. All promises by all parties are credible.
   Under these circumstances there are no complications in the

A-C transaction. *A*'s bargaining power is $(Ax + Cx) — (AZ + CZ)$, *A*'s EP is $AZ — Ax$, and *B*'s EP is $Cx — CZ$. *Z* is solely a service instrumental to the acquisition of *Y*, with no intrinsic reward value for *A*.

Under the assumptions of 3.27.21 the *C-B* relation is the same as the *A-B* except that *C* is negotiating instead of *A*. *C*'s bargaining power is the same as *A*'s: $(AX + BX) — (AY + BY)$. If tactics are employed in this negotiation there is still no difference between the *C-B* and the *A-B* relation except that of personality.

If strategies are employed by *B*, then by definition *A*'s EP may be changed during the negotiations. If *A* and *C* are in regular and accurate communication these strategic effects will be no different than if *A* were doing the negotiating. To the extent the communications from *C* to *A* are inaccurate, *A*'s strategic position might change more or less than if he were negotiating in his own behalf, depending on the nature of the inaccuracy. *C* will probably have some tactical advantage not possessed by *A*, since his statement that "*A* is adamant" is not directly subject to appeal, argument, or betrayal by gesture or tone of voice. But if *A* and *C* are out of communication, and if *A* has previously expressed to *C* an unequivocal but tactically understated limit to his EP, then strategies applied by *B* are of no consequence because *C* cannot act on them. This fact strengthens *A*'s bargaining power but weakens his power and may result in failure to consummate an advantageous transaction.

3.27.22   We now continue all assumptions of 3.27.21 except assumption **d**. Assume instead that the reward *x* is a fixed amount but is contingent on *C*'s acquiring *Y* from *B* for *A* at a price to *A* no greater than *A*'s EP. Assume also the following:

f. *C* can add inducements of his own to get *B* to give up *Y*.

g. The *A-C* transaction of agency is already completed and *C* has sufficient stake in his promise that he will serve as *A*'s agent.

Under these conditions, *C*'s EP for *Y* is $(AY — AX) + (Cx — Cz)$.

3.27.23   Now remove assumption g and assume instead that *C* has no commitment to serve as agent for *A* but will receive *x* only if he can get *Y* for *A* for no more than a specified price. *C*'s EP for *Y* is then $(AY — AX) + (Cx — Cz — CZ)$.

3.27.24   Now return to the conditions of 3.27.22 but change assumption **d** again so that, in addition to *x*, *C* will receive a bonus of *k* percent of any saving S by which the terms achieved are better for A than those of his EP. C's EP to complete the transaction and to receive x remains the same as in 3.27.22.

This conclusion may seem surprising. But it represents the same logic as that of A with reference to his own EP. A will be happy to get better terms from B—if he can. But his EP represents how much he will *give* if he cannot get Y for less. By introducing some percentage of the saving, kS, as a bonus for C, we also make him happy to get Y for less than A's EP—if he can. But we also leave him willing to give that much—if necessary—in order to get Y for A and hence x for himself. C might, of course, offer his bonus kS to B as an inducement to give more for X. But B would not thereby be willing to improve his offer by more than kS. In that case C's bonus would be $k \cdot kS$, which (k being less than 1) would be less than kS. The bonus therefore puts C in no better strategic position to improve the terms of the transaction for A. Thus under 3.27.24 C's power to acquire Y, and hence A's power to acquire it, remain unchanged from 3.27.22. The bonus (3.27.24) makes any terms "bad" in C's eyes that provide no saving for A. Under 3.26.6 (the effect of a "bad bargain" on bargaining power) C's bargaining power rises because of the bonus, since terms which would be otherwise satisfactory now become bad to C. The general principle of 3.26.6 about the bargaining power effect of a "bad bargain" is stated without regard to whether the measures of badness are objective or subjective. In this case they are objective in the amount of $k \cdot kS$— which, as we have seen, is insufficient to enable C to induce better terms from B. Hence 3.26.6 does not modify 3.27.24 under the present circumstances. A bonus larger than the saving ($k > 1$) could induce a saving but would not be offered by A.

A bonus like that of 3.27.24 is apparently useful to A only if we relax the latter part of assumption 3.27.21d (that clear criteria are available to determine whether C has performed satisfactorily). If we assume instead that C without the bonus would not bargain as effectively as he might and that A would have no way of knowing this, then the bonus might induce C to bargain more skillfully and achieve a saving for A. If there is a large overlap of EP's the gain to A from the bonus may be considerable; if there is little overlap the gain will be small.

3.27.25   If we relax assumption 3.27.21b (that the transaction is unique) we allow additional motives, such as that C wants to impress A in order to get future assignments or promotions. This relation is dealt with under multiple transactions (3.28). Many propositions of Section 3.28 can be applied to the principal-agent relation but need not be spelled out here.

3.27.3   *Competition*

We now return to the initial model of 3.21 but relax assumptions 3.21i and k (which assume that contact has already been established and that neither party considers the possibility of dealing with any other party). We now substitute the following assumptions:

a. There are two or more B's.

b. All B's are known to a single A.

c. All B's are perceived by A to have identical Y's they are willing to exchange for some quantity of X.

d. Transaction costs to A are the same for all B's.

e. There is no collusion among B's.

3.27.30   We also add two definitions. *Competition* is a conflict situation in which two or more parties seek to complete the same transaction. That is, if B's are competing, then $B_1, B_2, \ldots, B_n$ all want to be the B who receives X from A in return for some quantity of Y, and under the assumptions all B's are substitutes for one another in the eyes of A. Even if A engages in transactions with more than one B, a transaction with B_1 means the loss of that transaction to some other B. A *strong competitor* is a B who has a strong likelihood of consummating the transaction with A, compared to some other B with a small likelihood, who is a weak competitor. **3.27.31** Since A's power to acquire Y (3.22.3) and his bargaining power in acquiring it (3.22.5) are both direct functions of B's EP, A will seek to conclude his transaction with that B with the greatest EP for X, who is the strongest competitor and who has the greatest power to acquire X from A but the least bargaining power in doing so.

3.27.32   At this point the distinction between plain power and bargaining power (3.22.8) becomes crucial. Since the strongest competitor has the weakest bargaining power position, failure to separate the two concepts virtually guarantees failure to reach tight analysis. (I spent many frustrating days before convincing myself that the only analysis that made sense was one in which the *strongest* and *most powerful* competitor accepted the overtly *worst* terms. There was even greater frustration before stumbling into the necessity of separating the transactional from the pre- and posttransactional stages in strategic bads.)

In a unique transaction the B with the longest EP is the one with the strongest BX, the weakest BY (perhaps because he has lots of Y), or some combination. In a business firm a small BY represents a low cost of production.

3.27.33 The terms available from one B provide a floor under the bargaining power of A in dealing with any other B and a ceiling on the bargaining power of all other B's in dealing with A. That is, as soon as A knows the terms on which he can get Y from one B, we say that his EP for the Y held by any other B will not exceed the amount represented by those terms. By generalizing, we conclude that if there exist in contact with one another an indefinitely large number of A's and B's having identical X's and Y's to exchange, and no coalitions are formed, then over a period of time no EP for Y held by any A will exceed the terms available to other A's and no EP for X held by any B will exceed the terms available to other B's. All transactions of X and Y will proceed with identical terms. Any A's and B's whose EP's do not reach these terms will be unable to complete transactions (Kuhn, 1963, pp. 570ff).

3.27.4 *Consensus by Transaction*

In light of the above we define a *market consensus* as a single exchange ratio between X and Y among a multiplicity of mutually accessible and competing A's and B's, developed through the external availability of alternative transactions. Such a consensus is less than perfect to the extent that (a) the X's or Y's are not identical, (b) the terms of exchange are not identical (that is, there exists a range of ratios rather than a single one), or (c) the given terms prevail because they are felt to be "right" (that is, they reflect internalized preferences rather than overt opportunities). However, regardless of how a consensus is formed initially (communication, transaction, or coalition), once it is in existence and widely followed it has the same limiting effect on the external alternatives of an individual as if it were wholly market (transactional) in origin.

In the presence of a market consensus known to all A's and B's, there is nothing for any party on either side to gain from tactics or strategies against any other party. Unless there are other motives for using them they will disappear, whereupon single transactions can go forward with greater dispatch and efficiency than if there were no such consensus. Furthermore, if A is producing (or contemplating producing) X's for exchange he will know in advance how much Y he can get for them and can plan production accordingly. Obviously B's power and bargaining power are enhanced if he has no competitors, though such power is nevertheless limited by the strength of A's EP.

3.27.41   We now relax assumption 3.27.3b (that A is aware of all possible B's with whom he might conduct a transaction) and substitute imperfect information. Then among those B's known to A (or known equally well), the B with the greatest EP for X is the strongest competitor.   **3.27.42** We next relax assumption 3.27.3c—that all Y's possessed by all B's are perceived by A to be identical. Then B_1 improves his competitive position relative to B_2 if he can get A to perceive Y_1 as more desirable than Y_2—that is, to make A's EP for Y_1 greater than for Y_2—as through advertising or a good sales pitch. **3.27.43** Next relax assumption 3.21h (that the transaction has no costs) and substitute differences in the cost to A of transacting with different B's. Then, under 3.26.3, that B who can provide the lowest transaction costs to A is, *cet. par.*, the strongest competitor.

3.27.5   *Coalitions*

To discuss coalitions we return to the model of 3.27.3 but relax its assumption **e** (that there is no collusion among B's). Instead we substitute the possibility of coordinated action. Before getting into details we should note that although coalitions are in practice often mixed or overlapping in function, for theoretical purposes it becomes necessary to distinguish between (a) a coalition of B's designed to improve the terms of their individual or collective transactions with A's and (b) a coalition of only some B's designed to control the collective action taken by all B's. The first case is simply a *coalition* or *bargaining coalition* and is discussed immediately below. The second case is a *dominant coalition* and is discussed in connection with organizations (4.12.414). The discussion at this point deals only with the external relations of the bargaining coalition and a few repercussions of those external relations on its internal affairs. The strictly internal problems of the coalition coincide analytically with certain internal problems of organization as discussed in Chapter Twelve.

3.27.50   A *coalition* consists of coordinated action among actual or potential competitors (B's) to improve their power position in dealing with A's. A coalition can alter particular power, aggregate power, or bargaining power and influence transactions in either goods or bads.

If A's EP is given, then to raise plain power (particular or aggregate) of B's requires an extension of the EP's of B's. Since people are more likely to lack the ability than the will to acquire things, the way to extend the EP's of B's is usually that of raising the effective aspect. This may be done by pooling the powers of $B_1, B_2, \ldots, B_n$ into a larger aggregate than the separate B's can command so that an EP of A can be met which otherwise cannot be met at all. For example, by pooling equipment and know-how ten contractors together might be able to build a dam for A that none alone could undertake. Or competitors $B_1, B_2, \ldots, B_n$ might form a coalition with a joint EP larger than that of competitor B_0 so that the coalition could bargain X from A for themselves rather than have it go to B_0. For example, all NATO nations together might be able to offer France enough military security to keep her from concluding an alliance with Russia though no single NATO nation could do so.

To raise their *bargaining power,* competing B's must contract their EP's. This reduction is typically accomplished in either of two ways. In a *unilateral bargaining power coalition* the B's simply agree not to give up Y for less than certain specified terms—these terms may or may not be uniform for all B's or all A's, and the B's may or may not specify which B's shall deal with which A's. In a *negotiating bargaining power coalition* the B's jointly negotiate a settlement with A and specify the terms on which A may receive Y from any B if he desires it. The following discussion parallels that of N-person games.

Again the distinction between plain power and bargaining power is crucial. Whereas the coalition for plain power seeks to offer A more than he could get without the coalition, the bargaining power coalition seeks to offer him less by eliminating, or at least worsening, some of his alternatives. Although the nature and sources of power are generally the same for coalitions as for individuals and organizations, certain characteristics of coalitions seem worth spelling out. Of the many possible combinations we will deal with several simple cases.

3.27.51   Let us assume that two or more competing B's, with differing EP's for X, form a unilateral bargaining power coalition that agrees not to give Y to A for less than certain specified terms which are uniform for all B's. Then the coalition must include all B's whose separate EP's would be greater (stronger competitors) than the uniform specified terms; otherwise the coalition will not be able to enforce its terms. Those B's with shorter EP's (weaker competitors) would be of no value to the coalition since they could not undercut its terms in any event. Nor could they meet the coalition's terms, and so presumably would not join. Hence, if all B's are ranked in declining order of competitive strength, the coalition must include the strongest competitor and all successive ones down to the one that can just meet the coalition's terms. If it includes these competitors, the coalition's terms are no better, from A's point of view, than those of its weakest member. Thus the coalition raises the floor on the B's bargaining power and lowers that of A (3.27.33).

3.27.52   The strongest competitor in the coalition gives up the certainty of completing the transaction with A and receives instead the assurance that he will get more favorable terms if he completes the transaction. By contrast the weaker competitors stand only to gain from the coalition. With it they have at least a possibility of getting X whereas without the coalition they do not. The weaker competitors in the coalition will therefore be eager for it; the stronger ones will be restive and likely to break out and "go it alone." Hence unless some external stake is involved, the coalition is unstable and presumably will have to reorient its terms with A's whenever its membership changes.

3.27.53   Relax the assumption that there is only one A (3.27.3b and 3.27.5). Assume instead a multiplicity of A's with varying EP's for Y, no coalition, and no complete knowledge of all B's. Assume also a succession of identical Y's to be exchanged by each B. Under these conditions the effect of the coalition will be to reduce the number of Y's that the strongest competitors could otherwise exchange but to improve the terms of the exchanges. The effect of the coalition will also be to increase the number of Y's the weaker competitors could otherwise exchange, and it may improve the terms of the exchanges, depending on the length of their EP's. It can thus be seen that the coalition is nearly all gain for the weaker competitors, who will presumably want it. The loss to the strongest competitor is less incisive than when the sole transaction is assuredly his without the coalition; and the improvement of terms may or may not compensate for the loss of numbers. But the weaker competitors still stand to gain more from the coalition than the stronger ones.

      If the agreement of coalition is itself arrived at on the basis of transaction among its members, then in dealing with any particular B the stronger competitors will have more bargaining power than the weaker ones (from the 3.27.52 conclusion that the weaker competitors want the coalition more than the stronger ones do). The bargaining power coalition is thus partly cooperative (since it provides some benefit to all members) and partly conflictual (since it improves the position of some members at the expense of others). A coalition of B's, of course, is all cost and no benefit to the A's.

3.27.54   Remove the assumption that B's are indifferent to A. Assume instead that B's feel hostile toward A and wish to hurt him. If the coalition already exists for the benefit of the B's, the addition of the hostile motive further increases the bargaining power of the B's and decreases that of A (3.26.11 and .12). If the coalition does not already exist for the benefit of the B's, its formation by B's would have the same negative effect on A's power and bargaining power as would hostility by the single B in a unique transaction. Hostility held only by the stronger competitors among the B's could produce the same negative effects on A as a coalition. Hostility held only by the weaker competitors could not. Given multiple B's, the effective expression of hostility therefore requires a coalition that includes the stronger B's.

3.27.55   Some third party C may favor A over B, either in general or in a particular transaction. If so, C may exert tactical or strategic efforts on either A or B to improve A's position in the A-B relation. Although the range of possibilities is tremendous, the analytic nature of the event is the same. Examples include: (a) a parent C who always intervenes to prevent a neighbor boy B from picking on C's son A; (b) a group of townspeople C's who will snub a landlord B if he evicts an old lady A for not paying her rent; (c) a government C that will fine an employer B if he pays his employees less than a specified minimum wage; and (d) a policeman C who will arrest a thief B if he takes a wallet from man A.

      Much, perhaps all, government law enforcement and regulatory activity consists of putting the government in the role of a third party C in some transaction between A and B, so as to influence the outcome in favor of A or B. Typical government regulations are that certain goods are not to be exchanged in transactions (narcotics, machine guns) or that certain tactics or strategies are not to be used (fraudulent labeling, secondary boycotts, physical violence). In contract law the government provides a stake to make people keep their promises who might otherwise have insufficient stake. In civil law the government settles terms of disputed transactions by throwing its weight on one side or the other. Although the body of such activity is vast, the analysis is basically no more than that already covered. The main social question is how or why government chooses to support A's rather than B's. This is a problem of decision-making within organization, which is dealt with in Chapter Twelve. Implementation of these general decisions, although they have complicating social factors, is essentially a matter of logic, evidence, and identification: Which A's and which B's in the society fall within the categories government has decided to side with or against? Social norms can be enforced by the same process but without the government, as is discussed in connection with semiformal organization in Chapter Fifteen.

3.27.6   *Collective Bargaining*

Collective bargaining is coordinated action among B's who conduct transactions with A to improve their power position in dealing with A, but whose transactions with A stand in the relation of complements rather than substitutes. That is (in contrast to the coalition), the B's in collective bargains are not competitors, and A's transaction with B_2 is made in addition to, not instead of, his transaction with B_1. Common instances are the unionized employees of an organization, the relations between parents and their children or between an automobile manufacturer and his dealers. Although the relations among B's may be competitive, for the purpose of collective bargaining, as defined, it is assumed that transactions with *all* relevant B's will go forward. As with the coalition, collective action can alter particular power, total power, or bargaining power.

To raise plain power (particular or aggregate) requires that the collectivity be able to offer A more than its separate members, so as to induce or enable A to give an X, or more X, than he would otherwise give. This is the base for what Walton and McKersie (1965) call "integrative bargaining."

To raise their bargaining power the collectivity must be able to withhold something from A that he would otherwise be able to get. If complete, the collective action can withhold all Y's from A. There can also be *unilateral collective action,* in which case the B's simply agree not to provide Y for less than some specified terms. Or a *collective bargain* can take place in which the B's jointly negotiate a settlement with A, the labor union being the most conspicuous case. In either the unilateral or the bargained collective action the B's increase their bargaining power by being able to withhold more Y's, or all Y's, simultaneously.

3.27.61   Since the participants of collective action stand in the relation of complements, not substitutes, the generalizations about coalitions in Section 3.27.5 do not apply. The collective bargain is treated instead like any noncollective transaction involving multiple goals. In the most conspicuous case, the union-management relation, both sides are themselves organizations and both bargain through agents. In consequence the total relationship involves intraorganizational decision processes (Section 4.12.4 below) and the problems of agents (Section 3.27.2).[1]

Although union-management bargaining is a conspicuous feature of contemporary Western society, its analysis apparently requires no special generalizations. Virtually all the analysis of 3.2 is applicable, particularly that of bads (3.25) and of dynamics (3.24) in connection with the strike. The third-party relations (3.27.1) can be applied to the intervention of government or public pressures on one or both parties.

3.28   INTERRELATED TRANSACTIONS

Two transactions are *interrelated* if the terms of one are affected by the occurrence or expectation of the other.   3.28.1 Since the terms of one transaction depend on the initial EP's of the parties, on the changes of EP induced by strategy, and on tactics, one transaction can affect another by altering its initial EP's or its strategic or tactical elements. Except for transactions involving insignificant values, the completion of a transaction alters the selector states and hence the EP's of both parties, at least with respect to the things exchanged. That is, after the transaction each party has something he did not have before and also lacks something he did have before (or, in the case of time and energy, could have applied to other purposes). Among other changes, once A has acquired Y he has less X left to give for other Y's. And having Y, A will now presumably have a lower EP for other things that are substitutes for it and a higher desire for other things that complement it.

3.28.2   If A engages in a sequence of transactions to acquire the same or similar Y's, the terms on which he completes one such transaction communicate that his EP extends to at least those terms; and they also reduce the credibility of a tactical allegation of a shorter EP in subsequent similar transactions. A will therefore attempt to avoid giving more in one transaction than he expects to have to give in

[1] Walton and McKersie (1965) handle these two aspects under the heading of "Intraorganizational Bargaining," and in far more detail than this survey can encompass. Their approach seems highly compatible with the present one and, like it, is allegedly applicable to a much wider variety of situations than labor-management relations—as they suggest.

subsequent similar ones, even though he would be willing to give more if the transaction were unique.

3.28.3   If *A* engages in successive transactions with *B*, or with multiple *B*'s who communicate among themselves, the credibility of his tactics in one transaction will depend on the truth of his tactical representations in preceding transactions. Hence if *A* wishes his tactical representations to be believed he must either avoid misrepresenting his true EP in his transactions under continuing or repeated relationships or else avoid making concessions after a position has been stated, unless there are grounds, understood by *B*, why his EP should have changed —even though refusal to make concessions may result in losing a transaction that would have been desired. Such questions are particularly important in threat transactions, the outcome of which depends entirely on the credibility of the tactical statements—though present circumstances and anticipated future developments can be of importance equal to or greater than past performance in assessing the credibility of a threat. In transactions involving goods or bads, if *A*'s tactical statements are not credible on their face a modified credibility may be established if he consistently misrepresents his true position to about the same degree.

3.28.4   If *A*'s behavior in transactions causes him to be liked or disliked by *B*'s, their subsequent transactions with him will be modified in a generous or hostile direction, respectively.   **3.28.41** In continuing or repeated relationships *A* will therefore have more power and more bargaining power if he is liked than if he is disliked by *B*'s.   **3.28.42** Those who accept the validity of the preceding statements (and most of us probably would) prefer, *cet. par.*, to be liked rather than disliked by those with whom they anticipate future relations.   **3.28.43** To be disapproved, and hence to some degree presumably disliked, by someone with whom one anticipates no future relationships, and who will not communicate with anyone with whom one anticipates future relationships, constitutes a sunk cost and is irrelevant to the transaction which generated disapproval. But given frequent benefits from approval and low assurance that disapproval will not be communicated, the desire to be approved is useful and hence likely to acquire strong intrinsic reward value.

3.28.44   The propositions of 3.28 thus far indicate a developmental change in the relationship between parties. That is, their system states (EP's toward and information about one another) after a given transactional stage differ from those before. Hence the relation between them cannot be precisely duplicated but begins from a new base that incorporates the accumulated effect of all previous ones.

3.28.45   Although it does not seem necessary to introduce the term formally, it is clear that interrelated transactions mean that *expectations* of future transactions play an important part in the way the parties behave in a transaction. Among persons who interact repeatedly or continuously over a long period these expectations, or long-run factors, may dominate the short-run factors inherent in the EP's of the transaction at hand. The logic of such interaction is contained in the present analysis. The content, like that of all transactions, depends on the time, place, and parties and cannot be specified in the model. It is essentially an idiographic problem, subject to the observations of Chapter Sixteen.

3.28.5   *Main and Subsidiary Transactions*

A *main transaction (major bargain)* is an explicit or implicit agreement that there shall be a continuing or repeated relationship of a particular sort. A *subsidiary transaction (bargain)* is a transaction over some detail of a continuing relationship, on the assumption that the main transaction continues.

3.28.51   Transactions sometimes come in hierarchical sets—main transactions, subtransactions, and perhaps sub-subtransactions. Typical are contracts of employment, membership in an organization, marriage, continuing purchase contracts, union-management relations, and understandings that two persons will be friends or "go steady." Since the main agreement cannot anticipate every detail that will arise during the relationship, these points must be settled as they arise. If agreement is reached on a general policy, the negotiation of that policy constitutes a subtransaction under the main one and negotiations over applying it in a particular case constitute a sub-subtransaction. There is no necessary limit to the number of levels in such a hierarchy. But any one level may be subsidiary to the one just above it and main with respect to the one just below. Since the logic of the power relation is the same between any two adjacent levels, to simplify discussion we will speak of only two levels, main and subsidiary.

The subsidiary transaction differs from the main one (and from most others) in that some agreement about it must be reached, at least in an overt sense. That is, if two parties to a main or unique transaction do not reach agreement they go their separate ways; no relationship then exists. But if either party in a continuing relationship makes a proposal about a subsidiary part of it, either he gets his way or the other party gets *his*—or some compromise. Unlike the mutual termination of a main transaction, simple inaction or nonresponse to one party's proposal determines the terms of a continuing relationship. Terms can also, of course, be determined by response, negotiation, and explicit agreement. In the overt, operational sense we say that whatever is done constitutes the agreement, even if assent to it is not explicit.

If the terms of a subsidiary transaction, or the cumulative effect of a number of them, are unsatisfactory to A and he cannot negotiate more satisfactory terms from B, he can nevertheless escape those unsatisfactory conditions if he terminates the main agreement. Whether he will do this depends on whether the benefits from continuing the main relationship outweigh the costs of dissatisfaction with some of its details.

So long as details are satisfactory enough that the parties do not contemplate terminating the main relationship, transactions over details give rise to no special rules. As parts of a continuing relationship, however, propositions about interrelated transactions are strongly applicable. Particularly pertinent are 3.28.3, which suggests the pointlessness of tactical misrepresentations, and 3.28.41–.43 about the importance of being liked. Since tactics can be time-consuming, their reduction or elimination in continuing relationships can greatly increase the efficiency of negotiations over details. There is always the possibility, however, that in order to be liked one party may switch from selfish tactics to the tactics of generous transactions with no net gain in efficiency. The conditions under which one or the other tactic might dominate need not be detailed here. But if terms regarding details are so unsatisfactory as to call the whole relationship into question, the desires concerning the main relationship become relevant to the terms on which details are settled. More explicitly, the EP's for the main relationship are reflected in those for its details.

3.28.52   If either party is willing to terminate the main transaction rather than accept unsatisfactory terms on a subsidiary one, then in negotiating the terms of the subsidiary transaction the EP of each party consists of the sum of his EP regarding the detail and his EP for the whole relation, the latter possibly modified by some probability function.   **3.28.521** The party with the lesser desire to continue the main relationship (that is, the one who has the lesser stake in it) will add less to his EP, or subtract more, in the subsidiary transaction and hence enhance his bargaining power in it.   **3.28.522** The party with the greater stake in the main relation can, of course, increase the likelihood of continuing it (that is, the likelihood that the two EP's for it will continue to overlap) if he can raise the other party's EP.   **3.28.523** Since the satisfactoriness of the main relation depends on the satisfactoriness of the details, the EP for the main relation will rise as satisfaction with details rises. Hence one party can raise the other's EP in the major bargain by making more concessions (contributing more or taking less) in the details.   **3.28.524** Hence if continuance of the main relation is in question, the party with the greater stake in continuing it suffers a relative bargaining power disadvantage in the details. **3.28.525** However, if the detail is important enough to bring the major bargain into question, the party with the larger stake in the subsidiary transaction suffers a relative disadvantage in bargaining over terms of continuance of the major bargain.

3.28.53   Although established in the context of an explicit continuing relationship, the generalization in 3.28.521 can be applied to any case in which one or both parties contemplate the possibility of future transactions with one another.   **3.28.531** For example, if A learns in one transaction that B is a strong competitor and gives A good terms, A will be more likely to return to B for subsequent transactions than if A receives bad terms.   **3.28.532** In any cases to which B sees 3.28.531 to apply, his EP in the present transaction will increase with his estimated gain in such future transactions with A as are contingent on A's receiving good terms in the present one.   **3.28.533** In such a case the party with the greater stake in subsequent transactions will suffer the greater loss of bargaining power in the present transaction.

3.28.6   We have now traced two main axes of relationships among transactions. One axis is involved when two parties to a transaction are affected by the fact that other parties also seek or conclude transactions in the same or similar goods. The limiting case on that axis is the perfectly competitive market, in which the terms between any pair are uniquely determined by the net effect of all other transacters in the same good. The other axis is involved when parties to one transaction anticipate subsequent transactions with the same persons and the outcome of one transaction affects subsequent ones. In addition third parties, C's, often learn about the behavior of A or B in their transactions, which knowledge influences future transactional relations of A or B with C. Since these connections are often highly personal, this axis shows no limiting case comparable to the perfect market.

These two axes connecting transactions through goods and through persons have been identified by White (1959, pp. 242ff) and are traced independently here. In reality the two are not separate streams but a vast network of intersections of the two kinds of relations. Connections through goods are largely external and objective; they are studied primarily by economists. Connections between persons are largely internal and subjective; they are studied mainly by sociologists. Although we need not attempt to trace the intersections of the two streams here, the next two chapters describe part of their content.

3.28.7   *Implicit and Open-end Transactions*

An *implicit transaction* is one in which an exchange of two goods is in fact contingent but in which there are no negotiations whatsoever in the normal sense of the word. Such transactions are usually generous at the moment they occur—a favor performed now in the expectation of a return favor later. There can be implicit transactions involving bads, as when I stop annoying you and you then stop annoying me. An *open-end transaction* is one in which *A* does something for *B* in the expectation not of a return favor to himself but that *B* will similarly do something for some third party when the occasion arises. The most conspicuous case of an open-end transaction is that between generations—the children are expected to repay the costs of their upbringing not to their parents but to their own children in turn. Although these relations are widespread and deserve mention, they need not be dealt with explicitly except in connection with aggregate power, status, and related matters in the next two chapters.

3.29   FREE AND FREEDOM

The present approach to social science does not hinge on the concept of freedom. The term and the concept are nevertheless so important to feelings, behavior, and legitimacy that it seems worth a short detour here to see how freedom appears when dealt with in this framework. First, the customary definitions. A *free good* is any good or satisfaction achievable without cost; *freedom* is the inverse, or relative absence, of cost. *Freedom to* is the absence of cost in carrying out one's own actions; *freedom from* is the absence of cost imposed on one without his consent by actions of others. A *sense of freedom* is a condition in which some desired thing is perceived as having less cost than under some other condition to which it is compared: hence a reduction in cost is properly sensed as an increase in freedom.

In a cooperative relation between two goods (see 2.50p), if one alone justifies its cost the second may be said to be free. If I drive the Pennsylvania Turnpike for the transportation alone, the views along the way are free goods. Any wanted gift is free to the receiver if it brings no handling cost, obligation to reciprocate, or other cost.

In a transactional relation, one aspect of A's freedom is the condition in which B cannot impose costs on A without his consent. To illustrate, take the noise of B's power mower at six o'clock Sunday morning. A has freedom if B does not run his mower (does not impose this cost on A) or if A can choose between B's not running the mower and his being paid enough by B to compensate him for accepting the noise. Other costs, such as those of air or water pollution, can be treated in parallel fashion.

Another transactional aspect of freedom is the assurance that A will not have to give up some valued thing without his consent or at least without reasonable compensation. The freedom embodied in the concept of due process is of this sort. So is freedom from invasion of privacy. In this connection we have noted the tactical and strategic importance to B of certain information about A, and A's power position in many transactions may be stronger if certain facts about him are not known. Such information is therefore a matter of considerable value—positive to A's opponents and negative to A. In addition to its other facets, an invasion of privacy therefore takes away a valued thing without one's consent—a clear transactional phenomenon and a form of theft.

As the comparative absence of cost, freedom is highly visible when the comparison is strong, as when the cost of some desired thing drops dramatically. It is perhaps most conspicuous when the cost drops from a clearly unattainable level to a readily attainable one. Under traditional colonial rule, for local leaders to divide public land among natives contrary to the wishes of the rulers would involve not only the normal costs of such distribution but also jailing or even execution. If the normal costs could readily be borne but the penalty costs could not, the ending of colonial control would eliminate the penalty costs and constitute conspicuous new freedom. Similarly, if parents remove the penalties on their daughter for staying out past midnight the daughter has more freedom—that is, a lower cost.

Since A's power and bargaining power vary directly with B's EP, any extension of B's EP that achieves or increases overlap constitutes an increase in A's actual or potential freedom. If the Turnpike Commission extends its EP so as to reduce its tolls, drivers have increased freedom to use the road. Or if the commission gives drivers an additional traffic lane that reduces their time and frustration costs, it also increases their freedom.

In actions not involving other persons, freedom may be thought of figuratively as related to "nature's EP" extending toward us. Nature has a long EP, so to speak, in providing coconuts and bread-fruit in the tropics—and a dreadfully short one in providing much of anything in the Antarctic. The longer nature's EP in such cases, the lower the cost of wanted things and the greater the freedom in acquiring them. Conversely in all these cases a shrinkage of B's or nature's EP, or any other increase in cost, represents a reduction in freedom.

We normally do not think of a situation as involving freedom except when it changes—when freedom increases or decreases compared to expectations. If our expectations rise and reality does not match them, the gap is perceived as a lack of freedom. If the costs of some Y rise when we do not expect them to, we sense a loss of freedom. But once we are accustomed to a certain cost we no longer think of it as a problem in freedom. The sense of diminished freedom apparently occurs under the same dissonant circumstances (2.41.4) that give rise to anger or frustration. We feel frustration and even anger when some task or acquisition requires more time, effort, patience, or money than we anticipated; but the same costs would not have created frustration and anger if they had been expected. Conversely a sense of increased freedom is apparently related to circumstances that relieve anger or frustration.

3.29.1   *Methodological Note on Power and Freedom*

One can define power so that power and freedom are related in the same supplemental fashion as are the EP's of A and B. That is, we can define A's power as residing in his own EP and define his freedom as residing in the EP of the other party—B or nature. We could then say that for A to acquire Y, the more power he has the less freedom he needs; and the more freedom he has the less power he needs. This would be the equivalent of saying that the greater his own EP, the less need be that of B; and the greater B's EP, the less need be that of A. Similarly the two versions would be identical if we also said that freedom is the relative absence of costs (a long EP of B) and power is the ability to meet such costs as are present (A's own EP).

Perhaps unfortunately, it does not seem feasible to define power in such a way as to provide this identity, as by relating it solely to A's EP. We have said, for example, that A's power varies directly with B's EP (3.22.3). Part of the content of freedom as formulated in the preceding paragraph is thus subsumed within the concept of power, thereby eliminating the conceptual parallel between the two pairs of items—power and freedom on the one hand and A's and B's EP's on the other. The need to distinguish plain from bargaining power, as well as particular from aggregate power, seems to preclude this simple parallel.

Hence although the complementary relation of power and freedom might be useful for relatively simple situations, that formulation does not seem satisfactory for the varied and complex requirements of broad social analysis. It is more useful for slogans than for logic.

These complications arise, however, solely from the formulation of the concept of power. It seems fully feasible to maintain an analytically tight concept of freedom as the inverse of cost, and when thus viewed the looser, popular concept of freedom seems compatible with it. Hence the definitions in 3.29 seem tenable—which means that the *free* in *free goods* is the same as the *free* in *freedom*.

Chapter 10

Aggregate Power

We have defined interpersonal power as the ability to acquire wanted things from others. The transaction has been analyzed as the avenue through which a single wanted thing can be acquired, without regard to the magnitude of its value. The main questions were *whether* the thing could be acquired, a matter of particular, plain power, and the *terms* on which it could be acquired, a matter of bargaining power. In this chapter we move from a single wanted thing to the total of wanted things one can acquire from others. This is a matter of aggregate interpersonal power, which was defined as the ability to acquire a series of wanted things, $Y + Y_1 + \ldots + Y_n$, from others. In its simplest form, why do some people have more power than others?

Since any valued receipt from others constitutes a transaction, the study of aggregate power does not abandon transactional analysis. Instead it shifts attention to aggregations of successive transactions by particular parties, along with the facilitating communications. Large aggregate interpersonal power is thus seen as the ability to consummate transactions of large total value—a few of large value, many of small value, or some combination.

Many propositions in this chapter, particularly the later ones, include combinations of transactional and communicational elements. They thereby anticipate the discussion in Chapter Eleven of the relations between communications and transactions and may therefore be considered out of logical order. And yet it seemed more sensible to have the discussion of aggregate power follow immediately the two chapters on plain and bargaining power. The necessary definitions and assumptions have, I trust, been spelled out so that this discussion can stand on its own feet where it is.

Probably all really large power is acquired through or with the help of organization, a matter we have not yet discussed. The present sequence of discussion is feasible because the organization acquires power by the same fundamental techniques as do individuals. Hence for this chapter we can continue to think of the parties as either persons or organizations. The main difference is that whereas the individual is limited by his biological and mental capacities there are no such limits to the amount of skill, energy, information, man-hours, or time span the organization can mobilize. The organization, of course, faces problems of internal coordination that do not plague the individual. But these problems are essentially the same whether the organization's aim is accumulating power or feeding the poor and thus require no special discussion.

The techniques for exercising power—that is, for getting others to do as one wants—are transactions and the relevant communications. The principles are independent of the magnitude of their content; hence no new concepts are needed to deal with larger power. Nevertheless the problem of aggregate power does require an analytic shift of a different sort. The cross-sectional approach dealt with thus far takes effective preferences as given and examines their consequences. By contrast we are now concerned with the growth of EP's over time, especially the effective aspect that determines the magnitude of their transactional effect. This is a developmental process in which each transaction depends on the outcome of those that preceded it. Each such sequence is unique; we cannot predict how it will go. But since the underlying phenomena remain transactional, we nevertheless can state certain conditions that must prevail within and between successive transactions if power is to accumulate.

It was indicated (3.22.8) that a single transaction is a flow and aggregate power a stock of wanted things. To the large extent that it depends on transactions, that stock can grow only if the value of inflows exceeds that of outflows. It can also rise if values are generated inside the system. The relative importance of the two sources varies with the type of system (individual, firm, church, nation) and with the individual specimen. The propositions in this chapter are confined almost entirely to those that apply to all types. The problem of this chapter, then, is primarily that of stock, not flow.

If one wanted thing can be acquired through one transaction, then multiple wanted things can be acquired through multiple
transactions. Although large aggregate power can be defined as the
sum of things acquirable through a few large or many small transactions, the underlying factors are the same for both, and we may arbitrarily regard the concept of aggregate power as a function of the
number of transactions a party can consummate. *Large* and *small* refer,
of course, to valuations, not physical magnitudes.

3.31   MODEL OF AGGREGATE POWER

3.31.0   The model of aggregate power consists of the following
assumptions:
a. There is no specific limit to A's total desires.
b. The preference functions (EP's) of all other parties (B's) are given.
c. The actions used by A to acquire power are not subject to legal
   restraint or other third-party sanction.
d. A does not acquire his power by inheritance or happenstance. However pleasant windfalls may be as sources of power, they are analytically uninteresting. Hence for any power that is transferable we are
   concerned solely with its initial accumulation before transfer.

3.31.1 A's aggregate power is thus a direct function of his ability
to overlap numerous EP's of B's.   **3.31.2** A must therefore have multiple X's to give in exchange for the numerous Y's he wants to acquire.
These X's may be as various as the goods and bads discussed in Chapters Eight and Nine; they may include the ability to create and relieve
stress as well as to provide goods. Given an existing set of EP's of B's,
the limit on A's aggregate power lies in the total X's he can muster.

3.31.3   One way A can acquire X's is by producing them himself.
His ability to do this is limited by the same factors that limit his ability
to satisfy his wants for himself: energy, strength, skill, willpower, tools,
materials. These factors provide an important base—an initial minimum necessary for an accumulation. As with personal (not interpersonal) power we refer to this capacity as A's *productivity* (2.50ee and
2.94). When we move from personal to interpersonal power in goods,
A's ability to produce is the pretransactional base from which transactions can be built—the effective half of his effective preference. A can
also acquire X's by transactions. This is the socially interesting half of
the process, on which our attention will focus. Depending on the situa-

tion, productivity and transactions may proceed in a repeated sequence as A alternately creates X's and then exchanges them.

3.31.4   Many valued X's that can be transacted can also be accumulated. Durable material things can obviously be accumulated; services cannot. But some intangibles, such as status and the ability to command votes or audiences, can also grow by accumulation. Depending on its nature, either the thing itself or its use in behalf of someone else may be the X exchanged for some Y. **3.31.41** One or both of a pair of items can be increased if A can conduct successive transactions with different B's who have different EP's for the items. If A can give one cow to B_1 in exchange for four goats and then get one cow from B_2 for three goats, he can accumulate one goat per pair of transactions. This process is referred to by Boulding (1958, p. 51) as the "successive transformation of assets," and it can appear in many variations. The successful business firm acquires inputs of productive factors, exchanges the output for enough to acquire even more inputs the second time around, and so on. The successful traditional politician gives favors for support. This support elevates him to a position from which he can grant more favors, which bring still more support, higher office, or greater weight, and so on from dogcatcher to president. The sequence can be roundabout: X may be exchanged for Y, which is exchanged for $K, L, M, \ldots , W$ in turn until eventually more X is acquired for W than was given for Y in the first place.

In economics Gresham's law, arbitrage, speculation, and comparative advantage in international trade are all variations over time, place, or markets of A's who accumulate a stock of value by dealing successively with different B's who hold different EP's for X. Although the principle is easiest to see in connection with the quantified money units of economics, it is the same for subjective values whenever there are discriminable differences in values and the valued thing can grow by accumulation. Here, as in economics and other transactional analysis, it is the desires for things, not the number of units, that are relevant to analysis—not the mere number of votes but their importance to the candidate.

3.31.42   The smaller the quantity of X that A must give in each transaction, the more he will have left for other transactions, the greater the number of transactions he can complete, and the greater his aggregate power. **3.31.43** Hence for a series of transactions A's aggregate power is a direct function of his bargaining power in the separate transactions.   **3.31.44** A transaction not consummated cannot add to A's aggregate power; thus he must consummate transactions if they are to add to his power.   **3.31.45** A's ability to consummate a transaction and hence add to his aggregate power is a direct function of his plain power in each transaction.

3.31.46   Since plain power is related directly (3.22.2), and bargaining power inversely (3.22.4), to A's EP, to the extent that each is dependent on A's EP (rather than B's), 3.31.43 and .45 push in opposite directions. In a transaction a large EP may contribute to A's aggregate power by ensuring that he will get the needed Y; but it may simultaneously detract from that power by dissipating his stock of X's. The model provides no basis for judging which effect will be greater.

3.31.5   All propositions from Chapters Eight and Nine are potentially applicable to the transactional steps in accumulating power. Among other details, to accumulate power A would need (a) to understand at least intuitively many of the propositions about power and bargaining power and to acquire skill in bargaining; (b) to discern when and where advantageous transactions are available, select their proper sequence, and complete them; and (c) luck.

3.32   The analysis of aggregate power requires several definitions. *Internal power factors* are those system states or characteristics of A (knowledge, skill, physical strength, will, beauty, friendliness, courage, bargaining skills, ruthlessness) that give him power. *External power factors* are factors or conditions outside A that give him power, whether or not he is able to control them. These factors are of two main sorts. First are *material external power factors*—money available to A, books he has published, pictures he has painted, recordings he has made, printed mention of his name or works, monuments to him, and the like. Second are *system states of others*—his reputation, awareness of him, attitudes of friendship or loyalty, sense of obligation to him, and so forth.

3.32.1   The power based on internal factors is a direct and immediate function of the state of those factors: power to win a boxing match disappears promptly with a broken arm; power based on ruthlessness declines when chicken-heartedness sets in.   **3.32.2** The power that resides in material external factors depends on the durability both of their substance and of their valuation by others. To store this kind of power, gold and diamonds may be A's best friend, though they may change in value. Money is perishable during inflation, property values during depression, published ideas during intellectual ferment, and paintings in flood or fire—though in other circumstances all may rise in value. **3.32.3** System states of others are often loosely and idiographically related to A's internal power factors and to the material external factors. A's reputation may rise faster or slower than his performance justifies or it may be largely unaffected. A's reputation as a good pianist may go unscathed through several poor performances; a firm friendship may survive extended neglect.

3.32.4   From 3.31.41 on successive transformation of assets it follows that, where successive transactions are so arranged as to provide an increasing stock of material, durable external power factors, the process is one of positive feedback or deviation amplification. That is, the farther upward A moves in one step the farther he is able to move in the next, since the absolute amount of the increment depends on the size of the current stock. In short, under these circumstances not only is power self-reinforcing—power begets power—but it rises at an accelerating absolute rate.

3.32.5   Given the conclusion of 3.32.3, other conclusions about power based on system states, internal or external, are of uncertain status. Positive feedback can nevertheless apply in some such cases, particularly in successive interactions between internal and external, as follows. **3.32.6** If A has sufficient internal power factors to perform well (as comedian, lecturer, actor, musician, athlete) and if his external power rests on the desires of the public to see him perform, his power is self-reinforcing in two ways. First, the oftener he appears in public, the better known he becomes; and the better known he becomes, the oftener he is asked to appear. Second, the more skilled he becomes, the oftener he is asked to appear; and the oftener he appears, the more skilled he becomes. **3.32.7** A third factor also makes power accumulation self-reinforcing: the larger the stock of X's already accumulated, the greater is A's power to acquire a Y instrumental to further accumulation—and hence the less likely it is that his further rise can be blocked by his not having it. **3.32.8** Within the areas covered by 3.32.4, .6, and .7 (power accumulations subject to positive feedback), if A gains an initial superiority in aggregate power over A_1 then, *cet. par.*, the absolute difference between them will increase.

3.32.9   Various types of power can be "invested" to differing degrees. The extent can be determined only by empirical investigation and probably differs from case to case within type. Different types of power also have different breadths of currency. That of money is extremely wide. But the boxer *A*, whose power depends on the *BX*'s of sports fans, derives no power from the *BX*'s of symphony buffs. And the dictator who rules a nation with an iron hand may be unable to get his eight-year-old son to bed on time. Through his newspapers William Randolph Hearst could influence readers' *EP*'s on many questions—but never enough to get himself elected to public office.

3.33   Power based on ideas must be distinguished from influence (3.22.8). As the ability to get something in return for giving, power presupposes the ability to withhold as well as to give (3.21m). A person has power if others want personal *presentation* of his ideas, as through his lectures or new articles, since he can withhold these things. But once his *ideas* are loose he cannot withhold them. They have influence, but do not necessarily give him power. The wider they are spread, however, the greater (*cet. par.*) will be his status. His status *is* external, in system states of others, and it may or may not be closely related to the value of the ideas themselves.

3.33.1   As normally conceived by economists, consuming and investing are mutually exclusive. Consuming depletes one's money stock whereas investing augments it—if successful. As broadly conceived here, the two uses are not necessarily exclusive. A cocktail party can simultaneously be an act of consumption (!) that satisfies current wants and also be an investment in contacts or goodwill that may later enlarge the bank balance. Political influence can also be spent, as when *A* gets *B* to repay an obligation incurred by *A*'s previous favors for *B*. Clearly *A*'s stock of obligations receivable is diminished. However, *B*'s actions may simultaneously increase *A*'s status, reputation, or list of supporters and hence be an investment. Whether a withdrawal from stock is expenditure, investment, or both can be ascertained only in the particular case.   **3.33.2** Whenever consumption and investment do not coincide, investment requires current sacrifice. This sacrifice in turn presupposes enough power to do without some of it for the present. Hence the propositions about positive feedback apply with particular force to stocks already above the level where there is power to spare.

3.34 AGGREGATE POWER AND BARGAINING POWER

3.34.0   Let us now assume that many kinds of *X* show diminishing utility to *A* and hence a successively lower *AX* for added units. The business firm is a special case, which presumably has no desire concerning the X's it sells except as they produce income. Its AX is the cost of producing X, and its supply curve represents the differing values of AX as output is altered. Hence for a business firm the cost curves should be substituted for the principle of diminishing utility. The following propositions apply to AX's that *are* subject diminishing utility (probably most AX's except those of business firms).   **3.34.1** The larger the number of X's held by A, then for any given unit of X the smaller will be AX in any given transaction (*cet. par.*).   **3.34.2** the greater will be his power to acquire Y but the less his bargaining power.   **3.34.3** Thus the more units of X held by A, the greater is B's power and bargaining power in acquiring X.   **3.34.4** Hence, too, (a) the greater will be B's desire to deal with A rather than with other A's of lesser power and (b) the greater will be the number of B's who will want to deal with A. [Note the corollary to (a) in 3.35.66.]

3.34.5   Generalizing from 3.34.3 and .4, B's have more power in dealing with A's of large power than with A's of small power and hence will prefer to do so. The relation is reciprocal: A's would rather deal with B's of large power than with those of small power. Continuing the overall assumption that people in fact select the more over the less advantageous, the powerful will deal mostly with the powerful. And although the less powerful would also prefer to deal with the more powerful, they will be precluded from doing so by the choices of the latter—except for those things not available, or inadequately available, from the more powerful.

3.34.6   By definition, larger transactions (involving more units or more valuable units) are possible with persons of large power than with those of small power. Hence the potential gain per transaction is greater, as is also the rate of accumulation. This statement elaborates that of 3.34.5 about the powerful preferring to deal with the powerful. The desire of parties of power to deal with others of power adds to the positive-feedback aspect of aggregate power seen in 3.32.4, .6, and .7.

3.34.7   For external power factors the sum of many transactions by A with people of individual small power may add up to a large total power (and hence be sought by A) and is conspicuous in business relationships. This situation is the exception stated at the end of 3.34.5; but this case does not violate 3.34.5 since that proposition states a first preference, and this a second. The relative usefulness of few large and many small transactions may differ between goods and bads, though

no generalizations come readily to hand. The weak may readily conduct many small transactions in goods with the strong, for example,
whereas stress or credible threat would seem beyond their capacity.
Yet many petty annoyances by the weak may not be worth massive retaliation by the strong and may thus be bought off with petty concessions.

3.34.8   According to 3.34.2, under certain common circumstances
large power by A reduces his bargaining power in a transaction. But
also, according to 3.28.533, A's bargaining power will be enhanced if
B has a greater stake than A in subsequent transactions between them—
which, as the weaker party, he is likely to have. The net effect is indeterminate unless we can measure the relative strengths of the opposing
forces, which we presumably cannot. But this indeterminacy allows the
possibility that A's greater aggregate power in a transaction may enhance both his particular power and his bargaining power. Since there
is no apparent intertransactional relation that operates on plain power
in such a way as to offset the intertransactional effect on bargaining
power, if A has the greater power to begin with the net balance is
apparently in his favor.

3.35   CLASS AND STATUS: STRATIFICATION

3.35.0   For discussion of this topic we introduce the terms *class,
status,* and *deference.* A *class* is here defined as the collection of people
in a society who have approximately equal aggregate power. Since
power can rest on several different bases, classes based on different ones
may not be comparable—for example, classes based on money, intellect,
artistic ability, athletic competence, and political position. By definition
there is a potential continuum of class levels from lowest to highest
within or among types. Hence any division into a given number of
stratification levels must either be arbitrary, geared to the purposes of
the investigator, or else reflect discontinuities in the real world.
          To the extent that various types of power can command
money, it is a common unit of power. But different types are by no
means equally convertible to money, to one another, or to any other
common unit. Hence it seems unlikely that we can create a single
analytic hierarchy of class levels based on different types of power;
Polsby (1963) gives considerable attention to the nontransferability of
different kinds of power. The phenomenon must apparently be treated
as idiographic—dependent on the society, type of power, and individuals concerned. By the same token, schemes for categorizing classes

would have to be empirically based, not derived from the model; and they might differ for every society, subsociety, or ethnic group. We can, however, at least identify two distinct problems in this area. One concerns differences in level of classes based on different types of power—such as artists versus nobility—and the degree to which mere membership in one class carries power in the other. The second problem concerns differences in rank within a class level based on one type of power and the degree to which rank within one such class carries power within a class level based on a different kind of power—such as the degree to which high nobility has more power among artists than does lower nobility; or the degree to which the high-ranking artist has more power among nobility than does the low-ranking artist. If the problem is idiographic there is no point in developing a general model for it, and thus none is introduced here.

     For certain problems we might usefully view a class as an organization (system), formal or informal. I do not see sufficient need for such a model to justify formulating one here. Briefly the reasons are (a) that although members of one class may interact as individuals with members of another, the classes themselves do not typically interact as units; and (b) that interactions between members of the same class presumably follow the same transactional and communicational logic as interactions elsewhere and thus require no special treatment. When explicit treatment is needed I would expect the models of informal and semi-informal organization (Chapters Fourteen and Fifteen) to be more applicable than the formal, though in times of sharp social conflict a class could be formally organized. We will therefore attend to special configurations of transactions and communications related to class and status but need not treat class as organization.

     Status is apparent, or perceived, aggregate power—the images of A's power in the heads of others. To the extent that it provides power, status is an external power factor consisting of system states of others. Material external power factors (such as books published or paintings hung in galleries) may help to communicate status, but the status itself exists only as system states in others. In pure form status is a phenomenon of the information-communication axis alone. Like many other such phenomena, however, it also appears in mixtures with transactional. However widespread one's status, only that status in the eyes of persons with whom A directly or indirectly interacts is analytically relevant to A's power. As a perceptual phenomenon, status is to aggregate power as tactics are to the transaction.

     To the extent that A's power factors exist independently of the images of them (that is, internal power factors or material external ones) his status may lie above or below his actual power. The term *actual aggregate power* designates such independent power. As noted (3.32.3), the system states of others that constitute A's status are only loosely related to his internal and material external power factors— which is another way of saying that the others' impression of A's aggregate power may differ considerably from what he can actually come up with if challenged. On the average and in the long run others' perceptions tend to approximate reality and status thus tends to coincide with the actual power factors. But such coincidence will not necessarily occur in any one case, even over a long period. At the same time, status *is* power so long as it lasts, and if it lasts through the critical periods it may help bring actual power up to match the status—as we shall see. *Deference* is behavior by B that communicates (symbolizes) his acknowledgment of status inferior to A's.

3.35.1   Status and actual aggregate power, and hence also status and class, coincide whenever A's power factors themselves consist of images in the heads of others—as with a reputation for brute strength or good artistic performance—and A's actual power factors are accurately perceived by others.

3.35.2   From the advantages of "dealing upward" (3.34.5) and the definition of *class* it follows that A will find it more advantageous to deal with members of his own or a higher class than with those of a lower class and will try to do so—subject to the second choices stated in 3.34.5 and .7. In addition, to the extent that approval increases power or bargaining power, approval of A by B's of high status will contribute more to A's power and bargaining power than will approval by B's of low status—and the same may be said of "being liked" (see Section 3.28.4). Hence to that extent A will be more eager to be approved and liked by high-status than by low-status persons. To the extent that A finds deference rewarding in itself, he can achieve it by associating with B's of significantly lower status than himself.

3.35.3   Since persons respond to their images, not to reality, it follows from 3.35.1 and .2 that A will find high status more advantageous than low. This statement implies nothing about the efficacy or acceptability of a particular type of status-seeking activity.   **3.35.31** If A's status is less than commensurate with his actual power, he can raise his status by communicating awareness of the actual power factors. **3.35.32** Conversely, if his status is more than commensurate with his actual power, maximum status depends on concealing the latter.

3.35.33   As a communicational and perceptual phenomenon, status can be influenced by symbols. In Western society these symbols include manner of speech, home and neighborhood, car, church, club, vacation, and furnishings. They also include informing others of one's own high-status activities, which in turn include associating with high-status people—a condition communicated by name-dropping. Given the human faculty for making conditioned associations between juxtaposed things, one may gain status by almost anything he has in common with those who already have high status.

3.35.34   Since A's status will be perceived as related to that of his associates, if A wishes to maintain or raise his status he will avoid relationships with B's of lower status than himself unless his status is communicated in other ways.   **3.35.35.** If B is of lower status than A, if B wishes to continue a relationship with A, and if A's superior status is not otherwise established, then A will find the relationships more acceptable if B shows deferential behavior.   **3.35.36** If the relation in 3.35.34 is understood by B, he may expect to be rebuffed if he attempts to open a relation with an A of distinctly higher status. Hence he will not make the attempt.   **3.35.37** Under these circumstances, if A nevertheless wants to be open for transactions with B's of lower status he must somehow remain approachable. One way he can do this without lowering his status in the main area is to deprecate his power in less important areas (see Blau, 1964, p. 53).   **3.35.38** Paragraph 3.35.2 indicates the advantage of trying to "trade upward"; 3.35.34 indicates that such attempts may be blocked. To the extent that both forces operate simultaneously, they tend to produce a closed class.

3.35.4   Being a good negotiator is an important ingredient in acquiring power (3.32). Bargaining ability will therefore be seen by those who want power as instrumentally rewarding. Since this skill is an important instrument for acquiring power, those who possess it will be perceived by others, *cet. par.*, as more powerful than those who do not. For A to be perceived by B as a good bargainer will therefore tend in itself to enhance A's status.   **3.35.5** If A is aware of this factor he will want (*cet. par.*) to appear to be a good bargainer and thus enhance his bargaining power (under 3.26.6).

3.35.6   ACTUAL VERSUS PERCEIVED POWER

Let us now assume (a) that in a transaction A's status in B's eyes is greater than justified by A's actual power and (b) that A's actual

EP is determined by A's actual power but that B's estimate of it is based on his perception of A's status.   **3.35.61** Under these circumstances B will overestimate A's EP. Assuming overlap, in tactical maneuvering B will then be less likely to make a concession than if he had correctly estimated A's EP.   **3.35.62** Hence in a unique transaction if A's status exceeds his actual power he will find it harder than otherwise to get terms of a given degree of satisfactoriness.

3.35.63   Given that the party with a greater stake in subsequent transactions will suffer the greater loss of bargaining power in the present one (3.28.533), then in future transactions this exaggerated status may increase his bargaining power as between this effect and that of 3.35.62. Although the net effect is indeterminate, it is probably a net positive force for A (3.34.8).   **3.35.64** The logic of all parts of 3.35.61–.63 also applies, with reverse effect, if A's status in B's eyes is less than is justified by his actual power.   **3.35.65** More broadly, the same logic applies to any estimate of status compared to any other, whether or not one of them is commensurate with actual power.

3.35.66   If we return to assumption 3.35.6a (that A's status in B's eyes is greater than A's actual power), then B will have a stronger desire to deal with the perceived A than he would with the actual A— under 3.34.4a. If B gets intrinsic or expected instrumental benefit from dealing with higher-status persons, then under the transaction benefits of 3.26.4 his EP will rise, as will A's particular power and bargaining power. Even if B expects no benefit from the mere fact of dealing with an A of higher status, the logic of this factor, *cet. par.,* will enlarge the number of B's competing to deal with A. Further, if B anticipates subsequent transactions with A, the net effect of the exaggerated status on A's bargaining power is indeterminate but probably positive (3.34.8).

3.35.67   The preceding propositions about status and aggregate power refer only to relative amounts—to the positions of the parties relative to one another, regardless of their positions relative to the larger society.

3.35.68   High status within the relevant type of power tends to attract relationships of others (3.35.2). But since status is relative, for A to raise his own status lowers the relative status of B's. That result is disliked by them and tends to generate hostility with its attendant loss of power and bargaining power (3.26.11 and .12). The effect of raising one's own status is therefore two-edged, and the net effect is indeterminate within the model.   **3.35.681** An A will respond to the expressed wishes of a B of higher status than his own for at least three reasons: (a) A sees the advantages in probable terms of the transaction with a higher-status B, (b) A thus keeps the higher-status B aware of him and his potential usefulness, and (c) A's own status with C's is enhanced by his dealing with high-status B's.   **3.35.682** Given this conclusion, for A to respond readily to the expressed wishes of a B shows that A is of lower status. But the undesirability of low status induces the opposite behavior. The balance between the two is presumably determined as follows. If A is trying to appear of equal status with B, he will avoid this deferential response. But if A is so conspicuously below B that even to deal with B at all enhances his status, A will give the deferential response. I take this effect to be the partial equivalent, at least, of the statement that status includes claims on services and resources of lower groups (see Buckley, 1967, p. 134). ✓

3.36   GIFTS AND AGGREGATE POWER

We define a *gift* as any valued X given by A to B without explicit contingency that B will give Y in exchange.   **3.36.1** At any level of power, the direct effect of a material gift of value from A to B is to diminish A's aggregate power by reducing his stock. If the gift is a service that consumes A's resources (including time and effort), it too diminishes A's power.

3.36.2   For this paragraph let us assume that the status of A and B in one another's eyes is dependent solely on the interactions described here.   **3.36.21** Assume that the relation of 3.36.1 is understood, at least intuitively by others. Then if A gives gifts of significant value, whether material or services, he thereby communicates that he has power to spare, and the gift raises his status.   **3.36.22** If B desires to avoid lower status than A, he will want to reciprocate A's gifts. If these are of equal value to those received, B thus avoids lower status.   **3.36.23** But assume that A also understands this relation. Then if B cannot return gifts of equal value, the fact of A's greater power is communicated to both and it will be clear to both that A has the higher status.[1]

[1] The above relations starting with 3.36.1 are found in Blau (1964). Blau, however, apparently construes the unmatched gift as the *source* of power whereas I construe it as only the *communication* of it. To understand the source one must first explain why A is able to give more gifts than B can match —a matter covered by 3.31.3 to 3.34.8 above.

3.36.3   To relax the preceding assumption (3.36.2), if A is already widely perceived as of high status he need not seize every available opportunity to communicate it. If status is itself valued, then so is an opportunity to communicate it. And if giving up something valued communicates the fact of power, then to give up an opportunity to communicate status, if observed by others, demonstrates even greater power—a demonstration that one possesses so many such opportunities that he can throw some away. (See also 3.36.63–.65 below regarding the status-communicating effect of leaving resources idle or destroying them.) Reciprocation of gifts thus has potential conflicting effects. Under 3.36.22 it demonstrates that B has sufficient power to reciprocate. But under 3.36.3 it may also demonstrate that he does not have sufficient status to throw away the opportunity to communicate it. For B to reciprocate regularly and promptly may thus bring him lower status than to seem unworried about reciprocating. Whether prompt reciprocation, or any reciprocation, raises or lowers B's status thus depends on whether he is already perceived as assuredly able to reciprocate (Blau, 1964, p. 99). It then follows that a person of conspicuously high status is free to leave numerous gifts to him completely unreciprocated.

3.36.4   Let us now return to the selfish assumption (a) that A will not continue to give valued things to B unless he receives valued things in return and add (b) that—given their wide and frequent instrumental usefulness in satisfying wants—power, status, and deferential behavior by B have all become intrinsically rewarding to A.[2] It then follows that transactions may be consummated between A and B in which Y consists solely of deferential behavior by B. Analysis of the transactions is presumably subject to the whole body of transactional analysis, in which X may have either a positive or negative valence; A can be indifferent, generous, or hostile toward B; the transaction may or may not involve transactional costs or benefits; other desires may or may not be involved in the transaction; third parties may or may not enter the scene; and the transaction may or may not be related to others. That is, all propositions of 3.20 through 3.28 presumably apply. To illustrate with one obvious case, A will not continue to give gifts to B unless B gives deferential behavior in return.

        [2] Although this item is introduced as an assumption, for reasons implicit in this whole section it seems likely that most persons will in fact acquire an intrinsic desire for status and power in some form and that many who deprecate status-seeking merely prefer more subtle forms or different audiences.

3.36.5   Let us now assume (a) that the gift by A is an instrumental investment toward raising his status and (b) that although a pure gift may be discussed in advance, it can also be presented without prior discussion.[3] In that case the transfer does not require the mutually contingent decisions stated in the introduction of 3.2. If X is a material good B might be able to return it intact, though to do so normally requires some action by A, which he might refuse to take. Or A might preclude the return by saying, "I can't use it, so if you don't want it just throw it away." The gift can also be a service already rendered ("I noticed your tire was flat so I changed it for you"). Other gift services may be difficult to avoid, as when A gives B information that he obviously needs. Contrasted to the mutual consent of the selfish exchange, a gift can be imposed unilaterally on B even if he does not want it and would like to avoid receiving it. Hence if A wants to use the gift to establish superior status, and if B cannot match it, B may be helpless to stop him.

3.36.6   Even if B could refuse or return A's gift, intertransactional relations may preclude his doing so for two reasons. First (a), except for prenegotiated gifts it is often difficult to avoid implicit criticism of A's judgment in giving X unless B expresses pleasure in receiving it. Further, criticism of this sort is unlikely to be of constructive use to A for any purpose other than in his dealings with B and thus may not be desired by A. Since unwanted criticism is unpleasant, the normal process of conditioned association may cause A to dislike B and thus worsen B's position in subsequent transactions with A (3.26.1). For this reason, too, B will tend to accept the gift and incur the necessity of matching it or losing status. Second (b), if both parties know that A is trying to establish superior status and that B is trying to avoid it, and if both parties know that X is wanted by or useful to B, then B will have no apparent reason for refusing X other than to avoid having to match it. B's refusal of X therefore communicates his inability to match X and establishes his inferiority.

3.36.61   Sociologists and anthropologists often deal with situations

---

[3] Perhaps contrary to expectation, the terms of a pure gift may be negotiated and a settlement reached in the face of conflicting offers. If a carpenter friend offers to build you a gazebo, there could be discussion about size, shape, timing, and location. And if he finds your ideas more grandiose than he had anticipated there could be negotiations and compromise over the quantity he will give.

similar to these, noting that many societies show the normative behavior in which a person who receives a gift *ought* to match it. Blau
(1964, p. 94) summarizes this phenomenon as "the ubiquitous norm of
reciprocity." The present analysis replaces, or at least parallels, those
statements with nonnormative propositions that the person who receives a gift *will* try to match it because failure to do so *will* lower his
status and because he *will* generally want to avoid that result. I suggest
that the circumstances in which unmatched gifts do not lead to loss of
status coincide with the ones in which the sociologist would report the
absence of obligation to match. For example, 3.36.6b could appear in
normative language as the "obligation to receive" a gift.

Articles in Firth (1967) suggest that at least some anthropologists now conclude that exchanges in nonmarket societies are more
transactionally rational and less mystically normative than had been
thought. Since we have already observed that selfish transactions are
more "advanced" than generous ones (3.26.25), we may wonder whether
some unconsciously self-attributed higher status was involved in the
earlier conclusions reported about the ubiquitousness of the gift in the
societies the anthropologists observed! The communicational (symbolic)
nature of status in this model certainly coincides with the insistence of
many sociologists and anthropologists that the crux of a transaction is
often not the value of the thing transferred but its *meaning* to the
parties. Additional meanings of gifts—expressing group bonds, trust,
or cohesion—are dealt with in the next chapter. This section illustrates
that there is no necessary conflict between a norm-oriented and an
advantage-oriented analysis if the norms themselves reflect what experience or intuition has revealed to be advantageous.

3.36.62   It follows from the logic of the use of gifts to establish
status that if A destroys X, or refrains from using X when it could
satisfy his wants, and if these facts are known to B, then A communicates the fact of his power to B as certainly as if he had given X to B as
a gift. Unless B can establish his status by other means, he must destroy
or leave idle items of approximately equal value or else accept status
inferior to A's. Under 3.36.6b, B will do the former if he can.

3.36.63   If A's superior status is definitely recognized by B, who
offers deferential behavior, A's status is enhanced if he "throws away"
this presumably valued thing by asking B to behave as an equal. Conversely his status is not thus enhanced, and may be diminished, if by
not ending B's deference A seems to need the status recognition it im-
parts.

We pause to note that this section attempts only to trace certain lines of logic. This logic does not imply that when real people follow similar behavior they are necessarily following the same motives. Some real A, for example, may stop B's deferential behavior because it violates his philosophy of life, makes him uneasy, or simply displeases him. I nevertheless suspect that this analysis clarifies why some people want to avoid deference toward themselves. Here, as elsewhere, it is impossible to spell out all the ramifications of a transactional element. Section 3.26, however, indicates the general effect of multiple goals on transactions, and following that pattern there should be no difficulty in adding the attitudes of both A and B toward deference as part of the EP of each. A's attitude could also be modified by what he learns about B's attitude and so on. This section might also partially account for such behavior as the potlatch. Other components in such ritual behaviors, such as expressions of friendship and solidarity, are dealt with later (Section 3.44, especially 3.44.3, .5, and .7). It is also possible that when averaged over the long run (and if destruction is not involved) the terms of exchange in such ritual gift-giving are about the same as they would be under selfish exchanges. Another dimension of such exchanges may be the same as that in a well managed family Christmas. Under the guise of gifts the family collectively buy or make themselves things they really cherish, but which colder calculation might deem imprudent. Less well managed Christmas giving can also bring the net loss to both parties described in 3.26.23. (That this paragraph was written on December 28 is purely coincidental.) Under other circumstances goods might be destroyed to enhance terms of trade—as in plowing under crops.

3.36.64   Just as superior status can become intrinsically rewarding, so can inferior status acquire intrinsic negative value. If it is perceived that unreciprocated gifts or favors signal inferior status, then to continue "under obligation" (even in the technical, not the normative, sense) will also, by conditioning, become intrinsically unpleasant and discharging the obligation will be rewarding. Under these conditions A can add to B's obligations by giving him gifts or favors he cannot soon reciprocate. This stock of obligations due to him constitute power for A since B is obligated to do things he *can* when asked by A. This sense of obligation adds to B's EP's in later transactions with A. Such action by A clearly constitutes an investment in which the time delay grows from the fact that a gift may lose its quality as a gift unless the reciprocating gift is delayed long enough that it does not seem explicitly contingent. Thus if B wants X to be construed as a gift he must delay reciprocating it. Otherwise it may look as if A gave X in order to get Y, and the transaction may seem selfish.

3.36.65   In closing this section on gifts we may note four points. First, although the effects of gifts on power and status have been viewed as *A-B* relationships, A's status can be communicated to anyone who observes his gifts, and the same is true of disuse or destruction. Veblen referred to this situation as *conspicuous consumption,* a communicational phenomenon. Second, the gift thus shares with stress and threat the property that with it A can impose power on B without B's consent, a property not shared by selfish transactions. Someone subject only to selfish transactions "has his independence," which only generous transactions and transactions in bads can take away. Third, this section has dealt with gifts only in connection with status. Their role in friendship and cohesive bonds is treated later. Whenever gifts are not subject to these constraints they produce different results, as when they arise from love or empathy or within the close and special relations of kinship systems. And fourth, even within this context societies can make numerous arrangements, as by setting aside certain classes of objects whose transfer constitutes a gift and gives rise to a corresponding obligation, even if the objects have no value in themselves.

3.37   **DOMINANCE AND SUBMISSION**

We define *dominance* (of A) and *submission* (of B) as a condition in a continuing relation between A and B, or in some aspect of that relation, in which A generally exerts the greater power. The relation is most explicit when it is mutually recognized, as by deferential behavior by B, but such recognition is not required by the definition. The greater power must appear at least in the overt terms of the relationship, which we here define as the operational criterion for identifying cases of dominance and submission. It is definitionally impossible for dominance to exist without submission and vice versa. By definition we also exclude a single interaction, for which communicational or transactional analysis alone is sufficient. Nothing in this definition implies that dominance and submission must characterize all aspects of a relationship; A can be dominant in one area and B in another while the two hold equal power in a third. Nor does the definition imply anything about the intensity of the difference or its duration (beyond the definitional requirement that it extend beyond a single interaction).

3.37.1   Following the general model we classify the types of dominance by A as communicational, transactional, and organizational. The examples of each type are not intended to be exhaustive.

Communication is the first basis of dominance. A is dominant if he modifies B's detector function (code or decoding processes) more frequently or in more fundamental ways than the reverse —that is, if A exercises the greater intellectual power. Moreover, A is dominant if he modifies B's selector function (values and motives) more frequently or in more fundamental ways than the reverse—that is, if A exercises the greater moral power. Finally, A is dominant if, without modifying his own detector or selector functions, B nevertheless accepts those of A as providing the conceptual or value assumptions of their communications—even if only because A thinks more quickly or speaks more clearly or interrupts when B talks.

Transaction is the second basis of dominance. A is dominant if B's EP in the major transaction (which in this case is an understanding that a continuing relationship exists) is markedly greater than A's, so that B's bargaining power is regularly disadvantaged in the subsidiary transactions (3.28.521). A is also dominant if B's interest in or desire for the X's provided by A is relatively intense while A is relatively indifferent to the Y's provided by B. Moreover, A is dominant if he is willing and able to give gifts or bads in obviously larger amount than B. Finally, A is dominant if he uses superior tactics or strategy, either because he is more skilled at them or because he has fewer compunctions about using them.

Organization is the third basis of dominance. (It is discussed here as the apparent best location, though the full meaning of some terms will not be clear until Chapters Twelve and Thirteen.) A is dominant if he has superior strength, skill, or knowledge for performing the transformations that A and B jointly execute as an organization. A may also be dominant if he contributes more of the time, tools, effort, and the like that A and B use in their joint transformations. Finally, A may be dominant if, in decisions about joint activities, he displays stronger preferences than B does. In a choice between activities m and n, for example, if A strongly likes m and dislikes n while B is indifferent, A's preferences for m will prevail.

3.37.2   Dominance and submission are not phenomena independent of those already discussed. They are simply a way of indicating that the relationship between A and B is determined more by A's system states and resources than by those of B. This does not mean that B's traits are irrelevant. If a boot is to imprint its shape in mud, the densities of boot and mud are equally relevant. But it is nevertheless the initial shape of the boot, not the mud, that determines the resulting pattern. Dominance-submission is an interaction between two specific persons. As a phenomenon of aggregate power it is also the outgrowth of a developmental process (see Chapter Sixteen and the introduction to this chapter). On both counts it is an idiographic problem, unique to the two persons involved. Many nomothetic propositions helpful for analyzing such relationships have nevertheless already been stated in connection with power, status, bargaining power, or communication, and others will appear later, particularly in Chapter Eleven. Several illustrations follow.

3.37.3   Let us now examine a relationship in which we assume that *A* is and wants to remain dominant and wants to have his dominance acknowledged by deferential behavior, and that *B* would like to become dominant rather than remain submissive. In this relationship we attend only to details with status content, any of which could be identified from the propositions of Sections 3.35 and 3.36.

Under these assumptions any behavior by *B* that is mutually perceived as nondeferential reduces *A*'s status and hence his apparent dominance. But *A* will not allow such nondeferential behavior if he has the power to prevent it. Nondeferential behavior itself means that *A* is not in fact dominant. If *A* understands these matters he will either insist that all *B*'s actions toward him show deference or else try to have nondeferential behavior interpreted as deferential. Hence *B* will have to display deference in every subsidiary transaction until and unless he has sufficient power to change the major transaction by establishing his own overall dominance or equality—or by ending the major relationship.

In addition to relatively arbitrary signs of deference (such as dress, language, priority access to desired things, body movements) deference can take the form of *B*'s providing wanted things for *A* without receiving equivalent return. Such behavior indicates lesser power under 3.28.524 (by indicating *B*'s greater stake in continuing the overall relationship), under 3.28.533 (by indicating *B*'s greater stake in future relationships), under 3.35.68 (by indicating that *B* wants to deal with *A*), and possibly under Section 3.25 (by indicating that *B* is unable to match *A*'s power in applying bads)—among others. Such behavior may also symbolize *A*'s formal authority, as will be seen in 4.12.10f and m.

3.37.4   Although the general relationships among the concepts and phenomena of power, satisfaction, status, and dominance suggest that dominance will be preferred to submission, any actual situation may differ from the general expectation. Among the many reasons why a person may prefer, or at least willingly accept, the submissive position in a relationship are the following. (a) He may be dominant in enough relationships with other persons so that one submissive role is not unsatisfactory. (b) He may prefer to save the costs that dominance entails—"He doesn't want the headaches of leadership." At the minimum these include decision costs. If his dominance were challenged, the investment and fixed costs of attaining and maintaining it might be high. (c) The nonstatus benefits from the relationship may exceed all its costs, including that of submission. (d) Earlier attempts to break out of a submissive relationship with other persons may have been unsuccessful and unpleasant so that acceptance of the submissive position is strongly conditioned. (e) Culture or experience may have taught a value structure in which submission is placed above dominance ("Power corrupts" or "The meek shall inherit the earth"). (f) B may achieve more status from submission to a high-status A than from dominance over a low-status A. (g) B may feel that A's satisfactions are more important than his own. (h) B may feel guilt over some past behavior and want the submissive role as expiation (see 3.46.12).

As indicated following the definition, A may be dominant in one aspect of a relationship and B in another. This observation also includes the possibility that although A may be dominant and get what he wants in the overt aspects of transactions, by superior tactics or strategy B may achieve results quite close to what *he* wants. Moreover, the relation may be satisfactory to both in that each thinks *he* is dominant.

3.38   **METHODOLOGICAL NOTE**

The major divisions of this book, as reflected in its single-digit headings 1 through 4, form a clear hierarchical structure. I feel reasonably confident that their arrangement here is logically sound. But for some topics in the volume there is no one best location. The discussion of aggregate power, for example, might well have come in the section on interrelations of communications and transactions (Section 3.4), after formal organization (Section 4.1) (because organization is a large ingredient in the aggregate power of most individuals who have much of it), as part of informal organization (because social stratification and class structures fit best there), or even as a simulation illustrating the idiographic approach in Chapter Sixteen. In any of those locations it could have been written to include much of the content of this chapter. Yet a different context would have given different contours to the discussion of aggregate power. The different context might have produced a different set of propositions, possibly some that would contradict some in this chapter. (I may experiment along these lines after this book goes to press; obviously not before!) The point is simply that the content of knowledge may depend on its location in a large framework of knowledge.

Chapter 11

3.4

Interrelationships of Communication and Transaction

3.40   **INTRODUCTION**

    The initial model of communication in Chapter Seven attended solely to the information content of the interaction—the communicational and linguistic elements—without regard to valuation. Culture was identified as communication, and relaxations of the model at the end of the chapter dealt with desires for information. The focus remained on information, however, and certainly was not transactional. Similarly, the initial model of the transaction dealt with the valuational content of the interaction without regard to its informational or communicational elements. Later relaxations, especially those regarding tactics, strategies, and status, dealt with communicational accompaniments and consequences of transactions. But the main focus remained on the valuational consequences of such information.

    The initial purity of both models is nevertheless already broken, and this chapter attends exclusively to the breach—the respects in which every real interaction is simultaneously communication and transaction. It also deals with the effects of the interactions on the actors' system states—a sticky analytic area—and their subsequent effects on the interactions. Since these subjects depend on preceding ones some overlap is unavoidable.

As the model becomes more complicated the difficulties of tracing all assumptions increase exponentially and probably would require a computer. An increasing number of the following propositions therefore shift from positive to probabilistic. They do not explicitly use the total set of assumptions behind them, but rather a simpler assumption that the forces immediately under discussion are dominant within the context of propositions already developed. *Will* is increasingly replaced by *may,* and the *ceteris paribus* qualification appears more regularly. Whether it is possible (or productive) to spell out all assumptions at every stage I cannot even guess; I certainly will not attempt it here.

Sir Henry Maine (1906) once observed that social bonds are gradually shifting toward contractually defined ones. For analytic purposes it is easier to reverse the order. That is, we can first study the contractually defined relations that are logically explicit and from them learn certain principles of behavior under simplified conditions. Once those principles have been clarified in an explicit model, gradual relaxations can accommodate the mixed and relatively amorphous relationships. There is a parallel here to the previous observation that this model reverses the order in which economic theory developed: this analysis begins with pure bilateral monopoly and moves toward a consensus situation.

3.41   MUTUAL MODIFICATION OF PERCEPTIONS AND VALUES

To introduce the combinations and mixtures of communication and transaction we use an initial model of a communication from B to A about some behavior K engaged in by C. The materials of this model are necessary for subsequent analysis since they systematically scan several cognitive and motivational effects on A, on the A-B relation, and on the A-C relation of both confirming and disconfirming reports by B. The reader who is interested in substituting these deduced propositions for empirical findings in certain areas of social psychology, or in comparing the two, may want to examine this section carefully. Others will find a quick survey sufficient.

3.41.1   The following assumptions constitute the model for perceptions of K by A and B.

a. B communicates to A his perception of action K performed by C. This perception may or may not include B's inferences about C's relevant system states.
b. A has already perceived K for himself.
c. Action K is important enough to justify significant conclusions about C's motives and judgment (selector and detector functions) and to affect A's relationship to C. That is, A is motivated to clarify uncertainty about C.
d. A and B both have reasonably definite images of C and of one another.
e. A and B both have reasonably definite likes or dislikes of C and of one another.
f. A and B have no a priori preference to agree or disagree about C.

For systematic scanning the next three sections (3.41.2, .3, and .4) are arranged as follows. All propositions concern the effect of a communication to A from B. One subsection of each section deals with confirming messages and the second subsection with disconfirming ones. Each subsection traces the effects of the message on A (labeled "Intra-A"), on the relation of A and B, and then on the relation of A and C. Section 3.41.2 traces the effects on the cognitive aspects of the interaction (detector aspects). Section 3.41.3 traces the effects on certain motives or values related to the cognitive (selector aspects of detector). Finally, Section 3.41.4 traces the effects on certain motivational states not related to the cognitive (nondetector aspects of selector).

3.41.2 *Cognitive Aspects*

3.41.21   For the cognitive (detector) aspects of the interaction we assume that B's report to A about K substantially confirms A's perception of K.

A. Intra-A: B's communication (1) confirms or reinforces A's perception of K—that is, it reinforces A's decoding of that perceptual input (3.17.62). It also (2) confirms A's conceptual set (the code itself) about behavior of type K, about C, and possibly about A's view of "human nature" (3.17.781). And finally (3) it raises A's confidence in his own information-processing.

B. Relation of A and B: B's communication (1) supports (by not disconfirming) a hypothesis that A and B process information similarly. It also (2) raises A's confidence in B's code and decoding (3.17.791). And finally (3) in A it raises B's credibility as a source of information (see 3.17.77).

C. Relation of A and C: B's communication reinforces any relationship based on or already reinforced by A's own prior perception of K.

3.41.22   We now replace assumption 3.41.21 with the assumption that B's report about K significantly disconfirms A's prior perception of K.

     A. Intra-A: B's communication (1) throws doubt on A's prior perception of K, that is, on his own decoding. It also (2) throws doubt on A's conceptual set (code) about behavior of type K, about C, and possibly about "human nature." And finally (3) it throws doubt on A's own information-processing.

     B. Relation of A and B: B's communication (1) disconfirms a hypothesis that A and B process information similarly. Moreover, (2) for A it throws doubt on B's code and decoding and (3) raises doubt in A about B's credibility as a source of information (3.17.79). And (4) if the discrepancy is large, it suggests an inability of A and B to communicate accurately since it presumably involves significant differences in their referents of given signs. And finally (5) A now possesses conflicting information about K. The outcome of this conflict is an idiographic problem about which no firm predictions can be made. Involved in it, however, would presumably be such factors as sufficiency of cues, A's prior confidence in his own perceptions and in B's, and motivational factors such as those listed below.

     C. Relation of A and C: B's communication potentially calls into question every aspect of the relation based on A's prior perception of K.

3.41.3   *Cognitive-Motivational Factors*

3.41.31   We now assume that B's report substantially confirms A's perception of K.

     A. Intra-A: B's communication (1) makes A's information problem easier by keeping down information-processing costs in time, effort, frustration, and the like. It also (2) avoids the unpleasantness of dissonance in A (2.41.4 and 3.17.89).

     B. Relation of A and B: Because of B's communication (1) B's information about K is more desired than before and (2) A will have a stronger desire hereafter for more information from B (3.17.82 and .86).

     C. Relation of A and C: Because of B's confirming communication A will feel more at ease regarding aspects of the relationship related to K because of his increased assurance that he has correctly interpreted K.

3.41.4   *Other Motivational Factors*

3.41.41   We now return to the assumption that B's report substantially confirms A's perception of K.

  A. Intra-A: B's communication will make A pleased with himself and his self-image will be improved (at least in contrast to a disconfirming report by B).

  B. Relation of A and B: B's communication will (1) assist in producing a nonconflictual relationship; (2) increase A's liking of B (3.17.89); and (3) increase A's pleasure in B's company and hence the likelihood that A will associate with B. That is, in the transactional aspects of the A-B relation, A's EP to be with B will be extended (see 3.42.4 below).

  C. Relation of A and C: A's liking for C (or dislike—depending on the valence of the relationship before B's communication) will be strengthened by B's report.

3.41.42   We now assume that B's report disconfirms A's perception of K.

  A. Intra-A: B's report will make A displeased with and uncertain about himself, and his self-image may deteriorate.

  B. Relation of A and B: Because of B's communication (1) the relation will be conflictual and unpleasant to some degree, and (2) A's liking for B will decline or his dislike increase, because B will be associated by conditioning with unpleasant information and with A's unpleasant feelings about himself.

  C. Relation of A and C: A's general feeling about C will be reversed in proportion to the strength of A's positive feelings toward B.

3.41.43   In summary of 3.41.1 through 3.41.4 it is evident that the values attached to confirming reports by B are overwhelmingly positive; they make A feel good about himself and B, and they allow his relation to C to continue without the costs of reexamining it. The values attached to disconfirming reports by B are overwhelmingly negative; they make A feel unpleasant about himself and B, and they require added costs in reassessing the A-C relationship.   **3.41.5** In addition, A's general preference for accurate over inaccurate information will be fulfilled in some degree by a confirming report by B since it will increase the number of independent observations on which his information about K is based (3.17.75). That preference will be frustrated by a disconfirming report. Hence, even to the extent that A prefers accurate rather than pleasant information, a confirming report by B is more rewarding than a disconfirming one. **3.41.51** This factor reinforces conclusion (3.41.43) that, *cet. par.*, A will prefer confirming to disconfirming reports from B.

3.41.52 Whatever A's previous net position regarding his desire for accurate information or his attitude toward himself, B, or C—and continuing assumption 3.41.1f (no a priori preference to agree or disagree with B)—that position will be shifted by 3.41.51 toward pleasure in a confirming and displeasure in a disconfirming report from B. Other things equal, this result will produce a liking of B by A. **3.41.53** Assuming further that B has at least an intuitive awareness of 3.41.51–.52 and that he prefers a pleasant to an unpleasant relation to A, then as soon as he learns of A's interpretation of K he will modify his report toward greater congruence with A's (see also 3.41.41b). **3.41.54** Except for propositions dealing with A-C relations, all those of 3.41.2 through 3.41.42 are reciprocal and reinforce the conclusion of 3.41.53. **3.41.55** Furthermore, if A and B discuss in detail any discrepant views about K, and if there are differences in the details of K they observed, they would then possess more cues about K in common and be more likely to reach a common perception.

3.41.6 We now remove the assumption that K is an act of C about which A is concerned. Instead we substitute that K is some object or event, animate or inanimate, about which A and B hold no significant motives. Then both the communicational and the transactional aspects of the relation make mutual confirmation of perceptions more rewarding than disconfirmation since the negative effects of motive reorientation are removed. **3.41.61** Under the continuing major assumption that rewarding behavior is more likely to be executed than nonrewarding or less rewarding behavior, **3.41.62** continued interactions between A and B will lead to more confirming and fewer disconfirming reports of perceptions and **3.41.63** A will be more likely to associate with B's who confirm his perceptions than with B's who disconfirm them. **3.41.64** To generalize from 3.41.62, to the extent that members of the same culture or subculture interact they will become more similar in their concepts. **3.41.65** To generalize from 3.41.63, individuals will prefer (and, *cet. par.*, will engage in more interactions in) that culture or subculture whose members confirm one another's perceptions, and the confirmatory values rise with the number of persons sharing them.

3.41.66   By joining 3.41.43, .52, and .63 we conclude that if A witnesses an event (play, film, or real event) in the company of a B who shares his perceptions and evaluations of the matters involved, then in addition to any satisfaction from B's presence and other inter-actions A will have the satisfaction of having his responses confirmed by B.   **3.41.67** By corollary, if B's responses differ significantly from A's (he laughs at different things, disapproves what A enjoys, and so on) A's satisfaction will be reduced by this lack of confirmation. These two propositions are not of course proposed as the whole explanation of this phenomenon but as one aspect of it.   **3.41.68** This process (3.41.1 through 3.41.65) is a means of reaching consensus about concepts and perceptions by a combination of communication and transaction. It affects and is affected by the detector and selector states of the parties involved and supplements the consensus processes already observed for communication and transaction taken separately (3.18.1–.4 and 3.27.33).

3.41.69   If we remove the assumption that K is an object or event and substitute that it is a value judgment (creeping socialism is bad, raw oysters are nauseating) and that no relations with a third party are involved, all other 3.41 propositions to this point also apply to such value judgments.   **3.41.7** Value judgments are probably more subject to this consensus effect than are factual judgments since they are less subject to empirical verification. Value judgments may be more strongly held, however, and their modification may require more modi-fication of related values than would be the case with perceptual judg-ments. Therefore this proposition cannot be applied between pairs of value and perceptual judgments.   **3.41.71** Assuming that persons attach intrinsic reward value to things conditionally associated with other rewards (2.24), one will tend to like others who confirm his per-ceptual and value judgments, especially the latter, and will enjoy their presence and interactions for their own sake. For transactional analysis we can then say that A has an EP for B's company.

3.41.72   If A wishes to be liked by B (and he almost certainly will want to be liked by some B's (3.28.42) and if he understands the factors in 3.41.71, he will feel compelled to agree with B in significant re-spects.   **3.41.73** And from 3.17.82 it follows that A will also be more credible to B if he has already agreed with him in some respects. **3.41.74** Hence if A wants to be liked, or believed, or both by B he will hesitate to disagree with B and may appear to agree with him even when he does not.   **3.41.75** Since, by conditioning, inaccurate information tends to produce both distrust and dislike of its source (3.17.79, 3.41.52, and 3.43.4), however, feigned agreement by *A* may produce an effect opposite of that intended if it is discovered or suspected.

3.41.76 If we join 3.41.74 with 3.35.2 (that *A* will be more eager for the approval of high-status persons) it follows that low-status persons will be more likely to feign agreement with high-status persons than the reverse. **3.41.77** By corollary, to persons who understand this relation, agreement (or other communication) from a person of higher status than oneself will be more credible than that from a person of lower status and agreement from lower-status persons will be suspect— in which case **3.41.78** for *A* to feign agreement with *B* is a cue that *A* has (or thinks he has) the lower status. **3.41.79** The mutually reinforcing processes of 3.41.65–.78 constitute positive feedback: more interaction brings more similarity and more similarity brings more interaction. The asymptotic limit of similarity is, of course, identity. The limit on the other variable, amount of interaction, is presumably the diminishing utility of additional interactions, which will at some point reach zero. It seems safe to conclude from empirical observation, even if not scientifically controlled, that the utility of further interaction reaches zero before similarity reaches identity. Between these two variables diminishing utility of further interaction is the probable determinant of the equilibrium position in this relationship.

3.41.8 Let us now return to the assumption that *K* is an act of *C*, about whom *A* cares. Although propositions 3.41.51 and .52 are apparently valid as stated, they indicate only a direction of pressure, not a distance of resulting movement. Hence if continuance of the existing relation with *C* is important to *A* he may resist any change in his own perceptions of *K* in the direction stated by *B*.

3.41.81 The dissonance costs to others of a person whose values and perceptions differ strongly on important issues are high. Those costs can be reduced dramatically if the "deviant" is recategorized as irrational, stupid, strange, or just "not one of us"—in short, a person whose views are therefore irrelevant.

3.42 COMMUNICATION AS TRANSACTION

To achieve analytic separation of communication and transaction the initial model of communication assumed that the content of a communication had no value, either positive or negative (3.16.0d), and that its truth or falsity was irrelevant to the analysis of communication as such (3.16.0b). We now shift to the situation in which a value is attached to the information, to its truthfulness, and perhaps to the communicational process—that is, a situation in which the communication is itself a good or bad. To the extent that the communication has positive or negative value the interaction is analyzable as a transaction.

3.42.1   The following assumptions constitute the model for communication as transaction:

a. Information possessed by B is valued by A.

b. Information imparted to A by B (not the medium, such as book or map) does not cease to be possessed by B; and, once transmitted, it cannot be taken away from A (see 3.16.0c).

c. B has no desire to prevent A from acquiring this information, even though transmitting it has significant costs to B.

d. The technical question of accuracy of communication is absent or ignored for the analysis.

3.42.2   To view this relation as a transaction, AY is A's desire for B's information and BY is B's cost of communicating it—in time, effort, inconvenience, and the like. X is whatever A gives B in return—a fee, drink, dinner, thank you, acknowledged state of obligation, inferior status, other information, or release of a twisted arm. The terms of the transaction depend, as usual, on the magnitudes of AX, AY, BX, and BY, which may include generous or hostile components, intertransactional relationships, and so on. That is, potentially the entire transactional analysis applies to the transaction in which X, Y, or both are communicated information. A few cases are now identified, but without elaborating propositions.

3.42.3   The communication aspect of the relationship may be simple, probabilistic, and implicit, as when A and B agree to lunch together in a mutual but unstated expectation that useful information will be exchanged.   **3.42.4** The transaction may also be rewarding in itself (3.26.4), as when A keeps B amused while B drops tidbits of information.   **3.42.5** An exchange may include information about one of the parties applicable to future transactions between them: A interviews B, who is an applicant for a job with A; or A, allegedly seeking other information, studies B for clues to his tactics in later dealings with A.   **3.42.6** We have assumed that information possessed by B is valued by A (3.42.1a). If B lies to A, the value of the Y received from B is considerably less than if the message were truthful and may, in fact, be negative. Thus the magnitude of AY declines with B's credibility, with all the effects on power, bargaining power, and other aspects of transactions that such decline entails. In a continued trusting relation the transaction is of the general sort, "I will tell you the truth, even if it hurts, if you will do the same for me."   **3.42.7** Many real cases are amorphous, however, and the ingredients may be difficult to identify. Inability to identify the ingredients or even the applicable analysis is in itself no criticism of the analysis—as with the mushy ice in the Preface.

3.43   *Effects of Transaction on Communication*

3.43.1   Let us now assume that A is aware of the nature of power and bargaining power and desires both. Then even without specifying the strength or net effect of conflicting motives (such as a concomitant desire for scrupulous honesty) a scanning of the major facets of transactions would lead to the following expectations about A's behavior (in paragraphs up to but not including 3.44).

3.43.2 A will be more candid in describing his preferences and opportunities to persons with whom he does not anticipate having transactions than with others. By corollary, if he does anticipate having transactions he will be more candid about those preferences and opportunities not likely to be involved in the transactions than about those that are, and/or he will modify his statements to avoid disadvantaging his anticipated transactions. Here, however, the possible gain from distortion aimed at particular transactions has the cost of risking credibility in future transactions and hence of worsening their terms.

3.43.3   In communicating with those with whom he expects transactional relations he will tend to use the communication as a tactical device to create impressions that will increase his power or bargaining power. At the most obvious level these tactics include overstating the quality of X and alternative sources of Y and understating the quality of Y and alternative sources of X.   **3.43.4** By corollary, if he understands this effect he will tend to discount communications from B involving matters about which they may have future transactions— unless he believes B's stake in accuracy is high enough to prevent his distorting these matters (3.24.3).   **3.43.41** And if he understands this effect, A will tend to overstate his stake in keeping his promises. **3.43.42** A, in fact, has a larger stake in keeping promises to those with whom he has repeated transactions than with those with whom he has unique transactions.   **3.43.43** Never trust the promises of passing strangers.

3.43.5   According to 3.28.42 *A* will recognize the value in future transactions of being liked or at least of not being disliked.   **3.43.51** Furthermore, a hard bargain driven by *A* will bring *B* less value than a softer bargain; and by conditioned association *B* will like *A* less or dislike him more, *cet. par.*, if he is a hard rather than soft bargainer. **3.43.52** Hence *A*'s bargaining will be less tough with *B*'s with whom he expects future relationships, *cet. par.*, than with others.   **3.43.53** Similarly, since the use of bads by *A* brings distress to *B* and, by conditioned association, dislike or hostility, *A* will be less likely to use bads in transactions if he expects subsequent transactions with *B*. **3.43.54** If bads are generally disapproved (as they are likely to be—see 3.25.4) this fact will reinforce 3.43.53.   **3.43.55** By corollary, *A* will tend to understate his willingness to use bads except where they are approved or would not adversely affect *B*.

3.43.6   A basic proposition about transactions is that *A*'s power (3.22.3) and his bargaining power (3.22.5) are both direct functions of *B*'s EP. Since for *A* the only positive ingredient in *B*'s EP is *BX*, *A*'s power and bargaining power depend on *A*'s having traits, capacities, or tangible goods desired by *B*'s. In layman's language, to be or to have something wanted or useful is an indispensable condition of power other than that based on bads. Since power is the ability to satisfy wants, and since most wants have to be satisfied through others, then (adding to assumption 3.43.1 that the nature of this relationship is grasped at least intuitively by most people and probably at an early age) the desire to feel wanted and useful will acquire strong and widespread generalized reinforcement. If we add 3.28.42 (the advantage of being liked), the transactional analysis predicts a widespread desire to be wanted, useful, generous, and liked. (This effect could, of course, be limited if a society limits the range of terms on which transactions may be conducted.)   **3.43.7** Hence persons who see themselves as not wanted or as not possessing wanted things will see bads as their only source of power. Whether they will actually use them depends, of course, on the relative strength of conflicting motives.

3.44   **TRANSACTION AS COMMUNICATION**

3.44.1   Both the fact and the terms of a transaction are determined by the power factors in it. Hence any transactional behavior has implicit in it information about the power positions of the parties and can give an observer information about those power positions—perhaps the system states of the parties, their aggregate power and status, or their bargaining power. In one sense we have already identified information content of a transaction in stating that the two EP's of the parties and the terms are functionally related so that knowledge of any two factors provides implicit information about the third. This section deals with less formal extractions of information as well.

3.44.2   If A alters his EP, tactics, or strategy to create an impression in B or C about either himself or B, this act constitutes a communication, as defined, or at least an attempted one. If B and C extract similar information without any such alteration by A, they get it by perception, not communication.   **3.44.3** Generalizing from 3.44.2, in a continuing relationship between A and B the repetition of any subsidiary transaction, or of any element in it, can symbolize its continuance; likewise a change in a subsidiary transaction may symbolize a change in that relationship. This is presumably what sociologists, psychiatrists, and anthropologists mean when they insist that many transactions have a *meaning* beyond that of the overt exchange. By corollary, a change indicates instability and uncertainty in the relationship. If continuance is desired, such uncertainty will be disliked in its own right as dissonance and may elicit transactional behaviors that communicate dislike. It may also bring insistence that the symbols of continuance be regularly repeated, on pain of hostile responses.

3.44.4   Since C can learn about A or the A-B relationship by observing a transaction between A and B, if A anticipates subsequent relations with C that might be altered by C's observations, A may modify his behavior in that transaction to improve his later position with C. That is, when C is present A behaves differently with B than when the two are alone.

3.44.5   In further pursuit of this approach virtually the entire transactional analysis could be reviewed to identify the information implicit at each step. But since in general the effect implies its cause, one need merely reverse the emphasis to extract the implicit information. To illustrate the conspicuous types of transactions, a long, generous extension of EP can say "I like you" and a conspicuous contraction "I hate you." Since long EP's ensure repeated transactions, and since liking produces long EP's, the gift can also say "We're still sticking together."

Shrinkage to or toward the point of boycott similarly says "Our relationship is on the rocks."

3.44.6   A gift is often an investment (3.26.24 and 3.36.64) that will not be made unless a return is expected (2.53). Such expectation assumes a stake by B in making such return, which is the basis for A's confidence that he will do so (3.24.3a) even if B's promise is only implicit. Under such circumstances (and they are many) the gift can communicate A's trust in B. When combined with 3.44.5 this means that, by signaling trust and continuance of the relationship, the gift is integrative (see 4.12.51).

3.44.7   Since A will get better terms from a B who likes him, a transaction that communicates B's liking will lead A, *cet. par.*, to like B. This is a positive-feedback relation, subject to an equilibrium only if some element is limited—such as time, resources, and the pressure of alternative commitments. A movement in the opposite direction is similarly subject to positive feedback. In the latter case the relationship terminates rather than reaching an equilibrium. Hostility is thus disintegrative.

3.44.8   Given all the propositions about transactions in the preceding sections, an EP may be of a certain magnitude for numerous and sometimes quite different reasons. Hence the information implicit in the transaction can be highly ambiguous, particularly if tactical misrepresentations are used or suspected—and they are no less relevant to generous than to indifferent or hostile transactions. Hence the communicational aspects of transactions and the transactional aspects of communications—particularly the former—provide a potential base for the mutual misapprehension of motives.

3.45   **ASPECTS OF COMMUNICATION
                    AND TRANSACTION**

3.45.1   The following assumptions are applicable for paragraphs 3.45.11 through 3.45.3:
a. A believes that those who have high power and status usually achieve it through superior adaptation.
b. To A this adaptation implies their superior awareness of their environment.
c. A assumes this to be true until he has reason to conclude otherwise.

3.45.11 Other things equal, A will place higher confidence in the judgment of a B with high status than in that of a B with low status. **3.45.12** By corollary, given an initial discrepancy, *cet. par.* A will modify his own concepts or perceptions more toward those of a high-status B than toward those of a low-status B. **3.45.2** Furthermore, given the perceptual nature of status (3.35.0) and the tendency of association with status itself to convey or enhance status (3.35.33) A will achieve higher status, *cet. par.,* if he adopts the views (values and perceptions) of high-status B's rather than those of low-status B's. **3.45.21** And again by corollary, proposition 3.45.12 is thereby reinforced—that is, the views of high-status persons tend to be copied for both communicational and transactional reasons. **3.45.22** Hence to accept and act on a communication from a low-status B may be seen by A as lowering his own status and tend to make him disregard or downgrade such communications. **3.45.3** Items 3.45.12, .21, and .22 are additive. For all three reasons, A will give greater weight to communications from high-status than from low-status B's.

3.45.4 Once A has accepted and acted on a communication from B, A's own status will be greater with C's if B appears to the C's as having high status. **3.45.41** By corollary, once A has received and acted on a communication from B his own status will be enhanced if he can defend or raise the status of B. **3.45.42** However, given the benefits of high status, statements perceived as attempting to raise one's own status will appear as self-serving and, *cet. par.,* detract from the status effects of 3.45.4 and .41. **3.45.43** Since power presumes ability in the area on which the power is based, a high-status person's statements about his own field will carry more weight with low-status people, *cet. par.,* than will his statements about other fields.

3.45.5 Status is to a considerable extent communicated by association. Thus if A and B associate regularly, a higher status of either in the eyes of C's helps raise the status of the other. Hence persons who want status will attempt, *cet. par.,* to raise the status of those they associate with. **3.45.51** But since status is relative (3.35.67), by corollary one's own status is enhanced if those to whom his own status is not tied have their status lowered. Hence persons who want status will be motivated to lower the status of those to whom they want to look superior. **3.45.6** Transferring this logic to group membership we conclude that one would rather belong to a high-status than a low-status group and once in it will prefer to enhance rather than detract from its status. **3.45.7** Low-status members within the group gain more status from membership than do high-status members since the status of each is affected by that of the others. Low-status members therefore have the greater EP for membership and less bargaining power in dealing with the group.   **3.45.71** Hence if the group wants conformity among members, including confirming reports about matters considered important to the group, it will have more bargaining power in getting it from low-status than from high-status members.   **3.45.8** Aside from direct status effects, under the assumption of 3.45.1 A would expect to achieve greater power and status from emulating high-status than low-status persons.

3.46   INTERPERSONAL AND INTRAPERSONAL

3.46.1   We have noted the partial analogy between decision processes within a person and transactions between them (introduction to Chapter Eight) in that any decision is a quasi exchange. The costs in a choice are the things given up; its benefits are the things received.

3.46.11   For A to receive some Y of value without giving any X in return often leaves him with inferior status, an obligation to B, or both, and the discharge of either condition will probably be rewarding (3.36.64). If A understands the full logic of such a position, he will want to eliminate it only if he perceives his position as likely to lead to some undesired situation later, the avoidance of which is worth the cost of discharging the obligation. But if A understands the situation only intuitively and partially, he may know only that the obligation makes him feel uneasy, and he may therefore want to discharge it whether or not it is objectively rational (2.50u) to do so. He may also feel that his pride (elimination of inferior status) requires it. With or without having been taught it explicitly, A can extract the behavioral rule: "I must not accept things from others unless I give equivalent value in return." And he may then experience unpleasant emotions when he violates the rule. We may identify this unpleasant emotional state as one form of guilt.

3.46.12   Given the partial analogy between transaction and decision, an intuitive understanding that is perhaps reinforced by the culture may provide the operational rule: "Every benefit has its cost." Under such subjective rationality A would expect that for any benefit he will sooner or later have to pay a roughly equal cost; and he will feel guilty until he has paid it. Since the basis of the feeling is subjectively rational only, the cost incurred to discharge the guilt need be only subjectively related to the benefit. That is, if the cost is related to the benefit in A's perceptions and feelings it can discharge the unpleasant feeling whether or not any connection between cost and benefit exists externally. If A holds this set of conditioned connections he will impose some cost on himself for any benefit received and not otherwise "paid for." In the language of psychiatry, he will punish himself.

3.46.2   By parallel subjective reasoning about bads, if A has imposed costs on B (such as pain, grief, inconvenience) A may feel guilty unless he gives B some compensating positive value or imposes some cost on himself that transactionally balances that caused to B.   **3.46.21** If relevant others disapprove of getting something for nothing, for A to impose such costs on himself may demonstrate to himself and those others that he violated the rule by accident and not for selfish gain. **3.46.22** But if A feels he has given much but received little in transactional relations, even those as broad as with "life" or "society," he can unilaterally achieve balance by theft, embezzlement, hostile transactions, or mayhem. (This is not to suggest that this is the only source of such behavior.)

3.46.3   On the basis of 3.46.11–.22 it can be generalized that certain objectively rational rules of behavior are not necessarily valid for all situations and that they become objectively irrational when, because of conditioned motivational connections, they guide behavior in the inapplicable instances. Although the matter cannot be pursued here, one wonders how much behavior of the emotionally disturbed represents subjective misapplications of overtly correct propositions rather than applications of essentially incorrect propositions. Depending on the circumstances, however, love might overcome pride and guilt and help avoid emotional disturbance. Or it might intensify it.

3.46.31   It was indicated (3.28.43) that being liked and approved has clear transactional payoffs and that the desire to be liked and approved is likely to achieve high intrinsic reward value. This desire is most strongly conditioned, however, to the desire for approval by those with whom one has continued an important relationship, such as immediate family. Being both strong and intrinsic it will tend to continue long after the relationship that gave rise to it has ceased. **3.46.32** If a person so conditioned later establishes important relationships with others who hold values sharply different from those of the persons whose approval he intrinsically desires, he faces a difficult decision. Through the combination of past memory and present fact he can acquire approval on one score only by denying himself approval on another. If the early conditioned response and the present situation both continue, the tension of unresolved decision can continue indefinitely. Thus transactional analysis and the associated communications provide an analytic base for emotionally disturbed behavior. (Note that the so-called transactional approach now rapidly spreading in psychiatry would be referred to here as interactional—it encompasses but does not distinguish the communications and transactions of the present analysis.)

3.46.4   According to 3.44.1 a transaction contains implicit information about the power factors that underlie it, and 3.44.5 generalized that any detail of any transaction in a continuing relationship might similarly contain cues about some aspect of the relationship. However, whenever the same terms could result from different sets of power factors, these cues are ambiguous (see 3.44.8). A might get good terms either because his power position is strong (B intensely wants X) or because B is generous (does not want X at all but wants to please A). Furthermore, if B really is generous but believes that A would be reluctant to accept his generosity, B's tactic is to overstate BX and understate BY. If believed, both tactics would lead A to overestimate his power. Or B might actually be selfish but prefer to appear generous and thus reverse the above. In unique transactions such misrepresentations have, by definition, no effect on the subsequent relationship. But in a continuing relationship they will have virtually inescapable consequences if discovered. If the forces of Section 3.36 prevail we might expect the following situations.

3.46.41   If B represented his position as selfish when it was in fact generous, several results may follow when A discovers this fact. First, if his own status is not satisfactorily established elsewhere A will have to match the generosity or accept lowered status—or at least his own lowered estimate of his status. This situation will tend to produce hostility toward B. Second, if his own status is satisfactory elsewhere or rests on other grounds he can feel and express appreciation without deference. This situation will tend to produce affection for B. Or third, B's status will be enhanced by the discovery.   **3.46.42** But if B represented his position as generous when it was in fact selfish, several different results may follow when A discovers this fact. First, A's estimate of his own power position and status will rise and B's status will decline in A's eyes. In this respect the discovery will be pleasant for A. But lies in support of a selfish position are large and red whereas those in support of generosity are small and white (3.26.24)—with the result that A's feelings toward B will move in the direction of hostility. Second, if the relation is one in which generosity is expected, as among family or friends, the whole relation is called into question (3.44.3). Doubt will be cast forward on allegedly generous transactions in the future and backward on those in the past. A will also question whether he should continue generous responses to B if B behaves selfishly in return. **3.46.43** An almost limitless number of questions can thus be reopened. Were the terms of that past relation really hostile rather than generous or indifferent? Was all that distress caused to A by B, and apologized for profusely, really accidental?

3.46.44 Let us generalize from 3.46.4–.43. Given the ambiguity of cues in many transactions, when motives are distrusted it may be difficult to establish their real nature. As a result a relationship may be broken because of a relatively small, or even imagined, breach of expected attitudes. And as each reinterpretation of B's behavior by A seems to justify another reinterpretation of still other behavior, A's perception of B may move farther and farther from its previous position in a positive-feedback reaction. It could also move farther from reality. If fear is introduced into the relation, its tendency to dominate (2.41.12) may bring a massive move to distrust. And because it tends to generate an avoidance response it may be difficult to extinguish once it has started. **3.46.45** For the same reasons that observed communications and transactions provide information about others, they also provide information about ourselves. But since the information is often ambiguous (3.46.4) it can produce widely varying conclusions. Given the general preferability of confirming over disconfirming communications (3.41.52 and .63), interpretations that confirm one's self-image will also be preferred.

3.47 NOMOTHETIC AND IDIOGRAPHIC APPROACHES: SCIENCE VERSUS THE CLINICIAN

As we have moved away from the strictly nomothetic analysis of the initial models, the propositions have become less positive and more probabilistic, less precise and more amorphous. This result is inescapable if we are to have the model deal at all satisfactorily with the complexities of reality. Importantly, however, no new main variables have been added; all new analytic constructs are combinations or special cases of the initial ones. This section demonstrates that a vast array of combinations can be derived from a few social variables and hence that a great variety of social phenomena do not necessarily require multiplicity of basic constructs.

We have categorized transactions as selfish, generous, and hostile. Any one of these three attitudes by A in a transaction could be accompanied by any one of the three by B. Nine basic situations are therefore possible:

A's Attitude	B's Attitude
Selfish	Selfish
Selfish	Generous
Selfish	Hostile
Generous	Selfish
Generous	Generous
Generous	Hostile
Hostile	Selfish
Hostile	Generous
Hostile	Hostile

But the parties are not necessarily aware of the real situation. Either may misinterpret his own motives, the other party's, or both. For each of the nine actual possibilities there are nine possible beliefs about the situation by A, and for each of those eighty-one there are nine possible beliefs by B. These 729 possibilities deal only with reality and the parties' beliefs about it. For each of these 729 there are nine possible beliefs by A, and for each of those there are nine possible beliefs by B as to what the situation *ought* to be—which may differ from both reality and the beliefs. Beyond what A thinks is or ought to be, his behavior will be influenced by what A believes that B believes about each item already listed and by the parallel set of B's beliefs about A's beliefs. Reflection on some of life's relationships reveals that these "wheels within wheels" are by no means unreal. Nor are they easily resolved, eliminated, or even significantly limited by the norms of the culture. Otherwise psychiatrists and marriage counselors would be in much less demand. To summarize all these possibilities:

Actual situation	9
A's beliefs about the actual situation	9
B's beliefs about the actual situation	9
A's beliefs about what the situation ought to be	9
B's beliefs about what the situation ought to be	9

A's beliefs about B's beliefs about all of the above   9
B's beliefs about A's beliefs about all of the above   9
    Total possible combinations 9^7

The relatively simple situation of two persons, each with three possible stances in a transaction, taken in conjunction with some beliefs and expectations about them thus gives us 9^7—nearly 4.5 million possible combinations. And we have not even introduced such mild complications as these: the parties might be concerned about the attitudes of some third parties, they might have varying intensities of hostility and generosity and varying degrees of confidence in their beliefs, bads as well as goods might enter the transaction, a communication between them might take place on an aspect of the transaction, and such a communication might or might not be accurate and might or might not be believed. Only a few more items of this sort would bring the number of possible configurations of one transaction into the order of magnitude of 10^{30}. This number of possible combinations of even a simple transactional relation clarifies, incidentally, why it is impossible for the most general A-level analytic tools (diagramed at the left of Figure 4) to work across the diagram by deduction to include all cases that might arise at level D. By the same token it indicates why the simulation model (see Chapter Sixteen) is apparently the only feasible way of dealing with complex reality. Why a transaction takes its unique form out of all these possibilities is an idiographic, developmental problem of the sort discussed in Chapter Sixteen. It cannot be predicted from the nomothetic analysis covered thus far. A splendid description of the multiple layers of ingredients similar to the above in close personal relations is entitled *Knots* (Laing, 1970).

3.5   DEDUCED VERSUS EMPIRICAL FINDINGS

Organization, which is discussed in the next chapter, is here construed as a higher-level system than its component members. We have now completed the main discussion of communications, transactions, and many of their mixtures—that is, the main discussion of *inter*-personal relationships. This seems an appropriate point to discuss the question of empirical verification of the propositions that have been deduced from models. For that purpose we will bypass the question of their logical status and focus only on their empirical status.

To be empirically testable, at least some points in a theoretical construction must be operational so that a given piece of reality can be identified as corresponding to a given piece of the construct. In probably no area of science, however, can all portions of the model and the relevant reality be matched (Margenau, 1966). I know of no general rule for specifying the number and types of correspondence that are "sufficient," but do believe this model meets any reasonable criterion of sufficiency. We will deal with this question directly by indicating some types of correspondence and indirectly by indicating similarities between this model and empirical findings reported elsewhere.

Regarding the direct approach with communication, re-searchers have already developed techniques for determining whether a given message means the same to receiver as to sender. These techniques include verbal responses, projective tests, overt performance of instructions, facial expressions, and the like. Tests of communicational accuracy are certainly possible, along with conditions in which accuracy can occur. Regarding other aspects of communication, the likelihood that an idea will be given a name varies directly with the frequency with which it is communicated (3.18.6). Although this proposition is readily subject to empirical testing in "scale-model" societies (see Chapter Fifteen), it is much more difficult in practice (though not necessarily in theory) to test it for the whole of a large society. Paragraph 3.16.631 (stating that named concepts are nomothetic and unnamed ones idiographic, relatively speaking) also seems readily testable. So do paragraphs 3.17.62 (that greater confidence can be placed in the average of ten perceptions of the same thing by different persons, which are then communicated, than in a single perception by one person) and 3.17.63d (that the informational value to an individual of such multiple observations varies inversely with the amount of redundancy in his own perceptual cues). Most propositions of Sections 3.17.7 through 3.18 also seem empirically testable.

Regarding transactions, the definitions of both power and bargaining power can readily be made operational. Power can be measured by overt observations of whether B does or does not do as A requests; bargaining power can be gauged by the overt terms on which a transaction takes place. Experiments in this area would presumably utilize *changes* in bargaining power factors (for example, an increase in A's desire for Y) or *differences* (for example, A's desire for Y as compared to that of A_1) rather than absolute magnitudes, and there should be no difficulty in creating experimental situations in which these factors could be measured. Observations could also be made of non-laboratory situations by using highly probable inferences about the effect of observable circumstances (that the United States government's EP for bombers was greater after Pearl Harbor than before or that one's EP for film is greater with a camera than without one). Regarding generous or hostile modifications of initial EP's, there is no reason to assume that introspection about such internal states is unreliable, and it is possible to arrange circumstances that will create such states. Nor should there be any difficulty in empirically distinguishing goods from bads, agents from principals, the presence or absence of third parties or of competition, and of other ingredients of the various modifications of transactions. And although it might be difficult to determine just how the transactions of everyday life are interrelated, there should be no difficulty in experimentally creating some transactional situations that are related to other transactions and others that are not. If these suggestions are correct, then most of the preceding propositions about transactions should be empirically testable, as will be those in Chapter Twelve about transactions in organizational contexts.

The indirect approach to empirical testing is to discover how the propositions deduced here compare with empirical conclusions extant elsewhere. If coincidence is found for a deduced proposition and the terminologies are compatible, we could say that the proposition has already been validated experimentally. It would obviously require tremendous time and space to attempt such comparisons for all the propositions, so we will sample only a few. For this purpose Berelson and Steiner (1964) present a notable collection of empirical findings. The works of March and Simon (1958), Hare (1962), and Pepitone (1964) are also useful compendia.

In this connection the reader might note that, although I was aware of many empirical findings before formulating deduced ones, the vast majority of these propositions were written with nothing before me except typewriter and paper; extant findings thus operated through general recollections rather than explicit notes. In any event many propositions did not occur to me until their logical antecedents had been spelled out, though in some cases I worked backward from known findings to build the logical structure from which they could be deduced—a not unusual procedure in theory-building. Furthermore I am reasonably sure that many of these propositions, perhaps most of them, have no currently available empirical counterparts. Among these propositions, at least those close enough to the initial models to be positive rather than probabilistic could provide cogent tests of the model. Any weaknesses of the present version are, I think, in principle remediable by other investigators and more sophisticated techniques.

To begin the comparison with Berelson and Steiner on broad terms, the present volume sharply distinguishes communication from transaction. In pure form the two are totally unrelated in the sense that no generalizations about one apply to the other. By contrast,

Berelson and Steiner (1964, p. 326) define *interaction* as "a generic
term for the exchange of meanings between people. . . . 'Interaction'
refers to communication in its broadest sense." They recognize that
"the term is somewhat awkward, but some single word is required to
cover all the various ways in which people can and do express themselves in face-to-face meetings." The illustrations that accompany their
definition include only items that I too would classify as communications—"gestures, glances, nods, . . . pats on the back, frowns, caresses
or slaps, and any other way in which meanings can be transmitted." In
short, in their conscious formulation *interactions* do not include what
we have called transactions.

Yet many of the interactions they report include such ingredients as support, security, rewards, punishments, expulsion from the
group (p. 337), services (p. 340), skills related to group activities (p. 343),
and task activities (p. 346)—all items one might hesitate to categorize as
communicational, even in the "broadest sense." Furthermore it is
evident that many interactions they describe take the forms they do
because of the valuations placed on the things exchanged, not solely
because of their meanings.

Their preoccupation with communication is not uncommon
among sociologists and social psychologists. Buckley (1967), for example, talks at numerous points as though all interactions are communications, despite his extensive and generally favorable discussions of the
transactional approaches of Blau, Homans, and others. At one point
(p. 82) he states that "the sociocultural system is to be viewed as a
set of elements linked almost entirely by way of the intercommunica-
tion of information (in the broad sense) rather than being energy- or
substance-linked as are physical or organismic systems." Among other
places, Buckley's emphasis on communication appears on pages 49,
54, 94, and 95. Along similar lines, Deutsch's cybernetic model of
government (1963) deals almost exclusively with the communicational
aspects of government organization and virtually ignores the transactional.

The fundamental difficulty throughout these illustrations is
the focus on information linkage to the exclusion of value linkage,
perhaps on the assumption that since values can be communicated the
interactions are presumed to have no value content distinct from their
information content. In sharp contrast I insist that concepts and
motives, factual judgments and value judgments, are not the same
just because both can be communicated and that the analyses of the
two are thoroughly distinct. In fact I am distinctly puzzled how any
sociologist or social psychologist can think of all human interactions as

communications when he lives in the midst of such notable transactional analysts as Homans (1961), Thibaut and Kelley (1959), Goffman (1966), Blau (1964), Emerson (1962), Pen (1952), Schelling (1956), Siegel and Fouraker (1960), Chamberlain (1955), Cartter (1959), Walton and McKersie (1965), and others.

Berelson and Steiner include perhaps several thousand empirical findings. In their conclusions they properly indicate that these generalizations are scientific findings that provide an impressive range of prediction. But overall—in their words (p. 660)—these generalizations must be viewed as "the stuff of theory: the material of which theories are built . . . the raw material [for] higher-order principles." The present volume suggests that communicational and transactional analyses are the higher-order principles and theory they speak of and that failure to separate them sharply is the main reason those findings remain "stuff" rather than "theory."

Many of the findings they report can be converted into the present language and, with appropriate translations of terms, can be deduced from the present model. Many others cannot, since they involve particular circumstances not included in the present model—as, for example, the relative proclivity of Catholics and Protestants to become employees rather than enterprisers. At the same time I propose that when differences between Catholics and Protestants are described as differential states of detector, selector, and effector such predictions could be made. We would also, of course, have to accept either the *ceteris paribus* constraint or the alternative constraint that the differences between Catholics and Protestants are the only significant factors involved and are uniform for all members of both groups. The former constraint is normally the more sensible. Many other findings reported in Berelson and Steiner deal with behavior at the psychological level and hence have no parallel in the present model, which is primarily social.

Let us now move to the messy business of relating deduced conclusions to empirical findings—messy because the conceptual structures are not entirely parallel and also because key terms of the empirical language are defined poorly if at all and the ordinary language referents of those terms are vague. Berelson and Steiner's findings are referred to by their code number and page; my deductions are noted by their paragraph numbers. For convenience we refer to these propositions as Berelson and Steiner's; actually those authors compiled the propositions but did not originate most of them.

We will first examine a straightforward transactional proposition from Berelson and Steiner (C-10, p. 352): "Both the effectiveness of a group and the satisfaction of its members are increased when the members see their personal goals as being advanced by the group's goals, i.e., when the two are perceived as being in harmony." To speak of the "effectiveness" of a group implies that it does something jointly, which means in present language that it is to that degree a formal organization. The transactional problem is that of terms of affiliation— contributions or withdrawals. This problem can be formulated by assuming that one member is A and the others B. If A sees his own goals as being advanced by achieving the group's goals, his EP for membership consists of his desire to achieve his own goals (AY_1) plus his desire that the group's goal be accomplished (AY_2). This means a longer EP than if he desired only one of these two things. Thus he will give more to the group than he otherwise would, whether of time, effort, money, materials, constructive thought, or acceptance of responsibility. Other things equal, larger inputs from A will accomplish the group's goals faster or more fully than smaller inputs. At the same time A's longer EP means a greater overlap, which also means a greater subjective gain from the transaction, or greater "satisfaction," to both A and B. Berelson and Steiner's statement speaks only of "the group's goals." For simplicity we have assumed that these goals are not divisible among the members; hence we refer only to terms of input, not output, transactions.

Proposition C-10 then adds that "when members push their own needs, both satisfaction and effectiveness decline." A would presumably "push" his own goals only if he was not interested in the group's goal or if it conflicted with his own. In either case his EP for membership and participation would be smaller and would reverse the effect described above. Although Berelson and Steiner implicitly omit transactions from their definition of interactions, their explanation in this case is wholly transactional.

A second finding we will try to translate appears under "Face to Face Relations in Small Groups" (A-1, p. 327): "The more people associate with one another under conditions of equality, the more they come to share values and norms and the more they come to like one another." Since the term is not defined we will assume that *equality* means the persons are not superior and subordinate in some organization, a relation which often involves withholding or distorting information for tactical purposes. But since authority is not the only relation that may lead to distortion (see Section 3.23 on tactics), perhaps equality should be construed to mean any situation not affected by actual or expected transactional relations in which the parties would be opponents. (I think that equality, unless explicitly defined, is too vague a term to serve as the base for scientific propositions about social behavior.)

If we accept this translation, the proposition is related to several deduced ones herein. Paragraph 3.41.55 indicates that with continued communication A and B will disagree less about a factual situation as both come to possess the same set of facts; 3.41.62 indicates that on the basis of advantage alone "continued interactions between A and B will lead to more confirming and fewer disconfirming reports of perceptions." In addition, according to 3.41.69 the same conclusion applies to both value and perceptual judgments. Joining the two kinds means that people in continued association will find it more rewarding to confirm one another's perceptual and value judgments than to disconfirm them and that (under the additional assumption that they will select the more rewarding) they will increasingly confirm. Paragraph 3.41.71 concludes that people will like others who confirm their perceptual and value judgments. Berelson and Steiner do not define their criteria for "sharing" values and norms, but we will assume that if A and B express consonant views about them they have filled the definitional requirements.

In comparing the two versions we find the deduced one more complicated and roundabout and hence less efficient for this situation. But the law of gravity applied to a batted baseball is also clumsier than a simple generalization that everything that goes up must come down. The overall advantage of the law of gravity is that it deals with matters the simpler generalization does not, including airplanes and balloons, orbiting satellites, and rates of fall. Similarly the deduced version is tied logically to other propositions whereas the empirical one stands more or less alone. An appropriate aftermath of having both versions, of course, would be to see whether the assumed conditions underlying the deduced proposition are in fact present in the empirical experiments.

To move now to a third translation, by combining 3.41.63, .64, .65, and .69 (which the reader can readily do for himself) we reach Berelson and Steiner's proposition (A-1.1, p. 328) that: "There is a tendency for people to gravitate into groups or subgroups with the effect [not necessarily the intent] of *maximizing* their shared values." [Emphasis added!] Furthermore paragraph 3.41.63 alone is very similar to A-1.1 except that it deals with perceptions instead of values. Berelson and Steiner refer also (p. 328) to a suggested "need for consensus" in human beings, and Sections 3.41.2–.4, leading to the summary paragraph 3.41.43, provide an analytic basis for it while 3.18.4 describes a technique.

Another finding (A-2, p. 330) reports that: "The larger the proportion of new members joining an established group within a given period of time (short of actually taking it over) the greater will be the resistance of the group to their assimilation." This conclusion follows in part from those above, and it is reinforced by others discussed below. If we accept the previous conclusions that continued interaction leads to both consensus and liking, then the newcomers (who, by definition, have not had continued interaction in the group) will be less likely to share its existing consensus and be liked. (The logical connection between this conclusion and the previous ones is mine, by the way, not Berelson and Steiner's.) To this Berelson and Steiner add the fear of being "overwhelmed" or "inundated." The logic of that fear is dealt with below (Section 4.12.4) in connection with decision processes in organizations. That section concludes that the ultimate decisional control of an organization resides in a coalition that is dominant in whatever type of power is relevant to that organization. If many newcomers then arrive, that coalition may no longer be dominant. Since its members presumably did not attain and maintain control without wanting it, they will resist losing control.

The next step of the comparison involves a terminological problem. Now I would be the first to grant that "inundated," "overwhelmed," and "taking it over" are more poetic language than mine. But if one tries to find precise scientific meanings for them he is inundated, overwhelmed, and taken over by alternative meanings. By contrast, my techniques for decisions in groups are precisely defined and readily operationalized terms.

To pursue this proposition further, even if decisions are made by communication or transaction (see also 4.12.4) the newcomers dilute the existing group's ability to control. And even if there is no joint decision and no control—as in a group that does nothing but chat —the rewards of chatting are reduced by the unknown detector and selector states of the new members, as we have seen. The effect might be offset, however, if the new members provide new and interesting grist for the conversational mill. Furthermore, if we view the group as a scale model of a society (see Chapter Fifteen) then a few new members follow the logic of the induction process and take on the traits of those already there (see *induction*, 4.12.510). But an influx of many new members parallels the fusion of cultures (3.17.0g) without advance assurance about which will survive.

Another proposition (A-2.4, p. 330) states that: "The more

eager an individual is to become a member of a small group, the more he will conform to its norms of behavior." Assuming that the group *wants* conformity, this proposition is an instance of the initial model of transaction. Let A be the individual, B the group, X conformity by A, and Y the benefits the group confers on A. Then the greater A's EP for Y, the more X he will be willing to give for it (3.22.7). And if A likes the B's he may by simple conditioned association like their behavior and imitate it (see also 3.45.71 above).

Proposition A-3 (p. 331) states that: "The more interaction or overlap there is between related groups, the more similar they become in their norms and values." Some ambiguity arises because "related" is not defined. In any event, however, this statement seems a simple extension to groups of what has already been said in A-1 (p. 327) about individuals and likewise presumably involves the additional assumption that no transactional conflicts hold the groups apart. Similarity develops between two groups for the same reason it develops between individuals. Both groups are composed of individuals, and the logic of earlier propositions is in no way softened if the interacting individuals are members of different groups. Further, any movement toward similarity presupposes interaction, and logically the greater the quantity of the interaction the sooner or more completely will the product of interaction be accomplished.

In such cases it might be useful to apply the extended version of the second law of thermodynamics directly: any closed system is subject to loss of differentiation. The proposition as stated by Berelson and Steiner may then be taken to assume implicitly that the system which encompasses the two groups is analytically closed. That means, for example, that group A's communications are not interpreted for group B by the Chamber of Commerce while B's communications are interpreted for A by the Communist Party. If such outside influences are allowed, the norms and values of the two groups presumably become more divergent instead of more similar. To make a strict scientific principle we must therefore say that if groups A and B interact with one another and only with one another (a closed system) the initial differences between them will drop asymptotically toward zero.

The thermodynamic law also assumes free interaction. Two contiguous bodies will not reach the same temperature if there is perfect insulation between them, and the freer and faster the thermal interaction the sooner will their temperatures coincide. Importantly, for both the physical and the social interaction, the principle seems the same whether we start with one body (group) or two. If the unit under study falls within a single boundary, if that boundary insulates it from other influences, and if internal barriers are removed, then for purposes of this principle the unit behaves as a single body. The mechanism is again that of 3.18.4.

The same paragraph from Berelson and Steiner adds that: "The less communication or interaction between them [If all interaction has already been identified as communication, why both terms?], the more tendency there is for conflict to arise between them. And vice versa; the more conflict, the less interaction." However it may appear on the surface, this statement is not a simple obverse of the preceding proposition. The present model suggests distinct reservations about it and raises doubt whether it is tenable except in some extremely vague sense. Berelson and Steiner (like many other behavioral scientists) apparently do not carefully define conflict between persons, presumably on the assumption that a sharp analytic definition is not necessary or that everybody already knows the definition. They do implicitly define internal conflict (p. 272) in connection with goals that "are at once satisfying and threatening, pleasant and painful, attractive and anxiety-arousing." That definition is compatible with the one used here as the predecision stage of costs—namely, that one goal can be satisfied only at the cost of denying another. In the present framework the definition of conflict remains the same between persons—namely, a situation in which *A* can reach his goal only by preventing *B* from reaching his. It is not clear whether Berelson and Steiner construe *conflict* to mean this or something else. If *conflict* means simply dislike or hostility (see Chapter Seventeen below) there may be some point to their proposition. The language is vague, however, even for that purpose, which we need not pursue further.

Moreover to say "the less communication . . . the more tendency there is for conflict" seems to mean that conflict varies inversely with communication and that absence of communication is a force which itself increases conflict. My reservation would be that, although communication (within a closed system) will reduce differentiation and increase people's liking for one another, nothing in that logic indicates that absence of communication will increase differentiation rather than leave it unchanged or random. I do not want to overemphasize the thermodynamic analogy; but whereas putting two objects in contact will assuredly reduce their temperature difference, leaving them out of contact will not assuredly increase it. In fact, unless we believe in action at a distance, then for either social or thermodynamic purposes two things out of contact will have *no* effect on one another.

By corollary a reduction in interaction will bring a reduction in effect, whatever the nature of the effect.[1]

But if *conflict* is used in the sense defined here the proposition seems unacceptable. If two groups are in conflict because both want to use the same meeting room at the same hour, for example, there is no a priori reason to believe either (a) that this conflict will reduce the amount of communication between them or (b) that increased communication will reduce the conflict. The opposite may quite as readily occur on both scores, and conflict may be the sole cause of communication.

I doubt whether Berelson and Steiner would disqualify disagreement over the terms of a selfish transaction as conflict or deny that two groups could have such a conflict. The difficulty here apparently stems from their earlier implicit denial of transactional relations since their proposition may make sense applied to communications alone. But the following quotation, which they cite from Broom and Selznick (1957, p. 33) in support of their own statement, indicates that their thinking does include transactional relations: "When labor-management conflict is intense, union leaders avoid all except the most official and circumscribed contact with their employers, partly because they would be exposed to criticism from their own members." Now certainly most union-management conflicts are transactional disputes involving the terms of a contract or the terms on which a strike will end. Although certain communications may decline during such conflict, particularly those not related to the conflict, other kinds rise. In fact, from my impressions as a labor economist I incline to the opposite hypothesis and suggest that communications between union and management are greatest during conspicuous conflict and that they typically dry up when there are no disagreements. Unions and managements were startled several decades ago when a noted mediator and arbitrator proposed that they communicate when *not* in conflict. His suggestion has not had many takers.

[1] For any reader inclined to react, "But B's total absence could have an enormous effect on A," it may be useful to reexamine the definitions and discussion of *mutuality* and *contingency* in 3.10d and e. As the terms are defined there, some aspect of A's behavior might well be contingent on B's absence and yet not be due to any current interaction. As noted in 3.10d, contingency nevertheless presupposes past interaction. We may also recall that absence of a message (nonsign) can be a prearranged sign and hence part of a communication and an interaction (3.16.42). Without such prearrangement, however, any conclusions reached by A from the absence of a message or of other action by B must be analyzed as a perception by A, not a communication (see 3.10.0j) and hence not an interaction.

In this model all selfish (indifferent) and hostile transactions involve conflict over terms, and so do some generous ones. All transactions also presuppose communication. Although I have not developed the intermediate logic to deduce the conclusion on this score, I think that for most transactions between the extremes of boycott and pure gift, which do not require explicit communication, the amount of communication would vary directly, not inversely, with the intensity of the conflict. If we reverse the dependent and independent variables I suspect that no reliable relationship could be established about the effect of communication on intensity of the conflict. The application stage of the stress transaction is a unilateral act (see 3.25.0g) and does not require communication. For that kind of conflict, including war, Berelson and Steiner's proposition may apply, though not necessarily. There is no reason to assume it would apply to the transactional stage, which might be replete with negotiatory discussions between parties, even those who had no previous contact at all.

Let us examine one more proposition from Berelson and Steiner (A-3.1, p. 331): "The less contact between members of different groups, the less will there be a mutually recognized, proper behavior for their relations. If such contact sharply increases, there will tend to be increased tension until the proper behavior is defined and established." This statement seems fully in accord with the present model, although some relevant parts are not stated until the later chapters on organization. To use the later language, if two groups are not in contact they cannot be components of the same formal organization or have defined roles specifying the content or terms of their communications or transactions and their respective authority and status. That is, uncertainty about the nature of these interactions can exist only when they do not fall within the model of formal organization. Any interactions (communications or transactions) that fall within the scope of this empirical finding thus fall, by definition, into the realm of informal or semiformal organization, or nonorganization. That means, in turn, that the roles are worked out by the interactions of the systems themselves on the basis of their separate goals and capacities. Given the demonstrated advantages of power and status (see Chapter Ten), each of the previously noninteracting groups will presumably want as much as it can get, *cet. par.,* in the new relation—a condition that meets the definition of conflict. As in any other continuing transactional relation, in due time the terms of the major bargain (3.28.5) will implicitly or explicitly define the roles of each, after which point subsidiary transactions will proceed with less conflict within the constraints specified by the roles.

It may be useful at this point to recall that under present definitions detector processes and communications can be inconsistent or contradictory but not conflicting. Conflict is confined by definition to selector functions and transactions. Conflict in connection with communications can occur only within the transactional aspects of a mixed communication-transaction, never in a pure case. Furthermore in this model hostile feelings by A toward B do not in themselves constitute conflict since they do not themselves provide satisfaction to either party at the expense of the other. Hostility moves into conflict only when a transactional relation is affected, as when A's EP is shortened by the hostility or A tries to get a law enacted to restrict B's behavior.

In summary, some deduced propositions in this model correspond closely with reported empirical findings, thereby suggesting empirical confirmation. Other correspondences appear (after translation to accommodate terminological differences) with the same effect. The sample is small, however, and is conspicuously biased by being limited to instances in which correspondences and translations were readily discernible—though one case of disagreement was included. The purpose of this section is not to test the model but merely to demonstrate that some propositions are at least partially testable with extant empirical findings. For reasons indicated earlier, social psychologists and other specialists should have little difficulty designing experiments to test additional deductions.

If the deduced and empirical conclusions do not agree, the nature of the next steps is reasonably straightforward even if their execution is complicated. The first is identification: to check the correspondence between the conditions of the empirical test and those of the model. Part of that task is to see whether some conditions not included in the model might account for the discrepancy. If the discrepancy remains when correspondence seems good, the alternative is to see where the model might need change. The change might be altering an assumption or two—a relatively simple task. Or it might require modifying one or more concepts, as by changing their definitions. There the difficulties are much greater since a definition may be tied to scores or hundreds of other propositions, which would then have different meanings or need to be restated. Adding a concept instead of modifying others is another possibility, whose consequences cannot easily be foreseen. However, a need for many such additions would run counter to the main purpose of the model, which is to reduce the number of independent concepts.

If social scientists do use the model, I expect that their main problem for some time will be that of identification, as follows. The model is penurious in basic concepts, rich in combinations that can be made of them—its alleged main virtue. The investigators' problem will be to select that configuration of pieces of the model that best matches the reality they are studying. If that configuration already appears in the book, so will some deductions about its consequences—which the investigators can confirm or negate. If not they will have to make their own deductions and test them. That problem, however, coincides with the making of a simulation model. We leave it for the moment and return to it in Section 6 of Chapter Sixteen.

Chapter 12

4.1   # Formal Organization:

General Theory

The first part of this volume dealt with systems and systemness; the second examined interactions between systems. Whenever two or more systems interact they fill the definition of an acting system in their own right—a new, higher-level, and emergent system relative to the initial ones. But in connection with communicational and transactional interactions we attended only to the interactions and how they reflected and affected the states of the interacting systems. Although we recognize that the interaction definitionally constitutes a system we focused on the parts, not on the systemness of the whole. The study of organization makes that shift from parts to whole, though without denying that the whole continues to consist of interacting parts.

We have distinguished controlled from uncontrolled systems: the former have a particular goal whereas the latter accept whatever state they happen to reach through interaction of their parts. For systems that consist of multiple interacting humans we make the parallel distinction between formal and informal organizations. The former have some goal as a unit and engage in coordinated actions to reach that goal whereas the parties involved in the latter seek merely their own goals, interacting through communications and transactions, and let the joint result fall where it will. The informal organization nevertheless produces a joint result distinct from the sum of its individual actions. This chapter deals with formal organization, which, as we define it, engages two or more parties in a joint action toward the *same* goal.

This definition does not require that all parties to the organization want to achieve the goal but merely that their joint activity be directed toward it. Nor does the definition specify the number of goals, although there must be at least one. The goals may be explicit or implicit, clearly formulated or vague. The analytic logic of formal organization does not depend on the clarity of the goals, although the efficiency in reaching them presumably does. As before, we assume for simplicity that the parties are persons; however, much of the logic is the same if some parties are themselves formal organizations.

However simple the joint action, formal organization requires a single decision regarding effectuation—whose end of the sofa will be carried through the door first, whether the A-frames for the cabin roof will be assembled in place or on the ground, whether the affair will go forward in apartment or motel. At a more complex level a decision that the organization will produce only totem poles precludes a decision that it will also produce storm windows. And a decision that all staff members will both paint and carve precludes a decision that some will only carve and others only paint.

Nothing about the concept of a single decision specifies how it shall be made—by unanimous consent, dictatorial decree, or some intermediate method—or that anyone in the organization necessarily "likes" it. Whatever technique makes the selection *is* the decision process (this process is examined later in the chapter). Nevertheless, since we have defined a decision as involving a degree of complexity that requires conscious comparisons, the model of formal organization assumes some conscious thought in its decisions. The single decision as the crucial distinguishing feature of formal organization contrasts with the communication, the analysis of which requires no attention at all to decisions; with the transaction, which involves mutually contingent but different decisions; and with informal organization, which makes no decision as a unit. As a controlled system the formal organization has detector, selector, and effector functions, and for purposes of definition we need not ask whether those functions are performed by the same or different persons.

Within an organization, by definition any relations between the persons who compose it are simultaneously intrasystem with respect to the organization and intersystem with respect to the individuals. Since analysis of organization in significant detail necessarily includes both the interaction of its parts and its behavior as a unit, that analysis must employ both the intrasystem and the intersystem approaches. The former is concerned with how the detector, selector, and effector functions are accomplished; the latter deals with the configurations of communications and transactions in the organizational context. Generally, intrasystem analysis resembles classic organization theory whereas intersystem analysis resembles the more recent behavioral approach.

4.10   RELEVANT DEFINITIONS

a. An *organization* is any interaction of two or more parties in which attention is focused on the overall process and effect rather than on the interaction itself or its effects on the separate parties. The attention to focus rather than to the nature of the entity parallels the earlier conclusion (1.20o and 1.20r) that whether something is or is not to be considered a system, and if so where it is to be bounded, depends on the investigator, not on the thing investigated. And to amplify the introduction to this chapter, when we are analyzing communications and transactions the interacting parties are viewed as essentially autonomous; but when we are analyzing organization they are viewed as interacting subsystems of a larger system. As we shall see, this shift does not alter the basic logic of communications and transactions, only their content.

b. A *formal organization* is the consciously coordinated action of two or more parties toward the joint effectuation of a goal. "Consciously coordinated" means awareness both of the goal and of the instrumental contribution of the various subsystems in achieving it. "Conscious coordination" of the instrumental contributions of the subsystems means, among other things, that the organization determines its own structure. And, as in our discussion of decision theory, we do not assume that all decisions on these matters are really conscious; rather we deal with them as if they were.

    In a simple sense the joint action may be thought of as a transformation of the environment, just as the simplest model of the effector of the individual may be thought of as such a transformation. But since the social analysis of the collectivity is essentially the same when the joint goal is communicational or transactional (rather than merely transformational), the definition speaks of the joint effectuation of a *goal* and does not limit the goal to joint transformations.

c. A *pure formal organization* is a formal organization, whether or not it is a subsystem of a larger organization, all of whose detector, selector, and effector functions, including those of all its subsystems, are directed toward the goals of the organization as an entity. Although it is more unreal than some other concepts introduced here, this concept seems a useful vehicle for certain fundamental ideas about organization.

d. A *member* is a person who has made a transaction of affiliation with an organization. This definition does not require that either the fact or the terms of the transaction be explicitly negotiated, that they occur in a single well-defined act, or even that they be consciously recognized. The initial model that follows, however, assumes conscious explicitness.

e. A *simple formal organization* is an organization, all details of which —such as contributions and withdrawals, goals, and techniques of making transformations—are settled in the major bargain. An alternative definition is that any details not settled in the major bargain are agreed unanimously as they arise by all members on the basis of communication alone, as elaborated in Section 4.12.4 below.

f. A *major bargain* (or transaction of affiliation) is an agreement that a person will work with others in or as an organization toward some joint goal. The organizational relation is a special case of the main-subsidiary transactional relation of 3.28.5, particularly as it appears in the complex organization.

4.11   SIMPLE ORGANIZATION

4.11.0   *Model and Assumptions*

The model of simple organization consists of the following assumptions:

a. Only two persons are involved, A and B, in a unique relation.

b. Both are selfish. That is, each wants to minimize his input contributions and maximize his withdrawals of or satisfaction from the other's output.

c. Each is indifferent to the position of the other, wishing neither to help nor to hurt him.

d. In operating the organization itself, neither has a motive—assuming input and output ratios to be determined by technology—except to maximize his own share of the joint output. And neither has a preference to work alone or together except insofar as the difference affects his cost/benefit position regarding inputs and outputs.

e. The costs to each party consist of his input contributions including the opportunity cost of not using them elsewhere.

f. The benefits to each party consist of his withdrawals of outputs (if the joint product is divisible) or of his satisfaction in the joint product (if it is not divisible). Indivisible products may be thought of as shared *values* when viewed on a small scale and as public goods when viewed on a large scale.

g. Each party's decision to make or continue affiliation is based on his best estimates of his advantage therein. Actual consequences are relevant only as they affect those estimates. (Actions are based on the images, not on the reality.)

h. The organization involves no contributions or accumulations of investment and has no fixed costs. As a result termination of the organization (which for a two-person organization is the same as either person's leaving it) brings no cost or benefit to either party other than that the organization no longer exists.

i. Only goods are used in transactions.

Other assumptions from the initial model of the transaction are relevant here but seem unnecessary to repeat.

4.11.1   Under these assumptions, if the joint product cannot be created alone (like procreation or the performance of a string quartet) or if the product cannot be divided between the parties (like clean windows or a neat lawn) neither party will join the organization (which then cannot exist) unless the benefit of the joint product to him exceeds or at least equals his cost. If viewed as a game the organization must be positive sum or at least zero sum. If B's cost were to exceed his benefit but A's benefit were to exceed his cost by an equal or greater amount, side payments from A to B might nevertheless bring the organization into operation.   4.11.2 If both inputs and outputs are continuously divisible between the parties, the organization will not be formed or continued unless each party's evaluation of his share of outputs equals or exceeds his valuation of his share of inputs. Again, under the assumptions the organization must be a zero-sum or positive-sum game.

4.11.3   We now define an *employee point of view* as residing in the individual's question whether or not he should belong to an organization in light of its costs and benefits to him. An *employer point of view* resides in the organization's question of whether or not it should have a certain person as a member in light of the costs and benefits of his membership to the organization.

4.11.31   If the outputs of the organization are of the same sort that A and B could produce for themselves from the same inputs, or if all inputs and outputs are convertible to equivalent units of some common type, then even the simple organization embodies both an employer and an employee point of view.   **4.11.32** Viewing himself as an employee, A will not be willing to join the organization unless it provides him a return equal to or greater than he could produce for himself from given inputs. His ability to produce alone thus constitutes the floor under his bargaining power.   **4.11.33** Viewing himself as an employer, A will not be willing to give B more than that portion of the joint product which would leave A at least as well off as if he had not accepted B as a member of the organization. That is, A will not "pay" B more than B's "marginal contribution" to what A could produce without him—a central tenet of economic theory.

4.11.34   To generalize from 4.11.32, A's EP to join the organization consists of the benefits he can get from the organization (AY) minus the benefits he could have by not joining (AX), which are the same as the costs of joining.   **4.11.35** Similarly to generalize from 4.11.33, A's EP to have B in the organization consists of the organization's output with B in it (AY) minus the organization's production without B (AX). **4.11.36** Under the assumptions of 4.11.0, in a two-person organization 4.11.34 and .35 coincide. That is, the total increase in output of the joint over the separate efforts is the same whether we view it as A's becoming an employee of B or B's becoming an employee of A. It is thus a matter of indifference whether A "speaks for himself" or "speaks for the organization." It makes no difference whether A views himself as an individual who may or may not join the organization or as the organization which may or may not accept B as a member.

4.11.37   If the joint product is not divisible (a shared value) then the party with the stronger EP for it has a relative disadvantage in bargaining power over allocating the costs or contributions. This proposition is reached by analogy with 3.28.521 and by treating the shared value as having the same effect as the continuing relationship in a major bargain. It is also in line with the identification in 4.10e of the organizational relation as a special case of the major-subsidiary transactional relation. When the output of the organization cannot be divided, the only possible transactional variation is obviously in the contributions.

4.11.4   We now relax assumption 4.11.0a (that only two persons are involved) and assume three or more parties. At least for countable out-

puts and in the absence of increasing returns to scale, the interchangeability of employer and employee points of view now disappears
—unless we treat one person as one party and all remaining persons as
a second party acting as a unit and of equal weight with the first.
Otherwise A's contribution potentially divided between the two parties
A and BC is different from that same contribution potentially divided
among the three parties A, B, and C. Similarly A's EP to hire the dual
party BC is not the same as his desire to hire B or C separately or to
hire either as a third person after the other is already hired. The basic
transactional alternatives of multiple-person organizations can be
delineated with three persons, assuming (a) that A, B, and C provide
inputs of equal quantity and quality, (b) that differences in output are
due solely to differences in the number of persons, and (c) that for given
inputs per person the total outputs are

One   person: 10 units
Two   persons: 24 units
Three   persons: 33 units

These figures illustrate increasing returns from adding one person but
diminishing returns from a third.

Under these assumptions A would rather join forces with B
than work alone so long as B will work for no more than 14 units of
output. If A and B are equally good negotiators, each will receive 12
units—two more than they could get by working separately. Starting
from that position, neither A nor B nor both together would be willing
to add C if he insisted on receiving more than 9 units since otherwise
the total available to A and B would decline. But C could receive 10
units by working alone and would do so. Under these circumstances
the number of persons taken into the simple organization is that which
will maximize the average return. The same conclusion holds even if
we relax the assumptions to allow contributions in the forms of invested capital or nonwithdrawn accumulations—so long as the contributions are equal.

4.11.5   We now change the assumptions so that some items essential
to production are available to A alone, that their use involves no more
cost to A than their nonuse, and that without them B or C working
alone could produce only 4 units each. A will again prefer to work with
B rather than alone so long as B does not insist on receiving more than
14 units. But whereas under the previous situation the EP's overlapped
between 10 and 14 units with mean terms at 12, they now overlap from
4 to 14 with midpoint terms of 9. Assuming A and B to be equally
good negotiators, they will receive 15 and 9 units respectively. Starting

from the preceding situation and with A making the decision, A would not add C if he insisted on receiving more than 9 units since that is all C adds to the total; to give him more would leave A with less than before. Between A and C the EP's overlap between 4 and 9 with midpoint terms at 6.5. Assuming that A and C are equally good negotiators, A will now receive 17.5 units and C 6.5 while B continues to receive 9. Even if C is willing to work for 6.5 units, A will nevertheless be better off to continue to give B any amount up to but not in excess of 9 units since by eliminating B he would lose the 9 units of output attributable to the third man.

4.11.6   The previous paragraph (4.11.5) provides the basis for a marginal productivity theory of wages if we add the assumption of a perfectly competitive labor market, and for a bargaining power theory of wages if we do not. In the competitive labor market both B and C would receive equal wages of 9 units. Under the bargaining power theory B could receive anywhere from 4 to 14 units and C anywhere from 4 to 9, depending on the tactical and strategic factors in each case. To make a different contrast we can say that data of the kind given in 4.11.5 provide the basis for wage theory when the goods involved are objective and explicit and the basis for attraction, attachment, and participation when the goods are subjective and not readily measured.

If we ignore other factors of production, by comparing this conclusion with that of 4.11.4, we may presume that a producer cooperative owned by its employees would employ the number of people that would maximize the average wage, whereas a profit employer would employ the number that would equate the wage and the marginal product. The former amount would presumably be higher. The conclusion might or might not be valid if there are costs other than labor, depending on their magnitude. A consumer cooperative, however, would presumably act the same as a selfish profit employer and would continue to add people as long as the wage did not exceed the marginal product of the last one. (I hope someone fairly soon will try fitting some materials from small-group analysis into this context.)

To generalize from paragraphs 4.11.3–.5, if neither inputs nor outputs are divisible in differing ratios between the parties, a person can join or not join an organization but there can be no question of terms. If inputs, outputs, or both are divisible, however, the terms of affiliation can be construed as a transaction whose ingredients are as follows. AX is A's desire not to make his contributions—the cost to him of making them. BY is the parallel item for B. AY is A's desire for the product of the joint effectuation. BX is the parallel item for B. If various assumptions of 4.11.0 are relaxed and others from Section 3.2 are substituted—assumptions about such factors as multiple motives, bads, generosity and hostility, interrelated transactions, satisfaction or dissatisfaction with the relationship itself or with one another's company, and the like—the entire transactional analysis of Section 3.2 is potentially applicable to the transaction that forms the simple organization.[1] These assumptions should provide a good model for most real cases of affiliation with a simple organization.

4.12   COMPLEX ORGANIZATION

4.12.0   *Initial Model*

  In contrast to simple organization, the model of complex organization is characterized by the following assumed conditions:

a. There are more than two persons. This is a possible condition for simple organization but a necessary one for complex organization. This means (as in 4.11.4) that the terms of a member's affiliation have a different effect on the organization as a unit than on any single member or (except under constant returns) on the average of members. By the same token, the consequences of a person's joining or not joining are not the same to the organization as to the potential member.

b. There is a transaction of affiliation, or major bargain between organization and member, that specifies certain aspects of the relationship but leaves others to be settled by later subsidiary transactions. Among other things the transaction specifies the role to be filled by the member. Whether explicit or implicit, the understanding about such subsidiary transactions is that the member will provide certain inputs when and as instructed by the organization in return for remaining a member and receiving outputs. Typically the member's input is to perform tasks as instructed, though it could also include contributions of money, materials, or simple physical presence.

c. Main and subsidiary transactions use positive goods only.

---

[1] Some propositions that result from these substitutions, along with additional detail of illustration, appear in Kuhn (1963, chap. 21). Since they merely spell out what happens when the various assumptions of Section 3.2 are applied to the organizational transaction—which the reader should have no difficulty doing for himself—they are not included here.

d. At least one party's receipt of outputs depends on the performance of the organization as a unit rather than on a prearranged input-withdrawal ratio. All members could occupy this position, but the model assumes that they do not.

e. The conditions that give rise to the subsidiary transactions dealt with in assumption **b** are not known in advance. Decisions regarding them will be made as the occasions arise and without a requirement of unanimous consent of members or even of those directly affected. When added to assumption **b** this condition means that a member's agreement to carry out instructions is not contingent on his necessary agreement to each instruction. As will be seen in connection with legitimacy, a requirement that the instruction fall within an agreed category to be accepted is not the same as a requirement that the instruction itself be agreed to if it clearly falls within that category. This condition clarifies the earlier indication that the formal organization requires a single decision regarding each act of effectuation whether or not all parties involved agree to it. By the same token the member agrees in advance in the major bargain to accept certain subsequent decisions whether or not he agrees with them when they occur. Whereas the major bargain involves the mutual contingency of decisions found in nonorganizational transactions, the subsidiary ones do not. Continuance of the major transaction, however, is contingent on completion of the subsidiary ones.

f. The organization's work is sufficiently complex to require subdivision of its tasks (1) horizontally so that there are at least two different tasks to be performed and (2) vertically so that there is a higher-level act of coordinating the lower-level tasks. The multiple levels constitute the hierarchical arrangement that characterizes complex organization.

g. A significant amount of initial investment is required for the organization's transformations to go forward; and a significant amount of fixed cost is needed if that investment is not to deteriorate.

We have identified the organization as a system whose behaviors as a unit can be analyzed with respect to its detector, selector, and effector processes. But whereas we have ignored the internal processes of the individual human's maintenance system (in contrast to his control system) as primarily biological, many in the organization are social and thus command the attention of social science. We have similarly ignored the content of the individual's inputs and outputs and have attended only to the social interactions by which they were transferred. This approach can be followed for the organization, but with an important exception. In all his communications and transactions the individual human acts as a unit—even if with vacillations, contradictions, and unconscious components. By contrast, external interactions of an organization can be carried on simultaneously by different levels of the system—as if, by analogy, people's livers could communicate or transact independently with other persons or their livers at the same time the whole persons were interacting. Whereas livers and circulatory systems are unable to use semantic communications and value-based transactions (see Chapter One), the components of an organization are whole humans or suborganizations that can do these things.

Following this pattern the examination of organization attends separately to (a) the main, or whole, system level and the subsystem level and to (b) the intrasystem and the intersystem aspects at each level. Since we have already delineated the basic areas of social analysis in previous chapters, the following pages present less new analysis and propositions and more definitions and identification of applicable analysis.

4.12.1   *Structure*

4.12.10   *Relevant Definitions.* To talk about structure, we need the following definitions:

a. *Parties to the organization* (categories relevant to the analysis of formal organization) are as follows: *Sponsors* are the person(s) whose motives concerning the organization constitute the goals of the organization as a unit. "Constitute" means effectively, not merely in form or on paper. The *staff* are those persons who effectuate the transformations that constitute the organization's function —that is, who are instrumental to the achievement of the organization's goals. This definition includes line as well as staff persons as those terms are used elsewhere. *Factor suppliers* are those parties who provide the organization's inputs. Staff are factor suppliers insofar as they provide the human contribution. But since their status as members requires separate analysis, we will treat them as a category separate from other factor suppliers. *Recipients* are those parties who receive the organization's outputs. These are the parties to, for, at, or upon whom the activities of the organization are directed.

b. A *role* is the set of system states and actions of a subsystem, of an organization, including its interactions with other systems or nonsystem elements.

To amplify, those other systems or nonsystem elements may be in or out of the organization, at the same or different levels. If the subsystem is an individual the role consists of only those system states and interactions that are specified for his part in the organization. In present terminology, the traits of the subsystem are its detector, selector, and effector states and processes; its interactions are its communications, transactions, and possible organizational connections. A complete statement of role specifies the position and function in the larger organization—though the definition does not require that all details of a role be explicitly stated or even understood (see also 4.12.3c). As indicated (4.12.0b), the role to be occupied by an individual is specified in the main bargain of affiliation.

Rights and *responsibilities* are presumed to be incorporated within the role specification as defined below. What sociologists and organization theorists often call *rules* are also part of the role specification, particularly the portions dealing with the kinds and terms of transactions. Certain rules that are parts of all roles within the organization can be stated in a general specification, or code, without having to be repeated for each separate role.

Since every role and its occupants are part of the environment of every other role, no two roles can have exactly the same environment or the same set of relationships with other roles. As indicated in Chapter One, to look downward from any level of the organization to subsystems and their roles is reductionist analysis whereas to inquire into the role of any such subsystem in the larger organization is functionalist analysis.

c. A *role specification (description)* is a statement or description of the content of a role. This definition does not require the specification to be explicit or in words; it could consist solely of images in people's heads. (In any event, in this model words do not affect actions until they become images.) In the language of the personnel officer, if they are reasonably complete a job description plus the job specification constitute a role specification.

d. A *role occupant* is the person or other acting system that effectuates the behavior specified in the role. Roles themselves are pattern systems and hence do not interact. (In this connection see J. G. Miller, 1965, p. 205.) And although roles can be said to be consistent or inconsistent with one another, they cannot be said to be in conflict (see 2.50q and Chapter Seventeen). That is, "role conflict" is *within* a person when he faces the requirements of inconsistent roles. The concepts of role and role occupant parallel those of office and officeholder or job and jobholder. In formal organization the role occupant is selected by the sponsors or their agents and is assigned to the role as part of the major bargain. Although we will think of role occupants as individuals, a role can also be occupied by an organization, such as a division or department.

e. *Structure* is the pattern of an organization described in terms of its subsystems and their roles. Since every subsystem down to but not including the individual is also an organization, any such subsystem also has a structure. Organization charts, flow diagrams, job titles, job descriptions, and the like are practical devices to describe the organization's structure.

f. *Authority* is the ability to grant or withhold rewards or punishments (sanctions) for the performance or nonperformance of instructions.

g. *Discipline* is the exercise of authority with special reference to the application of specific acts of sanction to specific violations of instructions.

h. *Sanctions* are rewards or punishments used in the exercise of authority.

i. *Responsibility* is the position of a party on whom authority is exercised. That is, a person holds responsibility *to* his superior if sanctions can be applied to him *for* performance or nonperformance of instructions.

j. *Delegation of authority:* Among three successive levels of systems— *A, B,* and *C,* with *A* highest—the delegation of authority is a transaction between *A* and *B* in which *A* agrees to provide sanctions to *C* directly or through *B,* when and as requested by *B,* in return for which *B* will request such release only as *C* follows instructions from *B* designed to effectuate goals specified by *A.* Any limits on the amounts thus to be released by *A* are part of the *A-B* bargain and necessarily constrain the *B-C* bargain within the same limits. If *B* by contrast were to provide his own goods to *C* for the same purpose, he would be exercising authority over *C*—but it would be his own, not authority delegated from *A.* For convenience, instead of releasing sanctions to *C* at *B*'s request, *A* may allow or instruct *B* to release them directly to *C* at his own discretion. This arrangement does not violate the definition since *A* can stop the flow as part of his authority over *B.* Any sanctions from *B* to *C* that *A* cannot stop are not delegated.

k. *Delegation of responsibility:* From the definitions of authority, responsibility, and delegation of authority, it can be deduced that *C* is now responsible to *B* to do work for whose performance *B* is responsible to *A.* One might say there is also a transactional relation between *A* and *C* in which *A* agrees to follow *B*'s instructions to pay *C* if *C* follows *B*'s instructions to perform work. Except for possible side effects—such as to assure *C* that *A* will pay him or to clarify that *B* is *C*'s boss—such a third transaction is analytically unnecessary if the *A-B* and *B-C* transactions are clear. Some organization theorists argue that responsibility, unlike authority, cannot be delegated. I think the difference between that view and the present definition is wholly semantic. Even if *B* remains "responsible to" *A* for *C*'s performance, matters can still be so arranged that *C* is responsible to *B* for that performance. Within the definitions used here we then say that the responsibility has been delegated.

l. *Management* is the sponsors or any of their agents or subordinates that are given the responsibility to see that actions by staff are directed toward the goals of the sponsors.

m. *Superior and subordinate* are a pair of persons, *A* and *B*, who hold authority and responsibility, respectively, in an organization under the same transaction of authority.

n. *Legitimate authority* is a situation in which instructions and sanctions from *A* to *B* fall within the categories and terms specified in the transaction of affiliation. For certain purposes, particularly in connection with government, *decision* may be substituted for instruction. This definition does not limit how categories themselves are defined. They could be defined by content (any work that can be run on a turret lathe), by quantity (up to ten round trips per day), by organizational structure or process that decides the instruction (any instruction acceptable to the board under Article ·X of the charter), by special criteria (any assignment that does not jeopardize parole), and the like.

o. *Illegitimate authority* is a situation in which instructions or sanctions from *A* to *B* fall outside the categories or terms specified in the major bargain. In this connection the real problem is not so much whether an instruction is legitimate as whether it is so viewed. For *B* to view an instruction as legitimate is to acknowledge that it falls within the categories he has agreed to accept; to view it as illegitimate is to deny that it falls within them.

p. *Effective authority* is a situation in which instructions are in fact followed. Such a result presupposes an overlap of EP's between supervisor and supervised, and adequate feedback information to the supervisor so that he can know whether his instructions were in fact followed. Unless indicated to the contrary, authority means effective authority.

Many of these terms are associated with employer-employee relationships and are adopted here because of their widespread currency. The terms and the analysis, however, are applied to any organization that meets the definition. If a bride promises literally to "love, honor, and obey" in the transaction consummated in the wedding ceremony and if the husband later employs sanctions for her obedience or lack of it, the two stand in the relation of superior and subordinate under a transaction of authority. The same can be said for a ski club, PTA, labor union, or trade association if a member can be denied privileges for failing to follow decisions of the executive committee.

4.12.2   *Authority.* It follows from the definitions that for B successfully to refuse an order previously construed as legitimate, or for A successfully to get B to accept an order previously construed as illegitimate, is to change the major bargain. By corollary, to attempt to do either is to try to change the major bargain. **4.12.21** Other things equal, it is advantageous to A to expand and for B to contract the scope of legitimate authority. **4.12.22** For organization members who are free to leave, to accept an order is in effect to acknowledge its legitimacy. If a member's bargaining power is weak in the major bargain, however, he may nevertheless perform what he considers an illegitimate instruction because his position is too precarious for him to complain. He will presumably force cessation of such instructions as soon as he has the power; meanwhile his situation is one of tension. **4.12.23** If the management makes a major change in the organization's structure or function, the number and type of roles and occupants will necessarily change—probably dramatically. Since role content and occupancy are part of the major bargain, many major bargains will have to be changed. Until the management successfully renegotiates those bargains with each affected member, some of its instructions will be illegitimate under prior bargains.

There are many possible outcomes and several ways of handling this situation. To illustrate, first, the new instructions may be more acceptable to members than the old ones or at least as acceptable. They will then probably be accepted and hence legitimized under 4.12.22. In that case the major bargain will have been altered. Subordinates may nevertheless resent a unilateral change in a transaction that was originally bilateral. If so, subsequent transactions may include a hostile ingredient on their side with consequences as noted in 3.26.10. Second, legitimacy of the new instructions may be negotiated separately with each member. The model does not predict the outcome, but bargaining power analysis is applicable. If EP's do not overlap in a particular case the member will leave. Third, if the new instructions are less satisfactory than the old, the members might form a coalition to improve their bargaining power. The outcome is again unpredictable, but the analysis of 3.27.5 applies. And fourth, management may implicitly renegotiate major bargains in advance of the change by describing its intended changes, soliciting and responding to suggestions or complaints, and only then putting the changes into effect. If subordinates at all levels freely express themselves to superiors, such a procedure can constitute a real negotiation and legitimize the new instructions. If subordinates are unwilling or unable to state their objections, the major bargain has not been renegotiated and the new instructions are not legitimate. If the members show mixed reactions, those who do not like the new arrangement may then leave, remain and feel hostile, remain and feel unhostile because they see no better alternative, or form a coalition—among other possible responses.

4.12.3   *Pure Static Model*

Although the organization theorist will need far more detail than is provided in this model or its partial elaboration later, to delineate the basics of the organizational problem we use a simplified pure static model with the following traits:

a. All detector and selector functions for the organization as a unit are performed by the sponsor. We assume for this purpose that the sponsor is either an individual or a group that agrees unanimously on all questions. Under these circumstances the sponsors and "top management" coincide. In relaxed models the sponsors or their agents remain the top management.

b. All effector functions are performed by the staff, in strict accordance with the instructions from sponsors.

c. The sponsors totally determine the organizational structure. They specify what subsystems are to exist and the DSE and interactional role traits of each as follows. *Detector:* The sponsors determine what information each subsystem is to possess or acquire and they provide information if necessary. *Selector:* The sponsors set the goals of the subsystem as a unit but not those of the individual who occupies a role. We assume that the individual has his own goals but is motivated by the transaction of membership and the subsequent applications of authority to provide the contributions wanted by the sponsors. *Effector:* The sponsors specify what transformational techniques the subsystem is to use and they select components, including individual persons, able to perform them. *Communications:* The sponsors specify the nature and quantity of information to flow from main to subsystem and the reverse, as well as among subsystems. The communications are assumed to be adequate for the sponsor to tell whether a subsystem is performing its role satisfactorily. *Transactions:* Among all subsystems and between main and subsystems, the sponsors specify all inputs to and outputs from each subsystem and the terms on which they are transferred. (Transactions with staff members are discussed below.)

d. All communications and transactions with the environment of the organization are performed by the sponsors or by agents who accurately reflect the relevant detector and selector states of the sponsors.

e. The relevant environmental parameters, including the human traits of staff, remain constant long enough for the organization to reach and retain a steady state.

f. Transactions with recipients over outputs depend on the goals of the organization (specified below).

g. Any positive or negative differences between the values of inputs and of outputs accrue to the sponsors.

h. Outputs are a function of inputs as modified by a technical co-efficient.

i. Values of inputs and outputs to the organization are determined by the transactions that acquire and dispose of them.

4.12.31   Under these circumstances a steady-state equilibrium is presumably possible for the organization, though not necessarily achievable, with the following characteristics: (a) The inputs and outputs of the whole organization flow at a constant rate of quantity and quality; the same is true for each subsystem. (b) Instructions from sponsors to staff consist solely of the latter's role descriptions, which are fully learned and accepted in the induction process (see 4.12.510). That is, the tasks of each subsystem are continuous or recurrent, in which case an initial set of instructions in the role description deals with all circumstances that arise. (c) The techniques of each subsystem are arranged to provide the most efficient transformations overall for this equilibrium level of output. Given different processes and techniques in each subsystem, however, with presumably differing curves of economy to scale there is no reason to expect that more than one subsystem could operate at its optimally efficient level. And (d) unless the output of the organization consists of information, no communications are necessary between subsystems, either directly with one another or indirectly through the sponsors. Movement of suboutputs from one subsystem to the next itself informs the latter that its inputs are ready and available.

4.12.32   If we change the static model to allow for moderate variations in quantity, quality, or value of inputs, outputs, or technique, but all within limits of competence of all subsystems, the organization could operate homeostatically much as in 4.12.31. Role descriptions could include procedures for accommodating such quantitative or qualitative changes. Communications to subsystems could be confined to informing them of changes in inputs or outputs with whatever lead time their technical processes and storage facilities require. Communications between subsystems would still not be needed if the sponsors informed each subsystem of the expected change to it or if each subsystem's role included means of calculating the effect on itself of any change in inputs and outputs of the whole system. Regarding efficiency, there would be no assurance that even a single subsystem could operate at its optimal level.

4.12.33   By observing the unreality of the assumptions of 4.12.31 we can see that the most obvious functions of management regarding the strictly internal aspects of the real organization are four. First, to design the initial structure of the organization. Second, to modify that design as needed to accommodate variations not encompassed within homeostatic processes specified in subsystem roles. Third, to replace or relocate component subsystems, human or other, when they cease to function adequately—that is, to change roles, role occupants, or both. And fourth, to maintain or modify the detector and selector states of staff through communications and transactions so that they will be able and willing to perform the subsystem goals assigned them.

Given that the initial design is a change compared to its absence and that maintenance is an attempt to prevent change, all these activities deal with some aspect of change. In short, given stability or moderate fluctuations the organization can "run itself." The task of top management is thus seen as solely to deal with changes of types or magnitudes that the homeostatic mechanisms of its subsystems cannot accommodate. The initial design of the organization creates a mechanism capable of homeostatic performance within certain limits and may or may not specify the level of a desired equilibrium. We may note in this connection that the bulk of economic theory deals with the homeostatic operations of the firm, not with those described here as managerial operations. If the model (4.12.31) inadvertently omits assumptions necessary to support these conclusions about the functions of top management, those assumptions can presumably be added when identified. The point of this model rises of course from the system view.

We begin with the notion of a system that is self-regulating in the face of certain environmental variations. Then we deal separately with responses to changes whose nature and magnitude fall beyond the capacities of the equilibrating mechanisms and hence may require modification of the system structure. In terms to be clarified in Chapter Sixteen, the task of top management may be described as to prevent the decay of the organization in the face of entropic and environmental change or to generate increasing order within it to improve its functioning in some respect.

The model makes no assumptions about centralization or decentralization of controls or whether a specific individual will make system-equilibrating or system-modifying decisions or both. The model does assume, however, that a hierarchy of organizational structure necessitates a hierarchy of decisions. In practice there is a continuum along this line, not a sharp division. An important question in practice, for example, concerns the degree to which a subsystem of the organization should design its own internal structure in light of the goals and constraints specified by the system level just above it. That problem must be dealt with by specialists elsewhere.

For purposes of social analysis we construe physical equipment, materials, and processes as falling in the environment of the organization's behavioral system in the same way and for the same reasons that we construed the biological system to fall outside the human's behavioral system. The purpose of such treatment at the individual level, it will be recalled, was to keep A's brushing his teeth on the same logical basis as his brushing his shoes; at the transactional level the purpose was to keep the physician's repairing A's broken bone on the same logical basis as the mechanic's repairing A's broken bicycle. The physical aspects of the organization not only constrain many of its decisions but also provide their subject matter. But for social analysis it is no more necessary to attend to these matters than to examine what the individual eats or his rate of digestion.

4.12.4   Group Decision Processes

The pure static model of complex formal organization (4.12.3) assumes that all decisions for the organization are made by the sponsors—either one person or a group that agrees unanimously. In relaxing that assumption we move to a distinctive characteristic of organization and an aspect of considerable interest to many political scientists. This is the question of how the sponsors or their agents who have conflicting overall or instrumental goals for the organization come to a decision about what the organization is to do and how. This situation can be contrasted to communication, which requires no decision in this sense, and to transaction, which requires mutually contingent but different decisions. It is the truly joint effort of formal organization which requires that two or more parties be bound by the same decision and which provides a crucial distinction between organization and communicational and transactional interactions.

4.12.41   *Model and Assumptions.* To analyze the group decision process we make the following assumptions:

a. A single decision must be reached on a problem on behalf of the organization for reasons delineated in the introduction to this chapter. The logic is the same for every subsystem, but for convenience we will speak of a decision by sponsors.

b. Assumption 4.12.3a (calling for unanimity) is relaxed, and the sponsors disagree at least initially on some matter.

c. All the rationale about the logic of decision-making discussed in Chapter Six applies to the decisions of an organization. That discussion need not be repeated here, however, where our concern is solely with the new problems added by initial disagreement among persons who must reach a single decision.

d. The problem to be solved is discussed among all members of the sponsor group. Moreover the decision is stated clearly enough for each sponsor to know whether his wishes have been accepted, rejected, or adopted in modified form. For example, the model excludes decisions by obfuscation or simple unchallenged action.

4.12.411   Agreement might be reached by communication alone—either through moral or intellectual influence by some members of the group or through successive communications among the group (see 3.18.4, 3.17.0c, and discussion of scale models in Chapter Fifteen) until they all perceive and evaluate the situation substantially the same. If this technique is completely successful the decision will be unanimous.

4.12.412 Besides agreement by communication the group may achieve agreement by transaction: one person or subgroup gives up something on one aspect of a decision if others reciprocate on another aspect. Although this arrangement incorporates the two mutually contingent decisions of transaction as seen at the level of the formerly disagreeing sponsors, it constitutes a single decision with respect to the level of organization. That is, to the organization only the resulting terms of this transaction are relevant, not the costs and benefits as seen by each sponsor. Since these transactions involve concessions on a decisional position we assume that they involve only goods. If bads are used it is their coalitional rather than their transactional aspect that is of interest in this context; therefore we discuss bads only in connection with the coalition.

4.12.413   An agreement reached by communication will be perceived, *cet. par.,* as preferable to one reached by transaction. In the former case everyone gets everything he wants whereas in the latter each must give up part of what he wants to get some other part. We may nevertheless note that if a person modifies his stated position after a discussion, it may be difficult in practice to know whether he did so for the communicational reason that he changed his mind, the transactional reason that he made concessions toward a position more acceptable to others, or some combination. The conceptual distinction is nevertheless clear, and the two reasons produce different consequences. The model does not presume that interactions proceed in the order listed here or that the parties involved understand what they are doing. The terms are those of the analyst-observer, not the participants.

4.12.414   Nothing about communication or transaction guarantees agreement, however, and the next question is how to get a single decision if they have failed. Presumably agreement must be reached by interaction. But this model posits only two types of interactions, strictly speaking, and both have failed. By elimination we therefore conclude that agreement is not possible. But to say that agreement is not possible is not to say that a decision is not. Any decision in such a situation must incorporate the desires of some and deny those of others. What, however, determines which persons will get what they want and which will not?

In its simplest terms the answer seems to be that some will do what they want and others will be helpless to do anything about it. To the next question—Who will be able to do what they want?—the answer seems to be: Those who have the power to. In every situation certain kinds of power are relevant, and the subgroup with the most of such power carries the decision. The power may be that of money, votes, muscle, bullets, applause, loudness of voice, or ability to stay awake longest. We need not ask why one kind is relevant to a specific situation and another is not.

4.12.42   For the analysis of this next step we add two definitions. A *dominant coalition* is the subgroup which holds the greatest power relevant to that group or its decision rules and which as a unit supports a single position on a decision. All the logic of coalition control is the same for a single individual who has more money, muscle, and so forth than any opposing coalition. Hence for this purpose we include the terminologically bizarre case of a dominant coalition of one person. The logic is the same for any type of party, but for convenience we speak of persons. A *decision rule* is some criterion for determining whose desires will become the group decision in the event of disagreement. Other aspects of decision rules are discussed in Chapter Thirteen.

Under these circumstances, if acceptance of the decision rule is made part of the bargain of membership and if each member prefers to accept a particular decision he disagrees with rather than leave the organization, then the decision that results from applying the rule will prevail. If the criterion of the decision rule is the largest number of a certain unit—votes, shares, dollars—then the coalition possessing the most units wins the decision. If the criterion of the decision rule is unrelated to the wishes of the members (suppose the decision will be determined by dice, senior member, color of the sunset, outside arbiter) no general conclusions can of course be made.

4.12.43   In the absence of a decision rule, *cet. par.*, a coalition that is dominant in the power relevant to the situation will win. Included in the *cet. par.* assumption would certainly be each coalition's awareness of its power, willingness to use it, knowledge of bargaining strategies, and determination to win. Suppose the five most popular actors in a local theater group refuse to continue under their present director and the remaining fifteen feel unable to command an audience without them. Then the fifteen will have to bow to the superior power of the five and accept a new director. Or suppose the local union's three experienced negotiators insist that a particular seniority clause is impossible to justify in negotiations. If the remaining six inexperienced bargainers on the team disagree but feel unable to handle the negotiations themselves, they will have to accept the choice of the three. In other cases where money is the critical variable, the coalition with the most money will control the decision. Where force is the criterion, the coalition that can defeat any other is dominant and will win the decision. If one person can beat up or out-duel any combination that tries to best him, he is the dominant "coalition." If force includes physical blockage or is otherwise great enough to be successfully coercive, it automatically becomes the relevant type of power. The amount of power that is dominant may also be thought of as "pivotal" (Riker, 1962).

The power that is relevant to a situation may be a marginal rather than a total amount. If a group strongly wants something that costs $100,000 and already possesses $90,000, the coalition that controls the only available source of the remaining $10,000 may be dominant for that situation. Similar power often resides in those who control a crucial "swing vote" in an election. As a bargaining power phenomenon, the power of the uncommitted lies in their ability to withhold (3.21m); the committed no longer have that power—though their plain and aggregate power may be greater.

4.12.44   Even in the presence of a decision rule, if a coalition is dominant in the power that is relevant to the group, if that coalition is willing to leave rather than accede to a decision they do not like, if their departure would effectually destroy the organization, and if the remaining members value the organization more than the rule, then the dominant coalition can get the rules changed or suspended so that their wishes will prevail. But as soon as the possibility of terminating the major bargain is opened, the bargaining power factors of 3.28.52 and .521 come into play and the question shifts from that of the decision process within the group to that of the transaction process between the group and its present and potential members—and hence to bargaining power factors. The problem of bargaining power in the context of a transaction between a group and this coalition of potential members should not be confused with the problem of making intragroup decisions after all possible transactional settlements have failed. Although the same persons are involved, the same analysis is not.

4.12.441   Without a decision rule virtually all the power relationships discussed in the chapters on transactions and aggregate power could come into play in a decision. A decision rule vastly narrows the range of possibilities and hence the range of possible outcomes. That is, an accepted decision rule greatly reduces uncertainty over methods and results; and by the same token it reduces the amount of time and effort needed to reach a decision. It may therefore be anticipated that an organization will normally adopt some decision rule, and unless specified to the contrary we will so assume. The advantages of simple uniformity may be so great as to make the content of the rule relatively unimportant—and the same may sometimes be said for other rules of the organization.

4.12.45   Since by definition the dominant coalition takes a single position, it faces the same problem of making an internal decision as does the organization. Hence we assume that the analysis of 4.12.4 thus far is also applicable within the dominant coalition—namely, that if the dominant coalition contains divergent interests, disagreements can be settled by communication, transaction, and/or dominant subcoalition. Like organization, the dominant coalition can be hierarchically structured.   **4.12.46** The question of when or whether it is advantageous for persons on the losing side to shift to a different subcoalition in the hope of making it dominant, although of obvious importance, is not treated in this volume. In this connection Riker (1962) and game theory are relevant.

4.12.47   Within any group (whole group or dominant coalition of it) it is uncertain whether decision by dominant coalition will produce more or less total satisfaction than settlement by transaction. Whereas the contrast between transaction and communication can be made within the preference function of each individual (4.12.413), in decisions by a dominant coalition the costs are borne by the losers and the benefits accrue to the winners. Since we rule out interpersonal comparisons of utility we cannot know whether the latter gain exceeds the former loss. It seems almost certain that decision by dominant coalition leaves more hostility than decision by transaction. If so it brings a decline of power of both winners and losers in subsequent transactions between them and a decline in the bargaining power of the winners (3.26.11–.12). Hence it may be presumed that, *cet. par.*, decision by transaction will be preferred to that by coalition.

4.12.48   Since we have already concluded that agreement by communication is preferred to that by transaction (4.12.413), we now add that, *cet. par.*, a group will prefer settlement techniques in the order of communication, transaction, and dominant coalition and will seek to achieve agreement by the higher-priority method before resorting to a lower priority. We may note here that March and Simon (1958, pp. 129–31) identify several processes for resolving conflict within a group —that is, processes of reaching a single decision among a group with disparate goals. Robinson and Majak (1967) refer to them as *"processes of decision-making."* These processes are problem-solving, persuasion, bargaining, and "politics." As I interpret it, the first process rearranges matters so that the conflict disappears—obviously a good solution if it can be managed (see 3.21i). The remaining three processes are ways of reaching a decision and seem to correspond with the techniques of communication, transaction, and dominant coalition, respectively. There seems to be no doubt about the correspondence of persuasion with communication and of bargaining with transaction. And when March and Simon (1958, p. 130) refer explicitly to "expand[ing] the relevant parties to include potential allies," one hardly hesitates to call it coalitional. Although I am not now prepared to provide the intermediate reasoning here, I do suspect that the macro-level parallels of these three methods would show the same order of preference in consensus by culture, market, and government decision.

Etzioni suggests three similar techniques for getting people to comply (1961, pp. 4–6; 1968, p. 371). His first technique is the normative; we will equate this with the culturally determined, which is communicational (3.17). His second is the utilitarian or remunerative, which involves payoffs and seems to be unequivocally transactional. His third technique, the coercive, does not directly parallel the third technique suggested here since coercion is also transactional in present language. But it certainly is similar in spirit since it invokes sheer dominance in a particular form of power, and I disagree with it only by suggesting a wider variety of types of relevant dominant power. All three models put the three techniques in the same order of acceptability. A leader of a group will prefer the same priority since an item high on the list will leave the members with a greater EP for membership than will a technique low on the list—and will thus enhance the leader's power and bargaining power.

4.12.49   *Group Preference.* The question arises whether a group decision can also be properly described as a group preference—a question raised (among others) by Arrow (1951), Downs (1957), and Rothenberg (1961). Regarding decision by communication, as here defined, the two are the same since the decision is by unanimous agreement.

Decision by transaction is more problematic. In a sense it incorporates and reflects the preferences of all participants. Yet it is difficult to say in what sense it represents a group preference beyond describing the process. This approach is formalized by Rothenberg (1961), who says in effect that the group preference *is* that which emerges from the group decision process. When the transactional decision incorporates only a single tradeoff between two well-defined positions it might be said to maximize some preference among the group. But if there are numerous positions on numerous aspects of the dispute and if each tradeoff is a compromise in several directions and at varying points in the overlaps, it is apparently impossible to say in what way the outcome reflects the preference functions of the individuals—except to say that all, with varying degrees of misgiving and reluctance, have "accepted" it. An Edgeworth box projected in multiple goods and multiple dimensions is probably more useful for this problem than for most other transactions, but it still does not answer the question (see Boulding, 1952). ∪

If we were to say that the status of group preference is somewhat uncertain in decision by transaction, we might say that it is unambiguously uncertain in decision by coalition. Let us look first at the problem of coalition itself by assuming that the dominant coalition wins straight off without first compromising its position through transaction. In this case the decision does not reflect the losers' preferences at all. Furthermore it does not give any clue to the intensity of their loss, which could range from mildly inconvenient to disastrous, or to their relative numbers, which could range from most of the group to only a tiny fraction. Whereas to describe the transactional process gives *some* notion of the sense in which it represents a group preference, to describe the coalition process suggests that there is *no* discernible sense in which it represents the preferences of the group. This statement about the pure coalition is strengthened if the dominant coalition's position is based on complex internal transactions and subcoalitions. It is probably weakened, but by no means invalidated, by a clear decision rule such as majority vote. The reasoning applies to a coalition defined in advance and stable over time, like a clique or cabal whose only problem is to settle its internal differences. It also applies significantly to a group like a political party, which is more concerned *that* it wins than about *what* it wins.

Three points are important in closing this discussion. First, there *is* a problem of "social welfare" associated with any technique of group decision, and the purpose of this summary is more to identify than to deal with it. Second, the group decision process is tied to the main framework by being identified as communication, transaction, aggregate power dominance, or some combination. And third, when the sponsors have divergent objectives the organization can be said to have a goal only in the formal sense that a decision has been reached about it.

4.12.5   *Internal Coordinating Processes*

4.12.51   *Cohesion, Attraction, Integration, Induction.* A crucial question of social organization is: What holds a group together? In an overall sense, authority is the main answer for formal organization. This section adds induction. But before discussing it we discuss cohesion, attraction, and integration—terms widely found in sociology though not used formally here.

Some important forces of this sort have already been discussed. We have seen, for example, why people generally prefer communications with others who hold similar concepts and values or who belong to the same culture or subculture (3.17.81 and Section 3.41). The expected gains from transactions, which means an overlap of EP's, also draws people together (3.22.6), and the anticipation of future transactions may hold them in contact (3.24.4). This anticipation also makes for terms more favorable to one of the parties and hence holds him in the relationship. Broadly, all transactions in goods, or expectations of them, are potentially cohesive, and the amount of the overlap of EP's is a measure of the amount of cohesiveness.

Liking induces generosity in transactions, and generosity induces liking—a positive-feedback relation (3.28.41 and 3.44.7). Hostility weakens the power of both parties to a potential transaction, and A's hostility weakens B's bargaining power as well (3.26.11–.12). Hence it is advantageous to avoid generating hostility in partners to future transactions. Gifts generate reciprocity and hence tie the receiver to the giver (3.36.64)—"You help me and I'll help you." Such gifts of "helping" are also an investment that ties giver to receiver (3.44.6).

We have also noted the great importance to transactions of confidence that a party to a transaction will fulfill his part of it (3.24.31). Such trust is therefore another key ingredient of cohesion. To the extent that accuracy of communication depends on the integrity of its source rather than on certain technical aspects of the communication process (as in 3.14), its effect on an interaction is also transactional (3.42.6). We have noted, too, that much deliberate lying consists of tactical misrepresentations in actual or anticipated transactions (3.23.3). Furthermore, because completed transactions themselves constitute communicational cues about the EP's of the parties (3.44.1), persons who have engaged in continuous and repeated relationships will have revealed so much about themselves that it may be difficult to misrepresent their positions successfully. Here trust based on any other grounds will be augmented by independent information of the parties about one another. In obvious contrast to the preceding observations, transactions that involve hostility or bads are disintegrative. Competition among B's is generally disintegrative in its direct effects on the B's—even if the benefits to A's and possibly to the B's indirectly (as by keeping them on their toes) makes it socially useful.

Cohesion is thus not seen here as a basic or unitary phenomenon in the sense implied by much social analysis. It is not a basic conceptual unit—like communication, transaction, and organization—but rather a mixture or series of special cases of many possible communicational and transactional patterns. It lies in the B-level superstructure of analytic concepts, not in the elements, and might conceivably be made into a C-level discipline of its own. Cohesiveness is a value-linked relationship through transactions or the value aspects of communications. To the rather large extent that cohesion is transactional, the degree of cohesion is a direct function of the overlap of EP's.

Organization is another element in cohesion. Its force is illustrated in the model of simple organization (4.11.0) and resides in the degree to which two or more parties are more productive together than apart. If changes in technology increase the productivity of the group relative to that of the individual, cohesiveness rises proportionately. Even without organization, if changing technology raises the productivity of individuals who specialize, or even if the mere fact of specialization raises productivity, the interpersonal power of all also rises. And so do the overlap of EP's and the degree of cohesion. Increased organizational complexity does not change the fundamentals. One aspect of cohesion is the amount of overlap in the transaction of affiliation. Another aspect, the ability of an organization to get members to work toward *its* goals rather than their own, is a function of the overlap in the ongoing transactions that effectuate authority. Increased complexity does increase the amount of specialization within an organization—and hence the dependence of all on the performance of each and hence the amount of cohesion.

Attraction is a related term that is not used formally here since it too is seen as a nonbasic as well as a relatively amorphous concept. In important respects attraction is not an interaction but a set of system states regarding interaction, actual or potential. When attraction is a factor in actual or anticipated transactions by A it lengthens his EP, as does friendship or liking for B or a strong desire for the Y he processes—the consequences of which factors have already been delineated. Similarly, if we view *integration* literally as the "oneness" of a group, in present language this means the degree to which the group constitutes an organization, formal or informal. Like cohesion and attraction, integration is a particular configuration of communication, transaction, and organization; it is not distinct. Although we do not use the term formally, we might define *integrative processes* as the total set of interactions between and contingent behaviors of the occupants of an organization's roles whose joint result is to fulfill the organization's goals.

4.12.510   By contrast *induction* is formally defined here as the communicational, transactional, and/or suborganizational relationships by which a new member of an organization, who is also necessarily the occupant of at least one role, is brought to acquire new system states and behaviors of his role—that is, he is transformed from an individual to a component subsystem. The term encompasses what is normally meant by indoctrination. For informal and semiformal organization (Chapters Fourteen and Fifteen) induction is the same as *socialization*. An *induction ceremony* is any collection of behaviors that provide a communication mutually recognized by the new role occupant and those who presumably need to know about it that a particular person has been installed in a particular role. The ceremony normally iterates or revises some important traits of the role. Moreover it communicates, and consummates, or both, the transaction of affiliation. A change of role, including exit from an organization by retirement or death, may be communicated by a role-change ceremony, usually designated more elegantly as a rite of passage. Such a ceremony may also communicate the addition or changed status of a role or subsystem itself rather than that of a person.   **4.12.511** If an induction ceremony is rationally related to its defined function, its magnitude—as measured by duration, money cost, number and organizational level of persons attending—is presumably proportional to the magnitude of change in role, measured by activities involved and numbers of persons affected. **4.12.512** As soon as a person is inducted into an organization and occupies a role he is tied into the network of communications and transactions which define that role. Furthermore, by the terms of the major bargain both types of flows to him are contingent on his fulfillment of the role. The same, of course, is simultaneously true for all other roles and their occupants.

In the language of much social science and psychology, *expectations* about the behavior of others are crucial in tying a society together. Certainly that view is fully accepted here, and it is only to keep many names from obscuring concept identity that we refer to such expectations of role occupants as their detector states. If the expectations have valences they have selector aspects as well—though analytically it seems wise to distinguish sharply between *what* one expects and whether he *likes* what he expects.

4.12.513   The definition of a person's role is determined and agreed in the major bargain (4.12.0b). Hence a change in role, as by transfer or promotion, changes the major bargain. Barring other factors, bargaining over the role involves the power relation of the major bargain, not subsidiary ones, although the content of those power factors will presumably have changed developmentally since the original bargain. By corollary the major bargain could itself include agreement about procedures for changing a person's role, in which case a role could be changed without changing the major bargain.

4.12.52   *Staff Modifications of Structure.* Let us now relax the assumption that the role is prescribed fully in the major bargain (4.12.0b) and that induction brings the new member into full conformity with it (4.12.3c). Instead we substitute that since no two persons are precisely alike and hence cannot fill a given role identically, each individual to some extent molds the role he occupies. Since role descriptions are by definition mutually interdependent, if one role is modified by its occupant all others with which it interacts must be altered in consequence and the amount of such alteration varies with the amount of interaction. To the extent that such mutual role modification occurs, the organization's structure is determined by the behavior of its staff, not by its sponsors. To a similar extent subsystem interactions within the organization must be construed as interactions of specific persons, not of roles.

4.12.53   To relax the model of authority somewhat let us now assume that the performance of instructions has more dimensions than simply "performed or not performed" and that the performance can also be large or small, good or poor, along multiple dimensions and to varying degrees. For present purposes the simple dichotomy of good and poor will suffice. We will ignore the fact that in some cases it is not possible or feasible to measure goodness of performance or to know what is rewarding to subordinates. Although for convenience this section speaks of persons, the systems to which the generalizations apply could also be departments, divisions, or other units of the organization.

If performance is graded and larger rewards are made contingent on better performance, staff members will be motivated to good rather than poor performance. Furthermore, if the goodness of performance by a person in one role depends on performance of persons in other roles, the exercise of authority on the former will motivate him to change the nature or performance of the latter roles in ways that will improve his own performance. Assuming individual differences in communicational and transactional skills, in direction and intensity of motivations, and in the power attached to various roles, the occupants of some roles will modify other roles more than their own role is modified by them. To the extent that occupant A of one role can affect the performance of occupant B of another role not formally subordinate to A, B's EP for certain behaviors by A will provide A with power and bargaining power in dealing with B. B, of course, will presumably resist such moves by A and engage in countermoves to increase or at least protect his own power position. One way of doing this is to seek more autonomy—to restructure the organization so that one's own performance is less dependent on that of others. To strengthen a previous point (4.12.52), to the extent that such modification occurs the structure of the organization is determined by specific members of the staff (not merely by "the staff") rather than by the sponsors. Unless we are to assume faultless decisions by sponsors (which we do not) role modifications made by staff will not necessarily perform the organization's work less effectively than the structure developed by the sponsors.

Once we have opened the possibility of role change at the initiative of individual staff members, many propositions about the interactions of communications and transactions, including those about aggregate power and status, become potentially applicable. If, for example, we assume that the B's whose roles have been modified by A have perceived that fact, they will further perceive A as a person of relatively large power who will then by definition acquire enhanced status. Other members of the organization will then have an increased desire to deal with A (3.35.2). As with other types of aggregate power, this effect is self-reinforcing. If the sponsors see the role modifications brought on by A as contributing to their goals they will presumably view A as competent in structuring the organization and increase that aspect of his role—that is, promote him into managerial levels. If his role modifications seem to obstruct their goals his superiors will presumably attempt to stop his role modifications.

4.12.54   To summarize 4.12.52 and .53, we can state that the inducements used by the formal organization to elicit good performance motivate the staff to transform the organizational structure in ways that are not planned by the sponsors and may or may not be useful to it. **4.12.55** Except in a steady-state equilibrium (4.12.3) persons occupying subsystem roles will have to adapt to changing conditions, and the size of their rewards will depend on the goodness of that adaptation. But their ability to perform their own role will depend in part on how well others do theirs. Hence, entirely aside from purely personal goals and without changing the role structure, there will exist both the occasion and the motivation for bargaining over matters related solely to the work of the organization. The salesman's ability to get a new customer, for example, may depend on the production department's willingness to commit itself to a delivery date, which may hinge on the purchasing department's procurement of materials and equipment, which may rest on the accounting department's willingness to expedite approval of the budget, which may require that management let them postpone a new accounting system, and so on. If we also assume that the major bargain does not legitimize instructions to do the impossible and that the subordinate will often know his own subsystem's capacities at least as well as his superior does, the superior will sometimes have to bargain with his subordinates over the instructions they will accept.

For all these staff-initiated changes of a role, potentially the entire transactional analysis of Section 3.2 applies. The problem in the specific case is to assess the content and strength of the EP's. These transactions, too, constitute an important part of the organization's internal coordinating and integrating processes.

4.12.6   *Structure, Organization, Power*

Differential power is inherent in any hierarchical structure that involves delegation of authority. Let us begin at the bottom of a three-level structure: C has nothing to grant to anyone below him. B has the ability to grant rewards to C for the performance of his instructions. A holds the rewards for both B and C, which are necessarily greater than those for B or C alone. If there are multiple subordinates at each level the power differential increases proportionately. The overall differential in aggregate power between two persons is thus a positive function of both the number of levels between them and the number of persons at each level. For such differences most of the Section 3.3 analysis of aggregate power, status, and class applies. In case anyone doubts it, it is obviously nonsensical to suggest that there could be "equal power" in all positions of a complex organization or, by extension, that there could be a "classless society" that includes complex organizations.

In addition to differences in level, different subsystems at the same level of the organization necessarily occupy different roles. No overall generalizations about them are readily apparent, but on a case-by-case basis differences of the following sort will presumably be found. First, different amounts of the organization's resources are channeled to them for their exercise of authority and for their rewards. Second, they have differing bargaining power based on the main system's relative dependency on (EP for) their outputs. Third, they undergo different types and degrees of interactions—perhaps subsystem A receives more information about subsystem B than the reverse and therefore has certain tactical advantages in transactions between A and B; perhaps A is allowed to engage in input or output transactions outside the organization if satisfactory ones cannot be achieved within the organization while B is not; or perhaps deferential behavior is a prescribed part of B's role. And fourth, they differ in the number or level of other systems with which they interact.

For the reasons given in the preceding two paragraphs the occupants of different roles will hold different amounts of aggregate power. If they want to augment their power the flexibility with which they can mold their own roles and indirectly modify other roles adds to this difference for those willing and able to use it. So does the general self-reinforcing nature of aggregate power (3.32.4, .6, and .7). Since the aggregate power of an individual is distinctly limited whereas that of the organization is not (introduction to Chapter Ten) and since top organization positions that control rewards necessarily hold more power than do lower positions, the largest power in any society or segment resides, *cet. par.,* in the top positions of its largest organizations.

Among things not equal in the preceding statement, moral and intellectual influence seem to be the most conspicuous possible exceptions—as exemplified in the charismatic leader. Charisma may be described as the ability of a person—usually in the face of some deep, widespread, but unmet desire—to generate long EP's in many people for the thing he can allegedly provide. The EP's generally are long enough to produce gifts for him or his designees. Even so, the charismatic leader may not have much power without organization although he may have the ability to build a powerful organization quickly. (See the discussion in Chapter Eleven—about Berelson and Steiner—regarding bargaining power of the organization when the goals of the members and those of the organization coincide.) In a large society the charismatic leader will also require organization to communicate his views even if the organization is not his own. His status may depend on the influence of his ideas, which he may or may not be able to convert into personal power (3.33).

4.12.7   *Taxonomy and External Relations*

4.12.71   *Types of Organization.* Except for the distinction between simple and complex, the preceding section did not provide a system for classifying organizations according to their internal characteristics. This omission does not imply there is no point to drawing such distinctions but rather that they do not seem necessary for this initial examination of organization. For complex organization, at any rate, the preceding formal model, which assumes generally selfish internal transactions, seems widely usable.

By contrast this section deals with the external relations of the organization—relations with individuals other than its staff or with other formal organizations. Within the present model these relations consist of communications and transactions over inputs and outputs. They could also include organizational relations in which two or more formal organizations join to produce a single effect. But since any such action would itself constitute a still higher-level organization we need not give it separate attention. Although computers and related machines may do part of such operations, we assume that all significant communications and transactions are decided on by human beings, either sponsors or their agents.

As to external communications, an organization may communicate about a vast number of matters. Moreover different organizations may communicate about vastly different matters. Nevertheless, although these differences may have intense interest at advanced analytic levels, for introductory social analysis there is no apparent need to make such distinctions or to distinguish the communicational problems of organizations from those of individuals. Most communications for most organizations are with the suppliers of its inputs and the recipients of its outputs. Many of those communications, other than those dealing with the sheer mechanics of the relationship, are tactical and strategic efforts to change the EP's of those with whom the organization conducts transactions. The relevant propositions are the same whether the parties are individuals or organizations and hence need not be restated here. The organizational transactions, however, would presumably require more use of principal-agent relationships (3.27.2) than would transactions between individuals.

As to external transactions, we need not deal in detail with transactions over inputs. Most organizations large enough to qualify as complex and to have a staff acquire most of their inputs through transactions that are largely selfish on the part of factor suppliers. Organizations in which this is not the case are generally cooperative or service types with special traits that are dealt with below. The distinguishing features of organizations that seem most interesting for introductory social analysis are the output transactions. On the assumption that the staff are merely agents for the sponsors and that output transactions are therefore really sponsor-recipient relations even if executed by the staff, the output transactions can be viewed in two ways: according to the type of transaction concerning outputs and according to the allocation of costs and benefits between sponsors and recipients. Table 4 lists four types of organization and indicates the defining characteristics of each. Although this classification scheme is here applied explicitly to formal complex organizations only, the simple organization also fills the definition of cooperative—two parties together doing

Table 4

Four Types of Complex Formal Organization

Type	Costs	Benefits	Transaction with Recipients
Cooperative	Recipients	Recipients	None (production for selves)
Profit	Recipients	Sponsors	Selfish
Service	Sponsors	Recipients	Generous
Pressure	Sponsors	Sponsors	Third-party strategies

something that can be done better jointly than singly, with costs and benefits going to the same persons. But since the goal of the cooperative is that its members receive its outputs, we assign the costs and benefits to them in their role as recipients rather than as sponsors.

In the simple organization all participants are simultaneously sponsors, staff, and recipients. If a simple (cooperative) organization becomes complex, however, these roles become differentiated, and it is important to know in what capacity its members continue their relationship. To speak of pure cases, they may shift their focus solely to their welfare as staff, selling their output to outside recipients and acquiring their nonhuman inputs from outside factor suppliers. In that case the staff double as sponsors; in conventional terminology the organization has become a *producer cooperative*. Although its decisional problems may be complicated by conflict between staff and sponsor roles within the same group and even within the same individual, this is a profit organization within the present classification scheme. That is, if its goal is really the welfare of its staff its output transactions will be selfish and the benefits will be received by the sponsors (who happen to be the staff) and the costs will be paid by the recipients through selfish transactions. To say that the recipients pay the costs does not mean that they do not receive equivalent value. It merely means that in return for receiving the outputs they give certain values in return that cover the costs of operating the organization. In fact the essence of selfish transactions is that each party does receive a benefit at least equal in his preference scale to the cost of what he gives up (3.21.03).

If in moving from a simple to a complex organization the members focus on their interests as recipients—and particularly if they hire a staff to do the work—the organization becomes a *consumer cooperative* in both conventional and present terminology. Since this is the only type of organization referred to as a cooperative here we will omit the adjective.[2]

4.12.72   *Definitions of Organizational Types.* With this background we can now define the pure types of complex formal organizations:

a. A *cooperative organization* is one whose sponsors are the recipients of its outputs and whose sponsor's goal is their own welfare as recipients. As noted, we assign the costs and benefits to them in their role as recipients, not as sponsors, since it is only to receive its outputs that they sponsor the organization. The definition accepts some nonsponsors as recipients but no nonrecipients as sponsors.

b. A *profit organization* is one run in the interests of its sponsors, who except incidentally are not the same people as the recipients and whose outputs go to recipients on the basis of selfish transactions.

c. A *service organization* is one whose outputs go to recipients on the basis of generous transactions. In the pure case, which is the only one discussed here, the outputs are pure gifts whose costs are borne solely by the sponsors.

d. A *pressure organization* is an organization C whose outputs go directly or indirectly to recipient B for the purpose of improving the position of A in a transactional relation with B. The sponsor A or his agent bears the cost z of the pressure activity Z, which is the output of the organization (see 3.27.1). The pressure organization embodies the pressure transaction raised to organizational level, and all analysis of the former is applicable to the latter.

e. A *mixed organization:* An organization can, and most real organizations do, engage in actions that fill the definition of more than one type, and some may be related only loosely to their main purpose. In the chemist's language, these are mixtures, not compounds. Their components can be put together in many different ratios and arrangements and are still separately identifiable after being mixed. They do not take on a specific and limited relationship that gives

---

[2] The term *cooperative* is adopted because it closely fits conventional meanings of the term. Although the term as applied to this type of organization is compatible with the broader meaning of cooperation defined in 2.50p, the two meanings are not coextensive and should perhaps be considered independent usages. For example, all organization, all selfish transactions in goods, and some aspects of decisions (among other possible things) are cooperative in the present broad sense.

the combination distinctive new characteristics. Thus to the extent that an organization engages in a certain relationship, the analysis of the relationship follows that for the comparable analytic model.

4.12.73   For initial statements about the external relations of organizations we assume that factor suppliers provide inputs on the basis of selfish transactions and that sponsors desire to maximize attainment of whatever goal they have set.

In a cooperative the benefits and costs by definition accrue to the same persons. Hence they can be evaluated within the selector systems of the same persons and do not require interpersonal comparisons of utility. Whether the costs and benefits are objectively or subjectively measurable, in the long run the sponsors will not continue the organization's existence unless the recipients want its outputs enough to cover their cost. By corollary, if the organization produces multiple types of outputs whose costs and benefit are separable, it will continue to produce those outputs whose benefits equal or exceed their cost and will discontinue all others.

In a profit organization the benefits to sponsors consist of receipts from sales (output transactions); the costs to sponsors consist of payments for factors (input transactions). Both accrue to the same persons and hence are measurable in the same value system. In the long run the sponsors will not continue the organization unless the recipients want its outputs enough to cover its costs. Each recipient, of course, also compares the costs and benefits of the output to him within *his* value system. The output transactions will therefore continue to take place only if benefits equal or exceed costs within the value systems of both parties (see 3.21.03–.04). The previous corollary about multiple outputs applies also to the profit organization.

In a service organization the direct benefits of outputs are evaluated by the recipients and the costs by sponsors. That is, the main cost/benefit relationship involves the value systems of two different parties. Unless we accept the validity of interpersonal comparisons of utility (which we do not) it is not possible to know whether the direct benefit is greater or less than the cost. However, the sponsors decide whether to provide the outputs generously to recipients. The decision therefore rests in a comparison between the costs to the sponsors and the satisfaction to the sponsors from having the outputs go to the recipients. Evaluation of the organization's performance thus presupposes some technique by which the sponsors can measure the amount of benefit to the recipient—a reasonably subjective task of indeterminate accuracy.

In a pressure organization it is irrelevant to the sponsors whether or not the direct recipients of the output want it or would be willing to pay its costs if required to. (Though some forms of pressure are negatively valued, for example, others consist of information or gifts that recipients might be delighted to pay for.) The ultimate benefits, however, accrue to the sponsors as improved power positions. Hence the costs and eventual benefits can be appraised within the same value system.

A service organization might indirectly provide transactional benefits to the sponsors, in which case it would resemble the pressure organization. We classify the relation as service if the sponsors do not take such benefits into consideration in making their decision or if such benefit to them is no more than some average benefit to the rest of society.

4.12.8   *Locus of Control.* Given the assumption of selfish transactions by factor suppliers (4.12.71), the organization will not receive the inputs necessary for its operation unless it can provide sufficient benefits to the factor suppliers. In other words, if the organization cannot cover the costs of its factors, the factor suppliers will force its contraction or demise by denying it inputs.   **4.12.81** But recipients' refusal to accept outputs cannot similarly force dissolution of the organization. Staff and other resources can remain idle if commodity or service outputs are rejected by recipients while the organizational structure remains intact. If the outputs are commodities they can be dumped or stored while the organization continues to operate. Storage space, like other inputs, can be as large as the ability to pay for it.   **4.12.82** It therefore follows that ability to consummate the necessary input transactions, not output transactions, is the crucial feature of the complex organization's ability to continue.

4.12.83   Ultimate control of the organization thus lies with those who decide whether or not to pay for inputs. In the pressure organization these persons are the sponsors who themselves provide the necessary resources; in the cooperative they are the recipients; in the service organization they are the contributors of its inputs or of the wherewithal to pay for them. These contributors may or may not participate directly in making such decisions or check on those who do. But if they do not contribute there is no organization. Furthermore, unless they want the services they will not contribute. Hence they fill the definition of sponsors in that their motives determine the organization's goals. At this level of the model we need not deal with fraud in acquiring or using contributions.

4.12.84   The profit organization is more complicated. Sponsors pay the costs until the organization is under way; then recipients meet the costs. We first examine the situation on the assumptions that the initial period is past and the organization is profitable. From then on its continuance, insofar as it depends on sponsor action, is subject not to the size of the sponsor contributions but to the size of their withdrawals. For example, the organization can grow merely because its profits are not withdrawn. Whereas a sponsor on whom the organization depends for continuing contributions can *exercise* power as a condition of contributing, a sponsor must already *possess* power on some other basis as a condition of making withdrawals. Although the details in specific cases may be messy indeed, in general it seems far easier in a profit organization than in other types for the staff to take control from the original sponsors under conditions that make it difficult for the sponsors to get it back. Once such a transition occurs it is the staff's goals that become the criteria for organization decisions—in which case the staff, by definition, occupy the sponsor role. That is, to say within our definitions we say that the staff have *become* sponsors, not that they have taken control from the sponsors. We are also speaking of limited-purpose organizations whose outputs are solely goods, not of government which may have unlimited purposes and use bads.

4.12.85   There are nevertheless conditions under which sponsor contributions to the profit organization may be needed—for the initial period before the organization has become profitable, for periods of loss to prevent contraction, or for expansion greater than can be financed out of profits. Assuming selfish transactions and uncertainty about the organization's expected profit or loss, we can trace the conditions for making such contributions in four steps. First, a supplier will not provide a particular unit of input unless he is paid (or promised, with adequate stake) a specified amount for it. Second, a supplier will not commit himself to specified contributions in the future unless he is promised specified withdrawals (payments) in return for them. Examples include loans in return for committed future interest payments or labor for specified future wage payments. In particular, a supplier will not commit himself to a specific future contribution in return for a promise to pay him some unspecified sum whose amount is contingent on what the organization is willing and able to pay. Third, by reciprocal reasoning the organization will not commit itself to a specified payment to a contributor without the promise of a specified contribution in return. And fourth, a person will not accept a contingent promise by the organization to pay him "whatever happens to be available" at some future time (a *residual withdrawal*) unless he can condition his own future contributions (*residual contributions*) on his estimate of how they will affect the size of that residual. Since all specified payments are tied to specified contributions, by elimination all residual withdrawals are available for residual contributors.

When sponsor contributions are needed, especially during the initial formation of the organization, the size of the residual contribution determines whether inputs can be acquired—and hence whether the organization can exist. Being contingent, those who provide the residual contribution can refuse to put it into the organization unless they think it will have effects acceptable to them on their residual withdrawals. In short, when residual contributors are necessary to the organization they can control it.

For reasons indicated in 4.12.84, once the organization can do without additional residual contributions it is free of *additional* residual control. How much control the earlier residual contributors can continue to exercise by virtue of the terms on which they made their initial contributions cannot be stated without additional assumptions. In the typical large American corporation their control is insignificant, and the technically residual payments to stockholders often become as committed as those to bondholders and carry no more control.

4.12.9   CONCLUSION

In this framework organization is one of the three intersystem relationships of humans. Hence in this view organization theory is a central aspect of social science. In this respect it coincides with the views of Etzioni (1969, p. v), Blau (1964), Blau and Scott (1962), and March and Simon (1958)—among others—and commands the attention of many sociologists and political scientists. This chapter has not tried to present a capsule version of organization theory, which (among other things) would certainly have to include much from the later chapter on semiformal organization as well. Its purpose is rather to suggest that such theory includes two distinct types of analysis—the system concepts to deal with the organization viewed as a unit and the intersystem concepts of communication and transaction to deal with the interactions of its parts and its own interactions as a unit with other systems. For this volume it is pointless to spell out numberless details. Instead the chapter has for the most part merely defined certain terms necessary for organizational analysis and then identified the applicable parts of the analyses already developed. For the same reasons as elsewhere, selfish motivations were assumed for the initial models except where definitionally precluded, as in the service organization. Where relevant, other assumptions can be substituted with consequences already indicated in the chapters on transactions.

Although other typologies of organizations abound, I suspect that for *social* analysis the distinction into cooperative, profit, service, and pressure is more useful, for example, than such categories as religious, military, educational, professional, and business organizations. A typology based on internal structures and their performance characteristics—which here means numbers and types of roles—is of obvious importance. Nevertheless this topic seems inappropriate at the introductory level. Much of the contemporary "human" and "behavioral" approach to organization is incorporated implicitly in an innocent phrase: the entire communicational and transactional analysis potentially applies here.

It was hinted (4.11.6) that the transactional analysis of membership in the simple formal organization coincides with a central aspect of the marginal productivity theory of wages. This aspect of economics, like others, is thus not distinct from the present framework but a special case of it. Furthermore, as the social science disciplines are currently structured much of this chapter could reasonably be said to fall within the bailiwicks of sociology, political science, organization theory, or even anthropology. I can think of no more cogent case for restructuring the disciplines to make organization theory an official social science. The ambivalence of social scientists who have kept organization theory as only a stepchild in the noble families of sociology, anthropology, and political science—while considering it fully legitimate only within the somewhat renegade family of business administration—needs prompt reform.

Chapter 13

4.13   Formal Organization

Government

This chapter discusses government and, to a lesser extent, controlled economies—both formal organizations covering the entire society. Its function is not to explicate political science but to identify some concepts and basic rationale about government as they appear in the present framework. Although the political scientist is not likely to learn anything new about government from this chapter, the explicitness of certain definitions—of legitimacy, sovereignty, the function of the judiciary, and the overall functions of government—might prove useful if pursued by specialists. The definitions in this chapter allow the analysis of prior chapters to be applied directly to government, and I suspect that the political scientist may find more immediate interest in the earlier discussions of power, bargaining power, authority, and other aspects of organizations than in this discussion of government. In any event the purpose here as elsewhere is to provide a unified, parsimonious, and largely deductive conceptual set across a broad range of social science, not to provide new knowledge. And I do think that to categorize governments as mainly cooperative or profit organizations, with certain service and pressure adjuncts, provides a more explicit basis for analysis than do such customary categories as democracy, dictatorship, oligarchy, and monarchy.

This chapter is the first to deal with a specific and

empirically identifiable system, a government, in contrast to the general principles about systems, interactions of parties, and formal organizations introduced in earlier chapters. In doing so this chapter demonstrates the advantages of a unified social science (if it works properly) —for merely to identify government as a formal organization means that much about it has already been stated in connection with the general theory of formal organizations. More explicitly, in this framework a theory of government starts with the theory of formal organizations, shows how government compares with the model, and then traces the consequences of its deviations from the general model. The study of government is thus not a separate discipline but a special case of certain broad concepts and principles. This volume identifies those relationships; it is up to specialists to determine whether they are of any use. In dealing with a particular kind of organization this chapter thus moves across the spectrum of Figure 4 to level C, which represents specialized disciplines.

Before launching into specifics we should point out that all the varieties of political theory covered in Easton's volume (1966) are incorporated in this book. The decision-making approach is covered in discussion of its general theory in Chapter Six and within the context of organizations in Chapter Twelve. The individualistic approach concerns the way in which each individual responds to his own situation and motivations and yet produces an overall result by interacting with others in particular contexts. This approach is dealt with not only in the previous chapters on communications, transactions, and aggregate power in formal organization but also in the following chapters on informal and semiformal organization. Politics as a phenomenon of power has its basis in the chapters on transactions and aggregate power; government as an aspect of organizational structure and process is dealt with in this and the preceding chapter. The systems approach is, needless to say, also dealt with here. As the platform for a general theory this volume does not deal with descriptive or historical approaches to government or with such details as statistical studies of voting behavior.

4.13.1   MODEL AND ASSUMPTIONS: SPECIAL PROPERTIES OF GOVERNMENT

The model of the special nature of government organization consists of the following assumptions:

a. Government is a complex, formal organization and as such is subject to all the preceding analysis of such organizations. In addition it has the following special characteristics.

b. Government covers the whole society. This assumption does not specify the size of the society. Theoretically a government could comprise as few as three persons (two would not be "complex"), but we are not interested in such small units.

c. Government has potentially unlimited scope in determining the subject matter of its decisions and their effects. The reasons for this assumption are twofold: the practical reason that there are few matters that some government somewhere has never attempted to regulate or may not in the future; and the logical reason that a government's inactions can be as important as its actions, and all actions plus all inactions constitute limitless scope. In the language of this volume unlimited scope can be described as authority to make and enforce decisions about any transformations, communications, transactions, or organizations within a geographic area—or to refrain therefrom. Although there are obvious imperatives for limiting the scope of actual governments, this assumption for the initial model seems necessary for clarifying the special problems of government.

d. Membership affiliation by individuals (citizens) is acquired automatically by birth and terminated by death; thus membership is totally involuntary to both citizen and government. Although nearly all real societies have immigration, emigration, expatriation, and the like, these factors are not included in the model.

e. Government exercises a monopoly on the legitimate use of force or its delegation. Use of force by private police, prizefighters, parents or teachers exerting discipline, duelers, and so forth could be prevented by the government's force (in the model) if the government chose to do so.

f. There are multiple sponsors of government who hold conflicting goals that could be affected by government action and who attach such markedly different importance to different issues that some conflicts cannot be resolved by communication or transaction. Furthermore each sponsor holds certain conflicts among his own goals concerning government action. These assumptions about conflicts apply also to citizens who are not sponsors.

Assumptions **a** and **f** apply to many organizations. The others apply only to government, and this model deals with the implications of those special traits.

4.13.10   *Relevant Definitions*

a. *Sovereignty* is the authority position of a formal organization (and conceivably of an individual such as a king) that is not a subsystem (member) of any larger organization whose authority over it or its subsystems includes legitimate use of force but whose own authority over its own subsystems does include legitimate force. This definition allows a sovereign entity to be a member of, and hence a subsystem subject to the authority of, some larger organization so long as its bargain of affiliation does not legitimize force as a sanction against it. Moreover this definition refers only to types, not amounts, of power exercised by or on a sovereign entity. A government can be sovereign and yet be powerless in its dealings with other nations or even its own subjects.

Assumptions 4.13.1a–e are incorporated implicitly or explicitly in this definition of sovereignty. The only sovereign units that concern us here are governments. Hence we are dealing with a complex formal organization (assumption **a**) possessing potentially unlimited scope of function (assumption **c**). The boundaries of such an organization can apparently be defined only geographically. The reasons seem intuitively clear but have been spelled out elsewhere (Kuhn, 1963, pp. 543f). If the boundaries are geographic, then all persons within them are subject to the government's authority whether they want to be or not (assumptions **b** and **d**). Limited exceptions such as extraterritoriality or diplomatic immunity need not be dealt with in this general model, which could easily be relaxed to accommodate them. The modified assumptions that might be necessary for analysis of multinational corporations are not dealt with here. Legitimate force (assumption **e**) is explicitly incorporated in the definition. With these clarifications we can say that the study of government is the study of formal organizations with attention to the special conditions of sovereignty.

b. *Government* is the formal and sovereign organization of a whole society. This definition does not presume that every society has a government: some relatively primitive societies do not. For this introductory purpose we need not extend the model to accommodate regional, municipal, functional, or other subdivisions of a government that do not possess full sovereignty—although there is no apparent analytic difficulty in doing so along the lines of the delegation of authority discussed in the preceding chapter. This definition of government is partly circular since we have already seen (3.17.0f) that the area covered by the government is arbitrarily construed to constitute the whole society. Since a geographic boundary is easily identified, even if contested, no significant problem arises from this circularity.

c. *Private goods* are goods that can be acquired and used by some parties whether or not others have the same goods.

d. *Public goods* are goods whose acquisition is necessarily the same for

all persons in a society or subdivision whether they individually
want them or not—such as national defense, the structure of government, or, more locally, clean air, paved streets, or low tax rates. A
shared value is the logical parallel of a public good at the level of a
small group such as a family. There are vast numbers of strictly
private goods including affection and peanut butter. Unequivocal
public goods are relatively few—even highways and police protection can be provided on a private basis for only those who want
them.

e. The *public* is all persons within the sovereign area of a government.
(This definition includes only the nonstaff aspects of government
staff.) The public consists of citizens plus foreigners within the
nation's territory. The entire public are construed to be recipients
of the government, even if the government never explicitly does
anything to or for some of them.

f. *Government staff* are those who do the government's work. Among
others these persons include elected and appointed officials, hired
employees, and consultants. The term is thus broader than the
customary meaning of "the bureaucracy."

g. *Citizens* are that portion of the public who are members of the
government organization with specified rights and obligations
whether or not they belong to the government staff. These specifications include an authority relation with government as superior
and citizen as supervised; they may also include the conditions
under which one is a citizen while residing abroad as part of the
public of a different government. This definition does not preclude
multiple classes of citizens, and the distinction between minor and
adult citizens seems inescapable. Citizens may or may not be sponsors, depending on the basic form of government organization (see
definition **i** below).

h. A *nation* is the whole of any society and culture—plus its territory—
that has a government. For certain aspects of international relations
the nation includes persons or goods outside its territory but under
its authority. These persons and goods need not be spelled out in
the definition.

i. *Sponsors (of government)* means the same here as in 4.12.10a. The
term is mentioned only to indicate that the sponsor groups may be
as varied as all citizens, a military junta, a dictator, a monarch, or a
hereditary landed aristocracy.

   The distinction between citizens and staff is like that
between the members of a union and the paid officers and other
staff of the union. The distinction between citizens and sponsors has

similar parallels. Sponsors may include all members, only some members, or possibly even nonmembers. There might, for example, be an explicit distinction between voting (sponsor) and nonvoting (nonsponsor) citizens or a de facto sponsor control by a limited group possibly including noncitizens. Examples of noncitizen sponsors of a government are General Douglas MacArthur in Japan after World War II or the English governors of the American colonies. The definitions allow for authority over members who are not also sponsors.

j. *Recipients (of government)* means the same as in 4.12.10a. The term is mentioned here primarily to clarify that taxes are paid in a recipient, not a sponsor, role. Whether or not the individual voted for a tax levy, the enforceable requirement that taxes be paid comes as an instruction from government to individual, not individual to government. It is a decisional output of government even if the ensuing movement of money is an inflow. I see no conceptual difficulty (or gain) if the government is viewed as "selling" its services to the public in return for tax payments, in which case the logical relation is the same as between a business firm and its recipient customers.

4.13.2   GOVERNMENT AND THE PUBLIC: EXISTENCE, AUTHORITY, LEGITIMACY

4.13.21   The authority of a government over a foreign citizen residing in its territory and free to leave is legitimized by his continued presence—as a corollary extension of the principle that acceptance of authority legitimizes it (4.12.22).   **4.13.22** By contrast with both the foreigner and the member of a limited-purpose organization, the citizen's inability to terminate membership in government (4.13.1d) means that any legitimacy of its authority to him depends on other factors. However, if (in contrast to assumption 4.13.1f) the whole public living at the moment a government is first formed were unanimously to accept its authority it would automatically be legitimate to all. This "first formation" might be by agreement of people who previously had no government, by merger of smaller governments, by successful revolt against a prior government, by reconstitution following unsuccessful revolt, or by other arrangement.

If a dominant coalition of the public accepts a government's authority it is legitimate immediately to them but not to the remaining public that does not want it. The remaining public is then in the same logical position as organization members in paragraph 4.12.23, whose alternatives we now examine seriatim. (a) If the new government's orders are satisfactory, the remaining public may then accept and thereby legitimize them. If the orders remain satisfactory, in due time (the model does not specify when) the remaining public may legitimize the government's sovereignty as contrasted to legitimizing its specific actions. (b) Negotiation with each individual is not a usable alternative (see 4.12.23) since under assumption 4.13.1d the member cannot leave if he does not agree. (c) Unless there is a shift in the relevant type of power, a coalition of the remaining public will not succeed since it is already assumed to be nondominant. (d) To speak of getting implicit acceptance in advance is also irrelevant to those who have not accepted the government already installed. Since they cannot leave, the only remaining alternatives are apparently to remain and feel hostile or to stop feeling hostile because there is nothing they can do about it. (e) If the remaining public is hostile and can form a coalition with power approaching that of the dominant coalition (which presumably includes government staff), the government's tenure will be unstable. To ensure a greater margin of stability the government can offer goods in transactions (such as concessions to the opposition to induce them into the dominant coalition), use bads of sufficient magnitude to coerce them in, reduce the numbers of the opposition or limit its actions by physical blockage, or apply some combination of these tactics. Thus, unless it can eliminate or buy off the opposition, even a newly formed government has little choice but to operate with some members hostile toward it while it nevertheless tries to exercise its authority over them.

4.13.23   Since a dominant coalition of force will gain control over one based on any other type of power (4.12.43), a government that desires to remain in power must use its own force when necessary to see that no opposition coalition grows large enough to challenge it. **4.13.24** In a government that continues through a turnover of its public through births and deaths it is not possible to have all persons accept it as legitimate as they might at the moment of its creation. To the extent that legitimacy is to be achieved it must be gained by inducing new members to support it through the induction process (political socialization). As before, the three techniques for achieving this end —in order of logical preference—are communication, transaction, and dominant coalition (4.12.48). Since communication involves virtually the whole society, it is the same as the process of culture and covers an important segment of cultural content. To the extent that this

technique is successful, the public *wants* the government. To use transactions in goods is to give the public enough value in return for its
contributions to ensure a substantial overlap of EP's for membership.
To the extent that this technique is successful, even those who do not
want the government per se will nevertheless find it to their advantage.
To use dominant coalition is to hold a large enough margin of relevant
power (namely, force) to ensure that no other coalition can become
dominant in the government. To the extent that this technique is
successful, those who neither want the government nor find it to their
advantage will nevertheless be unable to do anything about it.

4.13.25   Several conclusions about the continuance of a government
seem to follow. First, *cet. par.,* the longer the EP's of the public for the
government, the greater the likelihood they will enable it to consummate the input transactions necessary for its continuance (4.12.82)—
and, to the large extent that these inputs are acquired from the public,
the better will be their terms. Second, the stability of the government
varies directly with both the likelihood of such transactions and the
terms on which the government completes them. Third, the EP's of the
public for the government vary directly with their position in the above
preference ordering (4.13.24). That is, *cet. par.,* those who support the
government because they want it have longer EP's for its continuance
than those who support it merely because it is advantageous. (Under the
cet. par. assumption the government is equally advantageous to both
groups.) And fourth, the stability of the government therefore varies
directly with the relative proportion of its citizens whose support is
gained by the preferred techniques. The model provides no guide,
however, to the relative value to the government of numbers as contrasted to position—whether, for example, 50 percent in the top
position are worth more or worth less than 75 percent in the middle
position.

   Although the techniques fall into a clear trichotomy, *people*
do not; a person may support a government partly by persuasion,
partly because his payoffs are satisfactory, and partly because he sees
no chance of getting rid of it. The priorities thus also operate within
persons: the more a person's support is based on persuasion (belief,
conviction of its rightness, and so on), the stronger the government's
power with respect to him. Among these three techniques, the more
preferred technique the stronger the desire for the government and the
greater the government's legitimacy in the eyes of citizens. Hence we can
also say that the greater the legitimacy of government to its citizens,
the greater, *cet. par.,* its power with respect to them. Where relevant,

feelings of patriotism, awe, belief in divine right, inertia, fear of change, and similar attitudes have the same effect.

4.13.26   Thus a *stable government* is here defined as a condition in which neither the government nor the public makes any significant fraction of its actions contingent on the possibility that the government (as contrasted to particular staff members in it) will be replaced.

4.13.3   **GOVERNMENT AND SPONSORS**

4.13.31   A fundamental question concerning every complex formal organization is that of how the sponsors decide what they want the organization to do. We are for the moment still leaving open the question of who the sponsors in a particular case *are*—all citizens, property owners, a military junta, and so on. Whoever the sponsors are, we continue the assumption that they do not all agree on a particular government decision (4.13.1f). Such disagreement, however, does not preclude their agreement on decision rules (see Riker, 1962).

4.13.32   To examine the question of sponsor decisions we make two further assumptions. First, to operate a government with decision rules is obviously more satisfactory than without them; thus we assume that decision rules are adopted and accepted by the sponsors for reasons stated in 4.12.441 and because it seems easier to agree on such rules than on the decisions themselves. Second, we assume that no set of decision rules can be permanently acceptable under changing conditions.

4.13.33   We also add that acceptance of decision rules is an agreement to accept as legitimate any decision made in accordance with those rules, to accept sanctions for violating them, and to abide by those rules in one's attempt to influence decisions. Here the categories of legitimate decision are defined by procedure rather than by content (4.12.10n) although a decision rule may limit content as well—for example, a decision rule may specify that search without a warrant falls into an illegitimate category of content even if decided by an otherwise acceptable processes. A decision rule could also specify criteria for decisions—such as promotion of the general welfare. We need not specify how many parties must accept the rules, or how enthusiastically, for them to qualify as "accepted," though 4.13.24 and .25 are guides.

4.13.34   From the previous three paragraphs it follows not only that decisions made according to the rules will be accepted as legitimate even if disliked but also that the decisions themselves will in time become increasingly unsatisfactory as the decision rules become outmoded.

4.13.35   Let us now shift the preceding logic about decisions to the level of decision rules by assuming that any conceivable set of decision rules embodies power components which please some sponsors and not others and that complete agreement on changes of decision rules is not possible. These assumptions do not preclude agreement on rules for changing decision rules. We make no assumptions about the inclusion of such rules nor do we proceed to the matter of rules for changing rules for changing decision rules. Although situations involving such changes occur fairly often, they do not involve power forces different from the simple rules for changing rules. **4.13.36** In light of the above we define three degrees of legitimacy and specify some transactions concerning legitimacy:

a. A *fully legitimate* government uses accepted decision rules and rules for changing them.
b. A *conditionally legitimate* government uses accepted decision rules but has no accepted rules for changing them.
c. A *tentatively legitimate* government has no accepted decision rules, but its decisions are accepted.
d. The *major bargain between sponsors and government* is an explicit or implicit transaction over authority in which sponsors agree to accept government decisions if made in accordance with certain decision rules and the government agrees to abide by those rules. As noted in 4.12.10n the boundaries of "legitimate categories" can be defined by the organizational structure or process used in making decisions as well as by other criteria.
e. A *constitution* is the content of a major bargain between sponsors and government about decision rules, broadly construed. This definition does not preclude real documents called constitutions from containing other matter nor does it specify whether a constitution is graven in stone or hearts, explicitly negotiated or evolved through practice, clear or fuzzy, short or detailed. This definition of constitution, like that of government, does not imply that every society has one.

Given these definitions, and to sharpen certain additional questions, we arbitrarily say that if a government is both stable and fully or conditionally legitimate its authority over its entire public is legitimate, even over those who have not accepted it.

4.13.37   Whereas citizenship is acquired by birth and terminated by death, membership in government staff depends on the individual's acceptance or rejection of a transaction of affiliation as in any other organization. Within the model the sponsors then implement the constitution by making acceptance of it, through an oath of office, part of the major bargain with members of the government staff, subject to sanctions and termination of membership for violation. The major transaction between sponsors and government thus becomes a transaction between sponsors and each member of government staff—or at least those members who can significantly influence its decisions. **4.13.38** Thus, given continuity of sponsors (no revolution or foreign takeover) and freedom of staff to leave or of sponsors and their agents to remove them, a mechanism is available by which government staff can be held to legitimate exercise of authority. Whether the mechanism actually works in a particular case is, of course, a different matter. This conclusion about the mechanism is nevertheless valid for any sponsor—from a military junta to the whole citizenry—and is independent of the number of hierarchical levels in the staff.

4.13.4   FUNCTIONS (AS ORIGINS) OF GOVERNMENT

4.13.41   This section categorizes the functions of government within the conceptual framework already established. Since government is a controlled system established by human beings who are also controlled systems, within this model of formal organization we say that the goal of the sponsors is that the government perform its functions. Although our knowledge about the origins of real governments is murky, we pursue the introductory model of formal organization into this sphere by assuming that governments are formed *in order to* perform their functions. Hence the heading of this section. As we shall see shortly, the functions are also related to the kind of organization that is used for a government.

4.13.42   We have already seen the severe disadvantages of allowing force in private transactions (3.25.4). To these we add the following assumptions: that most people understand those disadvantages at least intuitively (detector); that most people prefer to avoid those disadvantages (selector); and that they therefore attempt to prevent such use of force (effector). Whether or not communications and positive goods transactions have failed, the remaining technique (in this model) of stopping such force is a bargaining coalition of greater force. If we assume that the availability of this technique is understood—which seems likely—those who want to eliminate force in transactions will gang up on the individual who uses it. If the membership of such a coalition remains more or less stable over a series of such incidents and the remaining people generally approve, there exists a legitimized monopoly of force and the nucleus, in fact and theory, of a government.

4.13.43   As noted (3.27.55), in such actions the government joins a coalition with the attacked against the attacker and brings a marked shift of power and bargaining power. Having clarified the principle, there is no particular limit to the number of areas to which it can be extended as the regulatory powers of government. Government can form a coalition with users of water against polluters, buyers of electricity against sellers, victims of discrimination against practitioners, property owners against trespassers, promisees against promisors, injured workers against careless employers, the libeled against the libelers —or the reverse. Whether or not the initial goal is to make coercive imbalances of power more nearly equal, the principle is the same whether the government joins a coalition of the strong against the weak or the reverse.

4.13.44   In joining such a coalition the government faces numerous practical difficulties, such as finding and identifying the trespasser or polluter, ascertaining his acts and their consequences, and then keeping track of him. To the extent that these difficulties can be overcome, the government's monopoly on force makes any coalition it joins dominant over any opposing coalition within the nation (4.12.43). By corollary one subset A of the public can gain tremendous power and bargaining power over other subsets B's if A can induce the government C to take its side in a pressure transaction or to join it in a bargaining coalition.

4.13.45   Power, however, can be intrapersonal as well as interpersonal. Intrapersonal power depends on one's productivity, which depends in turn upon his DSE states and the quality (state) of his environment. Hence government can raise the intrapersonal power of individuals, if it so chooses, by two main techniques. The first is to raise their transformational abilities, for which the chief devices are to increase their information (detector), strengthen their motives (selector), and improve their transformational skills (effector)—all of which may be encompassed under "education." The second technique is to improve the quality of their environment to increase the productivity of a given amount of transformational effort. Conservation activities in their broadest sense illustrate this category. Since Section 4.13.4 has thus far dealt with interpersonal and intrapersonal power, we can join the two under a heading of *allocating power*, the first major function of government.

4.13.46   Rather than merely affecting the power of others, a government can itself engage in transformations and thus fulfill the definition of organization rather than merely of coalition. Obviously it can produce public goods such as roads, irrigation, fire protection, a money supply, and postal service. Nothing about the nature of the function precludes it from producing private goods for sale or for free distribution if it sees fit. Producing goods, particularly public goods, is the second major function of government. In this model all governmental functions are thus categorized as the coalitional one of allocating power or the organizational one of producing goods. As noted elsewhere (Kuhn, 1963, p. 619), actions that allocate power at the micro level may be viewed as public goods at the macro level. Police and courts, for example, reallocate power between victim and thief at the micro level but constitute the public good of order and justice at the macro level. A school is designed to increase the power of the individual student at the micro level, but the educational system is a public good viewed at the macro level.

It is, of course, impossible here to scan the gamut of government's activities in order to justify this classification; but I feel reasonably confident that all can be handled thus without significant distortion while simultaneously identifying the theoretical analysis to be applied to each. Importantly, the techniques for making internal decisions are the same for coalition or organization (4.12.45) and involve communication, transaction, and dominant coalition. Further, the coalition *function* can be engaged in by an organization as a party (3.10a); hence the coalition function does not dilute the definition of government as an organization.

4.13.47   Since government dominance in force can vastly strengthen the position of A's by joining a coalition with them against B's, the distribution of power within a nation is strongly influenced by the government's selection of those it will support. By corollary it is also strongly influenced by the government's decisions whom *not* to support. This is the strongest sense in which a government's inaction in some areas can be quite as important as its actions in others (4.13.1c). Nothing need be said about cases where government intervention has never been suggested. But explicitly decided nonintervention has two effects: in the micro sense it leaves the power balance between A's and B's wherever they themselves have developed it; in the macro sense it lets the question of power balance be handled by the informal and semiformal organization of the society (see Chapters Fourteen and Fifteen). A government can hold a comprehensive sense of responsibility for all aspects of the nation, but by consciously doing nothing in some areas it may "assign" them to some other agency. By not producing the vast bulk of private goods the nonsocialist government assigns their production to the informal organization that is the market economy and to its formal subsystems of firms, banks, individuals, and so forth. When government leaves many functions to informal and semiformal organization and exerts pressure to influence their communications, transactions, or organizations, it helps determine the existence and content of roles in the informal and semiformal organizations.

4.13.5   TYPES OF GOVERNMENT ORGANIZATION

Since government is defined as a formal organization and formal organizations are classified into four main types, governments can likewise be classified as cooperative, profit, service, pressure, or some combination. After we identify the conditions that constitute each type of government, we can examine the principal logic of each type.

If the same persons are collectively both sponsors and recipients and bear the costs in order to receive the benefits, the government is a cooperative. However, the definition of cooperative (4.12.72) allows nonsponsors to be recipients but does not allow nonrecipients to be sponsors. The definitions of government, sovereignty, and public make the whole society recipients of its organization, but all need not be sponsors to qualify the government as cooperative.

If the sponsors are some subgroup of the society who receive benefits from a government whose costs are borne by some other subgroup in the role of recipients, that government is a profit organization. "Benefits from government" mean those from the organization itself, as an excess of its inflows over its outflows, and not from the government's help in a coalition against some other party. In its simplest (and fairly widespread) form this means output transactions in which the government collects more in taxes from the public than the costs of its outputs to recipients. The excess goes to the sponsors, who may or may not occupy important staff posts. In addition to public goods "sold" in return for taxes, outputs can include private goods sold to individuals.

If the sponsors are some subgroup of the society whose contributions to the government are given in generous transactions to some other subgroup, the government is a service organization.

And if the sponsors are some subgroup A of the society who use the government C to direct its outputs directly or indirectly to another subgroup B in order to improve the position of A in A-B relationships, the government is a pressure organization. It is also operating as a coalition partner with A.

Just as most limited-purpose organizations have some activity of all four types, the potentially all-purpose government almost inescapably does so. It is therefore difficult to imagine an actual government as a pure type; real governments can be classified only by their apparent dominant behavior.

4.13.51   GOVERNMENT AS COOPERATIVE ORGANIZATION

4.13.511   *Model and Assumptions*

The model of cooperative government consists of the following assumptions:

a. The government is operated as a cooperative organization.
b. All citizens are both sponsors and recipients. All remaining public are recipients but not sponsors—though this aspect of the assumption does not explicitly enter the analysis.
c. The functions of government are confined to allocating power and producing public goods; it produces no strictly private goods though some of its public goods and power allocations benefit some groups of citizens more than others.
d. The main decision rule is that sponsor decisions are by majority vote. The assumptions do not specify whether issues, top staff members, or both are decided on by the sponsors.
e. Additional rules proscribe coercion of sponsor choices by government staff or their agents. These rules provide political liberty as defined below.
f. All assumptions about the special properties of government as organization (4.13.1) are continued. But there is a specific addition to assumption 4.13.1f: although each sponsor accepts the decision rule, his vote and other actions to exert his sponsor control are exercised selfishly and with possible tactical misrepresentations rather than being exercised in the interest of the "public welfare."
g. As among sponsors of other organizations, decisions can be made by

communication, transaction, and dominant coalition. Although some decisions or aspects of decisions may be made by communication, we will attend specifically only to transaction and dominant coalition.

4.13.512 *Sponsor Decisions by Transaction*

Sponsors may be called on to make decisions about public goods or coalitional actions of government. Because both the costs and the benefits of coalitional actions are vague by comparison, we will examine the problem of making decisions about public goods and assume that each sponsor has a reasonably clear notion of his cost/benefit position with respect to each good. The transactional relations for making such sponsor decisions are of two main sorts. First, only those sponsors who want a particular public good contribute to its costs. Since "wanting" is private information, only those who voted for a measure can be charged its costs. And since we have assumed both selfish motivation and willingness to use tactical misrepresentation, a citizen who wants a particular good but thinks it will be voted in without his support will vote against it. Such a technique not only provides fewer public goods than people actually want but also shows at least five other difficulties. First, fewer people will vote for it than actually want it. Second, some goods wanted by a majority will be voted down by miscalculation. Third, the cost per person to those favoring and contemplating voting for it will be higher than if costs are assessed against the whole public. Fourth, the technique is incompatible with a secret ballot and lends itself to various pressures. And fifth, the technique requires that the costs of each public good be isolated from all others (an impossible task) and that decisions be made by referendum rather than by representatives (another impossible task in any but the simplest nation). Hence we conclude that, for these and perhaps other reasons, this transactional alternative is thoroughly unworkable and will not be used.

The second major type of transactional decision relation is that in which the tax liability of individuals is in no way related to whether they voted for or against a particular public good or did not vote at all. This technique is not subject to the above difficulties and has no apparent difficulties of its own; thus we assume it is adopted. With selfishness on the part of citizens we also conclude that the potential coercive power of government is employed when needed to collect taxes from the public. This means that the transactional relation between citizens and government over public goods is only of the general sort: the sponsors collectively pay for whatever goods they collectively authorize through whatever decision techniques they choose. More explicit consensus by transaction among citizens is subject to the same analysis as that among a dominant coalition and is discussed in the following section.

4.13.513   *Sponsor Decisions by Dominant Coalition*

The remaining technique of decision-making is the dominant coalition. Since the decision rule is the ballot, the dominant coalition of voters normally carries the decision. But we noted (4.12.44) that a coalition dominant in the power relevant to the organization could get the decision rule changed or suspended. We therefore pause to examine the possibility that a coalition dominant in some kind of power could get the decision rule of majority vote changed or ignored. Among the many possible kinds of power for this purpose we will examine the three most obvious—force, money, and mass communications.

Special conditions of government (in this model) limit the availability of these types of power. First, a dominant coalition of *force* among sponsors is ruled out under the assumption of government monopoly. Second, a dominant coalition seeking to withhold the necessary *money* is ruled out by the provision that each citizen's contribution is collected by the government's ability to use force to collect it (4.13.511f). And third, the ability to exert moral influence is indeed available through the *mass media*. But whereas force and the curtailment of money constitute power to stop government actions, persuasion has no such coercive effect except as it induces people to use some other type of power already at their disposal—one can gain no additional power over government by persuading to his cause those who have no power over government. Thus of the three most obvious types of power in a society, force and money cannot bring suspension of the decision rules for the reasons stated. Communication has power if it operates through the ballot; but it has no power to coerce elimination of the ballot as a decision rule. Effective communication directly to government staff can wield power, but at the moment we are concerned solely with the decision among sponsors. Hence we conclude that the sponsor decisions which constitute their instructions to government staff (4.12.3b and 4.13.511b) will continue to be made by a dominant coalition of votes—that is, by a majority.

4.13.513   *Problems of Sponsor Voting.* To say that decisions are made by majority vote, however, is only the beginning of the story, not the end. The simplest case is that in which voters attempt to decide only a single issue with only two alternatives—here a simple majority can reflect the (unweighted) preference of the sponsors. If some want one alternative strongly enough to try to entice the votes of others into their coalition, they can do so by communication or transaction. The former need not be detailed further here. The latter raises special problems, but these are essentially the same among sponsors as among their legislative agents and are dealt with in 4.13.515 below.

For the remaining discussion of government we assume that multiple issues and multiple positions on them are so involved that issues cannot be voted directly by sponsors, that decisions of government are made by agents of sponsors, and that citizens vote for agents (representatives) solely on the basis of the positions they take on issues. We assume also that each candidate for elective office wants to win. With this assumption we join the questions of how the sponsors make a decision and how a candidate gets elected, the election constituting the sponsor decision.

Given conflicts among and within sponsors (4.13.1f), no candidate can reflect the desires of all citizens; only by chance can he precisely reflect a citizen's desires. More strongly, it can be demonstrated that among a package of issues a "majority position" or "social preference" is not possible except by accident (4.12.49 and Arrow, 1951, p. 3). In fact, given that different voters attach markedly different importance to different issues (4.13.1f) a candidate can sometimes win by taking a minority position on every issue. To do so other than by chance requires explicit knowledge of voter sentiments; illustrations based on explicit assumptions appear elsewhere (Kuhn, 1963, pp. 665f, from Downs, 1957). Since the way a candidate goes about selecting a winning platform resembles the way a political party tries to build a winning coalition, we now shift to the latter problem.

4.13.514   *Political Parties*

We define a *political party* as a coalition of office seekers and their supporters who seek to increase their power over government decisions (which for this purpose range from broad policy to details of implementation). Rather than attempt to specify what is essentially a C-level conceptual structure of Figure 4 and beyond the scope of this volume, we will simply suggest an application of this framework to the problem of political parties, attending to types of power, hierarchical structure, and equilibrium coalition size within parties.

4.13.5141   *Types of Power.* To analyze the types of power within parties we assume that a party is not bound by the decision rules of the government and hence is open to more types of power manipulation. The power to achieve a dominant coalition within the party can therefore consist of money, status, time, effort, ability, popularity with voters, access to media, moral and intellectual influence, or virtually any other type of power except force.

As it directly concerns sponsor choice the main question is how a party selects its platform and candidates. For this question the present model would examine the same factors as before: the DSE factors within one individual or organization and the communications, transactions, and subcoalitions among them. A typical transactional question might concern how much the prospective candidate wants a wealthy contributor's money or a union's endorsement and how much those parties want the candidate.

To analyze these factors in a particular case would require a simulation model of the EP's of the factions within the political party; such a model would have to include the whole gamut of Sections 3.2, 3.3, and 3.4. As to communications, the model would include the detector states of senders and receivers of messages, their perceptual and linguistic codes, their confidence in their own perceptions, and so on as outlined in Section 3.1. To the extent that the party is also an organization, the conceptual apparatus of 4.1 through 4.12.9 could be added with particular attention to 4.12.414 on group decisions. Attention could also be directed to compatible and incompatible issues as in 4.13.515 below. Nevertheless the model of even the skilled politician astride the scene would necessarily be a highly oversimplified version of reality.

4.13.5142   *Hierarchical Structure.* We noted in 4.12.45 that a coalition, like an organization, can have a hierarchical structure with a dominant subcoalition deciding for the dominant coalition, a dominant sub-subcoalition deciding for the dominant subcoalition, and so on. An important question here, as in any other group decision involving multiple levels, is whether the lower-level coalition remains intact within the larger one or merges into it. The two arrangements are illustrated respectively by bloc voting (as when a state's total electoral vote goes for one presidential candidate) and proportional voting (as in a state's division of votes in a party convention). Bloc voting, of course, concentrates the greater power in a few hands. A simple case of such dominant coalitional hierarchy is a pyramid of corporate holding companies. Hierarchical amplification of power also occurs in electronic circuits and elsewhere (Kuhn, 1963, p. 7).

If there are only two parties, each tighly disciplined, the one that wins an election automatically constitutes the dominant coalition for controlling the government itself. If there are three or more tightly disciplined parties and none receives an absolute majority in a representative body, the government's decisions have to be made by a dominant coalition of parties as in certain European legislatures. To the extent that the parties are not tightly disciplined their representatives merge into a single large coalition rather than remaining intact, as is generally true in the U.S. Congress.

In short, although under assumption 4.13.1f it is not possible to get agreement of the entire sponsor group by communication plus transaction, it is possible to get agreement within the dominant coalition (or dominant subcoalition of it) by those two techniques.

4.13.5143 *Equilibrium Size of Coalition.* The size of a coalition is subject to equilibrium analysis. The benefit of increasing size is the increased likelihood of dominance. A plotted curve of this benefit would presumably rise slowly at first, rise steeply as the size approached dominance, and then fall sharply beyond a safe margin. The costs of increased size are, of course, dilution of objective. Assuming differences among sponsors, every new group brought into the coalition requires additional transactional tradeoffs by those already there. At the extremes, a coalition of all sponsors would have power to do anything it wanted but could not agree on what it wanted; a coalition of identical thinkers would know what it wanted but would be too small to have power. An equilibrium theoretically exists at a point that will not dilute the coalition's goals more than necessary to ensure dominance.

Parallel questions arise, but viewed through different eyes, for those outside the coalition who contemplate joining. How much will they improve their chances of winning if they join a particular coalition? At the cost of how much trading away of their goals? The bargaining power of coalitions and their prospective new members can be diagnosed by learning their respective EP's. Thus we can identify the equivalent of an employer point of view in the coalition's EP for more members and an employee view in the new member's EP to join. Because of their high marginal benefit, those voters who tip the balance between dominance and nondominance have considerably greater bargaining power with the party than those outside that range. They are the "swing" voters, as contrasted to the committed ones and those not worth going after (see 4.12.43). This equilibrium approach derives from Downes (1957), who certainly should be read by anyone interested in platform formulation.

4.13.5144 *Summary*. The coalitional nature of the political party has been examined in the context of a cooperative government, which corresponds closely with the notion of democracy (4.13.516a). This is the form in which multiple political parties are most likely to exist; hence it allows the fullest examination of the problem. If there is only one party, the dominant coalition within it makes the decisions of government. And in organizations that do not have a party at all, as with most organizations except governments (Michels, 1911; Lipset, Trow, and Coleman, 1956), the dominant coalition of the relevant power controls. If the organization has an accepted decision rule, the decision is carried by the coalition dominant in that criterion, even if the rule is only semiformal.

The general principles seem to be the same whether we are speaking of decisions among parties, within parties, within a legislature, between a legislature and executive department, and so on. Wherever there has to be a single decision on some matter it is made by communication, transaction, or dominant coalition of relevant power.

4.13.515   *Compatible and Incompatible Issues*

The decision process is not the same for all types of issues. Perhaps the broadest basic distinction is between compatible and incompatible ones. When goods X and Y show scarcity costs and there exists sufficient aggregate power to acquire both, we say that decisions in favor of both are *compatible*. Many, but not all, public goods are in theory compatible. By contrast, when goods X and Y are themselves incompatible so that the sacrifice of one is an inescapable cost of the other, decisions in favor of the two are *incompatible*. Some public goods and perhaps all coalitional power allocations are incompatible. Because transactions are both more possible and more readily analyzable in a legislature than in the whole public, we will examine this distinction briefly as it appears in a legislature whose members represent constituencies with goals that are strongly agreed internally but conflict externally with the goals of other constituencies.

With respect to a compatible decision, we assume that representative A's constituency wants a new harbor and B's wants a new irrigation dam. Neither constituency will benefit from the other's improvement and the cost of either will be spread across the whole nation. A introduces a bill for the harbor, B introduces one for the dam, and the question is whether a transaction over votes can be consummated between them. The transactional ingredients are as follows. AX is A's desire not to vote for B's dam. Since the cost of the dam will be spread across the whole nation, the cost to A's constituents will be small. Hence AX will be small. AY is A's desire for B's vote for the harbor. The benefit of the harbor will be concentrated on A's constituents while the cost will be spread across the nation. A's constituents will strongly want the harbor under these circumstances. Hence AY will be large. BX and BY are parallel to AY and AX, respectively, but with a reversal of items. With AY large and AX small, A will have a long EP for Y. Similarly, with BX large and BY small, B will have a long EP for X and the transaction can readily go through. Since the logic applies to all compatible issues and all representatives, within the limits of the total budget constraint A will get his harbor and B his dam. Furthermore, within these assumptions the motivation will be low to keep the budget constraint tight—though the same legislators may have other transactional relations creating pressure to tighten it. For this reason it may be expected that decisions on compatible issues will be primarily transactional among legislators.

On incompatible issues transactions are also possible, but they will be less likely and less direct. Suppose that A's constituency comprises textile producers who want higher textile tariffs while B's comprises heavy consumers of textiles who want lower tariffs. A and B introduce bills reflecting their district's desires with passage of one precluding the other. Here A and B cannot trade votes as before since either's vote for the other's bill will alienate his own constituency. And unlike the case with compatible issues, a representative may not have the chance to vote for both bills since passage of one may automatically kill the other. Although tradeoffs are therefore not possible between opposite sides of the same incompatible issue (or at least are highly improbable), they are nevertheless possible between two distinct and unrelated incompatible issues—such as higher tariffs and higher minimum wages. That is, two different incompatible issues may be compatible with one another. If we assume that an incompatible issue is of great importance to all constituencies, like a major aspect of foreign policy, and that each representative reflects his constituency rather than his own views, neither communications nor transactions among representatives will avail. The vote will fall where the power among various constituencies dictates.

4.13.516   *Relevant Definitions*

The model thus far has dealt with the basic concepts and processes of a cooperative government. Because some terms widely used in connection with government have not appeared here, this section defines them in the language of this framework and shows how they relate to this model.

a. A *democracy* is a cooperative government in which all citizens are sponsors. As noted (4.12.72a), in the cooperative sponsors are also recipients. Costs and benefits are assigned to the recipients since it is to provide themselves benefits that they pay the costs. To paraphrase Lincoln, the sponsor role makes it *of* the people and the recipient role makes it *for* them. In what sense it is *by* the people is less clear. In this model it is mostly *by* the staff, including its elected members.

This definition of democracy seems adequate. The problems lie in identifying the real conditions that fulfill the definition, and we can only sketch the task here. Sponsors have been defined as the "persons whose motives . . . constitute the goal of the organization." This means that the goals of citizens, not those of staff or government, are the ultimate end and goal of government. Such a definition puts citizens "over" government as well as "under" it. The definition also says that citizens are sponsors. This means that they exercise sponsor control actually and effectively, not in name only. If they do not actually exercise control, then other persons are the sponsors (under 4.12.84 and the definition of sponsor) and the organization is not cooperative. How responsive must government staff be to the wishes of citizens to justify classifying a real government as democratic? This is a problem of identification, not definition. We need not tackle that problem here except to insist that any difficulties on that score reflect the complexity of real situations, not difficulties peculiar to this definition.

This definition of a democracy says nothing about centralization or decentralization of government structure. Sponsor control might well be curtailed at either extreme. How far it can go either way without violating the definition is also a matter of identification. The definition also evades the question of how large a fraction of the public must be citizens to qualify the government as a democracy. Did the early United States disqualify as a democracy because females, slaves, and nonowners of property could not vote? Was fifth-century Athens a democracy? This definition deals only with the relation between a government and those who *are* citizens.

b. The *sponsor-staff relation* is that in which citizens select top staff officers of government and give them such instructions as their decision processes permit.

c. The *staff-recipient relation* is that in which government acts on citizens.

d. The *legislative function* is the process of making decisions on behalf of sponsors for execution by the staff. The legislative function is the essence of the sponsor-staff relationship.

e. The *executive function* is the process by which the staff effectuates the legislated instructions. Much, but by no means all, of the executive function consists of staff-recipient relations. Foreign policy is an important area of executive function in which citizens are only indirectly recipients.

Unsophisticated descriptions of American government often confuse offices or branches of government with functions. The functions are distinct in concept even though a function may be performed in any of the three branches. But they are continuous in practice because the hierarchies of government organization incorporate hierarchies of decision-making (see 4.12.0f). The relatively lowly decision to keep post offices open on Saturdays is in a sense legislative. Passing a law to stem pollution is in effect executing an instruction from sponsors who elected a candidate with a strong platform plank on pollution. It might be useful to go a step further and suggest that the distinction between legislative and executive is the same as that between policy and execution, which in turn is a matter of level of view. That is, what is viewed as a decision about execution at one level of a formal organization is received as a policy decision at the level below. Decisions effected at the second level become policy decisions for the next lower level, and so on.

f. *Political liberty* is the freedom of citizens to exercise their sponsor function in the sponsor-staff relation of a cooperative government.

g. *Civil liberty* consists of limitations on the power of staff to interfere through the staff-recipient relation with certain specified actions of citizens. A particular freedom may be important to both political and civil liberty—as is freedom of speech—and almost any liberty on one side could have indirect repercussions on the other. In a direct sense, however, access to the ballot is purely political and due process of the accused purely civil.

h. The *judicial function* is the process of deciding a dispute between two parties whether an agreement between them, or a decision that is legitimate under some agreement, is being executed according to its terms.

For obvious reasons the process is more fair and acceptable if the decision is made by an impartial third party. The facts, the terms of the relevant agreement or decision, or both may be involved, as in the following examples. Did contractor A fulfill his contract with homeowner B? In passing law X did the Congress violate the major bargain between citizens and government—the Constitution? Did the bottle labeled Papaya Juice, which A sold to B on August 13, actually contain orange juice with banana instead; and if so did A thereby violate the FTC's constitutionally legitimate regulation of labeling? Did government staff member A, in denying a loan to citizen B, violate or modify the congressional decision regarding government assistance to homebuilding embodied in the Housing Act? Did the president, in sending troops into area Z, violate the constitutional provision that only the Congress can declare war?

In practice no contract, law, or constitution can cover all possible contingencies. If a dispute arises over unclear or uncovered problems the judicial function will necessarily legislate in part if it decides the case. Although we need not thus broaden the definition, if we were to extend the concept of legitimacy to cover the execution of any agreement according to its terms, the judicial function could then be defined as deciding questions of legitimacy. Because there is a certain sense in which legitimacy does have this broader meaning, which also coincides closely with the concept of rights, and because it might make sense to adopt it formally here, we will briefly illustrate the logic of such an extension. If a law against stealing is enacted legitimately, then under 4.13.33 and 4.13.36d the sponsors have agreed to accept the law and to accept sanctions for violating it. And under 4.13.36 decisions that are otherwise legitimate are also legitimate to that segment of the public which does not accept the government's legitimacy. Under these circumstances the arrest and trial of an alleged thief is an attempt to find if he has in fact fulfilled the recipient's half of the major bargain.

The relation between sponsors and government in the cooperative organization can be diagramed as in Figure 15. In anthropology and political science such a diagram is sometimes called the pyramidal structure.

4.13.52   GOVERNMENT AS PROFIT ORGANIZATION

The model of government as profit organization consists of the following assumptions:

a. The government is operated as a profit organization.

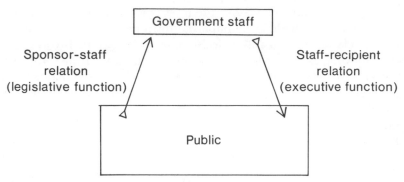

Figure 15. Government as cooperative organization.

b. The sponsors are a small minority of the public.
c. The goal of sponsors is to maximize their satisfactions from government—through receipts of money, travel, housing, and the like—with minimal contributions to it.
d. The whole public are recipients, notably in the role of taxpayers (see 4.13.10j).
e. Distribution of benefits from government among the sponsors is already agreed.
f. The government is firmly in power and stable.

4.13.521   Since the problem is one of maximization, equilibrium analysis is applicable. A detailed problem is to identify the rising costs and declining benefits (see Section 2.6 and 2.8). We have already identified the benefits as satisfactions to sponsors that decline after some point in accordance with the principle of diminishing utility. Incidentally, the agreement on distribution of benefits among sponsors in assumption **e** above removes the element of competitive display of power among them—a form of relative rather than absolute wants (Kuhn, 1963, p. 282) that might long postpone the operation of diminishing utility.

As to the cost of acquiring the benefits for sponsors, in a transactional relation the public will be willing to pay taxes if they receive a reasonable amount of goods in return. But the public will experience diminishing returns from more and more such goods while the opportunity cost of giving up more and more money will rise. Furthermore collection costs (as measured by, say, man-hours of collection time per unit collected) will rise for the same reason as in an extractive industry—because the types of persons and taxes that yield money most readily will be utilized first so that successive units of tax income will come from successively more difficult sources.

If bads are used in addition to goods (there seems to be no point in a model that provides *no* affirmative benefits), they too will meet diminishing returns, which is the same as increased cost per unit, for several reasons. First, negative actions against taxpayers' property will diminish the resources from which they pay. Second, hostility against the government and collectors will grow, raising the bargaining power of taxpayers and reducing that of the government (3.26.11) while reducing the (plain) power of both (3.26.12). Third, to use bads of more than minimal sorts will reduce the productivity of taxpayers. As with transactions in goods, the more units that are extracted from taxpayers the higher the utility of each unit they still retain and the greater the stress that will have to be applied to get more.

4.13.522   As in any other multiple-unit decision, the optimum position lies at the intersection of marginal cost and benefit curves as in Figure 16. This conclusion is independent of whether the preceding paragraph has correctly identified the ingredients of costs. Under these conditions the

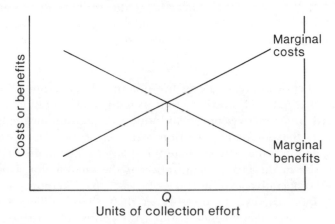

Figure 16. Cost/benefit equilibrium of profit government.

total units of collection effort are set at Q.   **4.13.523** If the profit government produces and sells private goods it must either hold some degree of monopoly or be more than normally efficient. Otherwise there would be no return over cost (there is no profit in the "representative firm" in a perfectly competitive market) and the activity would be incompatible with the assumed goals of government.   **4.13.524** If we relax the benefit-maximizing assumption and assume instead that the profit government also holds other motives (such as goodwill or hatred toward the people, desire to be good rulers, or desire to demonstrate power), the benefit curve shifts to the right or left in conformance with all the propositions of Section 3.26 about multiple goals in transactions. (See also 4.13.527 below.)

Without attempting to develop it here we can nevertheless suggest further analysis along the following lines. For any horizontal line drawn across Figure 16, there would be among a large public a corresponding pair of EP's—that of A, the government, extending from the left as far as the benefit curve and that of B, a member of the public, extending from the right to the cost curve. The Y's would be things the individual could provide for the government—notably money but perhaps also military service, deferential behavior, attendance at government-sponsored demonstrations, informing on dissidents, and general conformance with government regulation. The X's would include government services or transfer payments to individuals, public goods, possible sanctions, and the costs to the government of collecting or acquiring the Y's. All the pairs of EP's above the intersection would represent noncompletable transactions; all those below would be completable. If the government knew or could estimate the location of these EP's it would presumably maximize its position by setting the level of government services, taxes, and so forth at the point of intersection.[1]

4.13.525   Let us now relax the assumption that the government is firmly in power (4.13.52f) and substitute that its tenure is in question and that the rewards are attractive enough to induce competition for the sponsor role. Then the dominant coalition will decide who is to become sponsors, and thus they will become the sponsors. The main analysis of coalition size in the profit organization is the same as for a political party in a cooperative government (4.13.514); the differences lie in the content of the curves and possibly in their shape. Whereas in the cooperative government benefits and costs consist of achievement or dilution, respectively, of policy objectives about public goods and power allocations, in the profit government they consist of money and other perquisites. Given the relatively large volume of goods extract-

---

[1] The derivation of a pair of curves from a series of EP's for successive units of a good is found in Kuhn (1963, pp. 570–573). Although that description deals directly with a market for goods under competition, nothing about the underlying logic is affected if the EP's are modified by bads, additional motives, or monopoly. Such factors alter the content and length of particular EP's, not the nature of the analysis. The straight lines in Figure 16 are arbitrary; the cost and benefit curves could take any shape so long as their general direction of slope is not changed.

able from a whole society by a profit organization with dominant force, it seems unlikely that some group would not try to use it if they could. By the same token the public stands to gain by forcing an explicit or implicit major bargain on the government that the taxes they pay will be used mainly or exclusively for their own benefit. To achieve the latter presumably requires a cooperative government. To say which side will win in this struggle requires more assumptions than this basic model can provide. The profit organization can be diagramed as in Figure 17, a configuration known as the hierarchical.

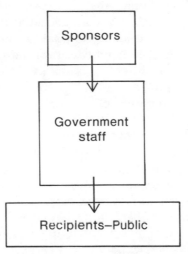

Figure 17. Government as profit organization.

4.13.526   With this background we now define a *dictatorship* as a profit or (less likely) a pressure organization of government that uses force to keep its coalition dominant—in contrast to using force merely to enforce decisions. Since most citizens are not sponsors (4.13.52b), the question of their political liberties is irrelevant and they have none. Civil liberties are not definitionally incompatible with dictatorship, however, and freedom of religion, due process in government pre-emption of private property, or jury trial in nonpolitical crimes might be compatible with dictatorship in practice. In fact any civil liberty not used to exercise sponsor control might be compatible with dictator-ship. Any such liberties would, of course, prevail at the discretion of the sponsors, though a stable dictatorship might constitutionally com-mit itself to some civil liberties. We might note in passing that since certain information about B is highly valuable to A in many aspects of transactions, it is valuable to B to keep it to himself and a loss of value if it is revealed. For these reasons an invasion of privacy by government is a seizure of a valued thing without due process. The purpose of this definition of dictatorship is simply to identify it as a profit organization. There is no need for further analysis here.

4.13.527   We now relax the assumption that the sponsors of a profit government seek to maximize their own satisfactions (4.13.52c). Instead we substitute that their goal is a restructured society, or ideological goal, not widely enough shared or desired by the public to allow democracy. We also reinstate the assumption that the government is firmly in power (4.13.52f). The cost/benefit analysis (4.13.522) would then apply, but the benefits and costs would be much more difficult to identify. The main benefits would consist of steps toward the ideological goal; the main costs would be setbacks in some areas that were necessary to achieve progress in other areas. Again there presumably exists an equilibrium position of optimal progress, but this stage of the model cannot specify how to find it.

Within the present categories a government with an ideological goal nevertheless classifies as profit so long as it works toward the goals of a limited group of sponsors while the costs are borne by the whole public, most of whom do not share that goal. If by successful induction the government persuades most of the public that its goals are desirable to them, its bargaining power with the public will rise and it will be able to get greater contributions of money and work than before. The government does not thereby become cooperative, however, unless the general citizenry are made effective sponsors. Their position is analogous to that of customers of a firm who think they are getting an excellent bargain and are delighted with the firm's performance although they exercise no sponsor control.

4.13.528   For the assumptions of 4.13.527 we now substitute that the sponsors A's want to maximize some selfish advantage that combines direct receipts of goods from government and government pressure on B's to improve the relation of A's with them. The basic equilibrium analysis of 4.13.522 would still apply, but the content of costs and benefits would be much more complicated and harder to identify. In contrast to maximizing a relatively simple sum of goods whose distribution among sponsors is assumed to be agreed, an important part of the composite that is maximized in this case—the changed relations between A's and B's—never passes through the hands of the government. The quantity received therefore cannot be measured nor can its allocation among sponsors be specified in the same way as are goods initially received by government. This situation might be too complex for even a C-level discipline (Figure 4); perhaps only a D-level simulation can deal with it—which would mean a separate model for each government studied. In any conflict between benefit maximization by a profit government in power and achievement of dominant coalition to stay in power, priority must obviously be given to the latter since sponsors cannot achieve benefits for themselves from a profit government if they cease to be sponsors.

4.13.53   GOVERNMENT AS SERVICE AND
PRESSURE ORGANIZATION

4.13.531   Service organization as a form of government consists of a portion of the population, the sponsors, giving resources to the government. With these resources the government provides public goods for the whole society, private goods to the remainder of the society, or both. Such an arrangement presupposes that the sponsors are either a limited group with extremely large resources or a large segment of the public helping a limited segment. The latter case approaches that of a cooperative government reallocating power; it would coincide with cooperative government if the specific recipients were also citizens with votes. The former case approaches that of a cooperative government in a nation with wide discrepancies in income distribution and progressive taxation; it would coincide if all citizens could vote. It seems likely that any such service government would be rare and short-lived. Among other things the generous motivation of sponsors might be inconsistent with their acquiring enough resources to continue the government on a generous basis. As noted in Section 3.3, generous transactions can demonstrate one's power but they cannot create it.

4.13.532   A government could fill the definition of a pure pressure organization only if it produces no public goods. This could be the case in an incipient government serving solely as a dominant coalition to prevent the use of force in individual transactions as in 4.13.4. To qualify as a pressure organization the government would have to exert pressure consistently on the same segment B of the public on behalf of some other segment A to improve A's position relative to B. The costs of government would be paid by A's, who would also be the sponsors. By contrast, if the costs were assessed against the whole public the situation would be a pressure activity of cooperative or profit government but not a basically pressure organization. This, too, seems an unlikely basic type of government.

Since service and pressure seem untenable as basic types of government organization, we will consider cooperative and profit types basic, but with admixtures of other activities. The profit form seems highly likely to arise because of its obvious attraction to potential sponsors. It is also the only type that can go on indefinitely without renewed contributions from sponsors (4.12.84). However, if the public had the power they would presumably try to make it cooperative. It could also evolve into a cooperative by gradual extension of the dominant coalition if the sponsors did not understand, or were not undilutedly selfish enough to use, the coalition-limiting logic of 4.13.525.

4.13.54   CONTROLLED ECONOMIES

Chapter Fourteen views a market economy as an uncontrolled system, or informal organization, and indicates the rationale of the categorization and how such a system operates. By contrast a centrally planned and controlled economy is a controlled system and a formal organization.

There is a distinct difference between the outputs of "normal" governmental productive acts and those of a planned economy. The former are generally public goods. They are not distributed among the population to be used only by those who want them. Decisions regarding public goods must be made according to the principles set down for formal organizations—that is, a single decision must be reached about them whether everyone agrees to it or not. An economy, however, produces private goods. Within the limits of his personal power and the technical capacity of the society to produce, an individual can have as much or as little of a good as he chooses. He can acquire and consume one package of goods while his neighbor gets an utterly different package. To speak in terms of two persons, A can have what he wants while B can also have what *he* wants. There is no need at any point for A and B to reach a joint decision about what they will have.

Let us deal with two pure, polar cases such as national defense and chocolate cookies. The government must reach a single decision on national defense. By contrast the government *can* make a single decision as to how many chocolate cookies each family is to receive but it obviously need not do so. In fact, no matter how dedicated a government is to the idea of central planning in principle it may well conclude that it is silly to make such a decision for the whole population from a central point. The present framework would suggest that markets provide techniques for such decisions but that the coalitional relationships do not—if we assume that consumer preferences have anything to do with the subject. But that is a long story that we need not go into here.

A second difference between an economy and government production of public goods is the sheer multiplicity of decisions required by the former. If we include component parts as well as finished products, an industrial economy requires decisions about billions of separate items—with the usual economic questions of what will be produced, in what quantity, by what techniques and with what materials, who will get them, and the ratios at which goods and factors will be exchanged.

The production of military goods is decided by the political process whether the economy as a whole is centrally controlled or not. Capital equipment, particularly for heavy, basic industries, can readily be treated as a public good if a nation is so inclined. But consumer goods inescapably show important aspects of a market system if consumers are paid in money and make their purchases with it since they will buy or not buy what they like.

This brief note on controlled economies is intended to make just two points. The first is to identify a centrally controlled economy as a formal organization. The second is to indicate that public goods require a single decision whereas private goods allow, but do not require, one. In conventional terminology we might say that politics deals with reaching a single decision among conflicting interests whereas economics deals with the mutually contingent but different decisions of transactions.

4.14   INTERNATIONAL RELATIONS

Relationships between nations can be viewed as intersystem if the emphasis is on the separate nations as entities and as intrasystem if the nations are viewed as subsystems of a larger system. As noted in the definition of sovereignty, a sovereign nation can be a member of a larger organization that is not itself sovereign; it can even accept sanctions from it other than legitimized force.

In the same sense that the theory of government is the theory of organization under the special condition of sovereignty (4.13.10a), the study of international politics is the study of communications, transactions, and nonsovereign organizations among sovereign parties. A model for interactions among governments would be complex. I would expect it to start with interactions among governments alone, then add interactions among subsystems of nations—international trade and tourism—not directly controlled by governments. Next would come effects of government actions on subsystem actions and the reverse, and government actions as agent of subsystems and the reverse. I do not develop any such model here but merely identify a small sample of effects on interactions of the fact that they occur between rather than within nations.

4.14.1   Whether between governments or any other units, communications between nations are subject to several difficulties besides those of communications in general. First, distances are greater on the average than for communications within nations. The relative importance of this factor depends on the medium used and the degree to which channels exist. Second, languages may differ. And third, the cultures may differ dramatically, in which case translation alone will not necessarily bring clear communication (3.17.81, .86, and .87). Since sovereignty is defined as an authority relation, which in turn is transactional, the analysis of communication, as such is not affected by the presence or absence of sovereignty.

4.14.2   Let us now assume that the relations between nations are strictly intersystem relations and not intrasystem interactions within some larger formal organization. Since the general transactional analysis of Sections 3.2 through 3.26 and the analysis of aggregate power in 3.3 are not limited by the assumption of sovereign governmental controls, since the absence of such limitations also characterizes relations between sovereign nations, and since governments can be parties as defined, those sections apply to the power and bargaining power relations among governments and constitute a large part of the basic theory of such relations. In fact the assumption of uniqueness in the basic model of transactions is the logical equivalent of sovereignty since it assumes that both parties are free of authority (or other transactional) constraints in dealing with one another. Problems of membership in organizations are also mainly transactional and reinforce the conclusion about the basically transactional nature of international politics, supplemented, of course, by bargaining coalitions. The reader can apply those sections to international relations for himself. The main task here is merely to identify a few salient aspects of transactions. The specialist might choose a different list.

     Since force involves much stronger motives than do transactions in goods (3.25.4) and physical blockage makes irrelevant any and all motives on one side (3.25.0c), the EP's of nations for goods will be dwarfed by their desires to avoid bads. In particular, war dwarfs other ingredients of transactional relations. Hence during peace the mutual assumption that peace will continue is the ongoing major bargain to which all others are subordinate.

An international transaction over goods (a bridge, fishing rights, terms of trade, airplane service, patent protection) whose value is not great enough to justify breaking the major bargain of continued peace is subject to transactional analysis in its own right. That is, the EP's in the main transaction do not become factors in the EP's over the particular transactions (3.28.51). However, if the desires involved in a subsidiary transaction are great enough to lead at least one party to question the major bargain of continued peace, the EP's of both parties in the major bargain must be added to those in the subsidiary transaction (3.28.52). It therefore follows that the nation with the greater stake in continuing the peace will suffer the greater loss of bargaining power in the subsidiary transaction (3.28.524). Moreover, being subject to the possible use of force, relations between nations are open to the instability of mutual threat relations (3.25.26). War is pretransactional, the stage of applying stress. In the overall sense its outcome depends on the relative transformational abilities of the two opponents. As such these abilities are not subject to social analysis although social processes within a side may influence the amount of such transformations it can produce. The transactional stage occurs after sufficient stress has been applied to block one side or to induce concessions and is subject to the general set of propositions about stress transactions (3.25.1). In this connection unconditional surrender reflects essentially complete physical blockage; conditional surrender constitutes a stress transaction.

4.14.3   Promises made by an individual spokesman on behalf of his country and concerning any important question of national power are rarely credible in their own right since an individual rarely, if ever, has sufficient personal stake to ensure performance of his promise (see Section 3.24.3). Although there are undoubtedly people who would not lie even to enhance their country's power position significantly, no other nation would be likely to risk its power position on their word. Promises made by a government as an entity may be more credible since a nation has stakes more nearly comparable to those of the individual. But the leadership of a government may change, perhaps dramatically, and new leaders may not honor promises of their predecessors. If the matters are important enough, no nation's or individual's promises will be credible. Who would risk anything important on the assumption that another would not lie to save his own life? Hence on matters of national survival, or matters that contribute significantly to it, trust is hard to come by since it is difficult to have stakes large and reliable enough to ensure performance of promises.

4.14.4   As agent for its own subsystems in dealing with other governments or their subsystems, a government is subject to the analysis of agent-principal relations detailed in 3.27.2 given the two following assumed special conditions. First, the government is not only agent for a multiplicity of its own subsystems whose goals are mutually conflicting but also a principal acting in its own behalf. Second, the government has a monopoly position on the agency function in numerous transactions on behalf of its subsystems. If the steel industry is dissatisfied with government negotiations on steel tariffs with another government, the steel industry cannot shift its "patronage" to some other government or private agency. The citizen who is unhappy about the government's slow service on his passport application cannot get one from a competing agency. The result is that the power and bargaining power of the government are higher, and those of the citizen or other subsystem are lower, than they otherwise would be (3.27.3).

4.14.5   In 3.22.6 we noted that the greater the difference in preference orderings between two parties, the greater the possible gain per unit from transactions between them. To the extent that different nations have different cultures with widely differing preference orderings, the possibilities for large gain from transactions is enhanced, particularly with respect to private goods. Such trade is limited, however, by transportation costs and by the possibility that the value systems are so different that persons in one nation simply do not want things available in the other. Americans do not want breadfruit and its producers do not want bronze valves. By contrast the per-unit gain in trade between nations of similar preference orderings is less, but the likelihood that they will want one another's products is greater.

4.14.6   A government may have many different kinds of interactions with another government; and each interaction affects its interactions with other governments. Given only a few governments and a few topics the number of possible combinations rises astronomically. The general logic is the same as in the earlier illustration about beliefs, expectations, beliefs about expectations, and so on (3.47), but it is even stronger now because of the larger number of parties. All we can do here is to identify the situation as as idiographic E-level problem calling for a simulation model.

4.15   Whenever two or more parties interact repeatedly it is presumably possible for them to handle their relationships through an organization if that seems advantageous. Thus nations can form limited-purpose organizations, and in doing so they are subject to the general analysis of formal organizations (Section 4.1). In particular, if the nation-member is free to leave when costs of affiliation exceed benefits, continued membership legitimizes the organization's authority over members. Similarly (but unlike a government's relation to its citizens), the organization can expel the member that does not accept its authority. By contrast, for a higher-level government unequivocally to outlaw force between regular governments (that is, to become sovereign over them) it must face the same imperative as any other government—namely, it must gather a coalition of force that will dominate any challengers. Without a formal organization, a bargaining coalition of sovereign governments might eliminate wars in a higher-level parallel of a police force among individuals (4.13.42). Such a result is in theory far simpler to arrange than through an organization of many nations with a decision rule of majority vote, though it is not necessarily easier in practice.

4.16   **CONCLUSION**

The long-standing preoccupation of political science with social and moral philosophy, "good government," social welfare, and related value-oriented notions is now a sufficiently dead horse within the profession to need no further kicking. To the extent that the discipline has been historical the observations about history in Chapter Seventeen apply.

It was suggested here that the study of government—and hence, presumably, political science—is the study of formal organizations with attention to the consequences of the special condition of sovereignty. We have already identified organization as involving a distinct problem of decision-making. As sociology deals with consensus by communication (culture) and economics with consensus by transaction (market), political science deals with consensus by dominant coalition. But consensus by dominant coalition is part of the study of organization and is implied in the suggestion that political science deals with organization. It is my impression that this delineation of the study of government is not a bad characterization of what many political scientists are now working toward.

Despite the difference in terminology, this chapter seems distinctly congruent with Easton's view (1965, pp. 51–54) of political science as dealing with political and parapolitical systems. To his indication that political systems differ from the parapolitical by covering the whole society and having exclusive use of force (corresponding to assumptions 4.13.1b and e here) this model adds involuntary membership and potentially unlimited function, though both are perhaps implicit in government's coverage of the whole society. I take exception, however, to Easton's "authoritative allocation of values" as a distinguishing feature of political interactions. As formulated here, any authority in any organization "grants or withholds rewards or punishments" and thereby "allocates values." Although the government's monopoly on legitimate force certainly can make its allocations *more* authoritative, authority without force is not therefore unauthoritative, and legitimate force itself seems to be the distinguishing feature. Or perhaps it is the dominant coalition that is authoritative (Easton's sense), in contrast to the consent involved in agreement by communication and transaction. But that, too, characterizes all formal organizations, not just governments.

It should be noted that the four-way categorization defined in Chapter Twelve and applied here to governments is relatively peripheral to the overall approach of this volume. Hence someone who dislikes this way of categorizing organizations in general or governments in particular can ignore it without necessarily affecting his use of other parts of the analytic structure. In any event the parts of this volume of most direct utility to political scientists are perhaps not so much the materials of this chapter (though I hope they make political sense) as the suggestions that "the whole of the preceding analysis applies." The most relevant "preceding analyses" are those of transactions in general, stress and threat transactions in particular, organization, authority, legitimacy, and dominant coalitions.

With reference to international relations I agree with Holsti (1967, p. viii) that "all governments seek to achieve or defend their objectives through basically similar means, employing threats, rewards, and punishments." And by putting such relations on a mainly transactional basis this chapter helps implement Holsti's suggestion. But Holsti's tendency to downgrade attention to "comparing capabilities" of nations I find untenable since it is precisely the aggregate magnitudes of the credible threats and sanctions a nation can muster that determine how much response they will bring.

Although the number of definitions and assumptions in this chapter may seem excessive, I am sure the list is incomplete. I would nevertheless insist that only by making explicit *all* significant assumptions underlying a statement will we achieve good science, even if we later drop some of them for workaday analysis. The effort, at least, I hope can be applauded, even if not the immediate result. And if this approach sometimes belabors the obvious I can only reiterate Jordan (1968, p. 15): "To be of any scientific use, the obvious must be reformulated into a conceptual system." Even if this approach to political analysis is not seen as an improvement over other approaches (or even if it is only slightly worse), it nevertheless deserves some attention because of its direct tie to other social sciences and the broad systems base.

Perhaps I should close with a reminder to those not fully accustomed to working with theoretical models. One can legitimately conclude that this model is of little use to the political scientist, if he so finds it. And one can also legitimately criticize a conclusion if it does not follow from the definitions and assumptions. But it is not necessarily a criticism of a statement to argue that it is not in accord with reality.

Chapter 14

4.2 Informal Organization

The preceding two chapters introduced a model of pure formal organization and then examined a conspicuous example— government. This chapter introduces a model of its opposite, the pure informal organization, and then moves to its most conspicuous (or at least its most theorized about) example—the market economy. This treatment is followed by a discussion of informal groupings in areas usually studied by sociologists, particularly ethnic groupings and stratification. At this level of system, which encompasses more or less the whole breadth of society, informal organization means human ecology as defined in this chapter. The model of pure informal organization, defined as an uncontrolled system of controlled human subsystems, requires total lack of concern by individuals about the whole system (compare Hawley, 1950, p. 179). Because we can never feel quite sure about those complicated critters known as humans but are nevertheless confident that the deer and fern neither understand nor seek to manage the total system of which they are a part, we use the ecological system of the forest or pond as a prototype.

Strictly speaking, a dyad constitutes an informal organization whenever they communicate or transact along the lines of previous chapters without pursuing a joint goal in the sense discussed in connection with simple formal organization. To the extent that continued dyadic interactions produce such outcomes as a private language used only between its members or some differentiation of role and function, they constitute a system worth analyzing in its own right—

368

that is, there is a significant sense in which the interaction produces a result that is more than the sum of its parts. For purposes of this volume, however, we confine attention to informal organizations of much larger size.

Organizations between pure formal and pure informal are dealt with in the next chapter. On the whole they are probably of more interest than the pure informal ones but cannot be properly analyzed until we have identified the two polar cases. Except for the following set of assumptions, the discussion of informal organization and ecology largely departs from the formal models with their definitions, axioms, and deduced propositions. Discussion of the general model focuses on two aspects of informal organization: the application of certain concepts from formal organization and the meaning of equilibrium in informal organization. I hope that my occasional shifts to the biological level of ecology clarify rather than confuse the issue.

**4.21   GENERAL CONCEPT OF
INFORMAL ORGANIZATION (AND ECOLOGY)**

4.21.1   *Model and Assumptions*

The following assumptions constitute the model of pure informal organization. With the exception of assumption **h**, which is irrelevant rather than wrong for that purpose, these assumptions should also be valid for nonhuman ecological systems (ecosystems) such as the forest, field, or lake.

a. There is no detector, selector, or effector for the system as a unit. The whole system has no goal and makes no decisions. The whole (or main) system neither determines what subsystems it will contain nor defines their roles or goals whether consciously, as in formal organization, or otherwise, as through the genes of an organism. Like many other traits of the system, its structure merely happens as an outgrowth of the interactions of its parts. Except in the sense that the whole system is environment to each of its subsystems and that the environment constitutes a set of parametric constraints for a system, the whole of an informal organization places no constraints by directed action on which subsystems may interact with which other subsystems or on the type or frequency of the interaction. In particular, the informal organization does not restrict the use of bads in transactions.

b. At least some components of the system are controlled systems, which do have goals. The essence of ecological systems and informal organizations is that they consist of the uncontrolled interactions of controlled systems. "At least some" means, for example, that the biologist might bound his ecosystem to include air, ambient temperature, and soil nutrients or he might focus solely on the interacting organisms. The human ecologist might similarly bound his system to exclude climate, space, machines, or natural resources. Since we are not interested in servomechanisms here, "controlled systems" means only organisms and formal organizations.

c. Every controlled-system component is oriented solely toward its own welfare; it is wholly indifferent to the welfare of the whole system. Except to the extent that it may be interested in helping or hurting other subsystems which are of direct concern to its own welfare (the bear cares for its cubs; the housewife maligns the grocer who overcharged her) each subsystem is also indifferent to the welfare of all other subsystems.

d. Either matter-energy or information may be exchanged among the controlled-system components. If those components are humans, their interactions may include communications, transactions, and suborganizations.

e. The outputs of some component systems, and sometimes the systems themselves, constitute essential inputs to other components, though the degree of direct dependence between a pair of systems may be insignificant.

f. The system is closed. No new component systems (or species of systems, if we are considering an extended period) are introduced by evolution or migration, and no significant aspects of the environment (such as climate, topography, or rainfall) change significantly during the period under consideration. This assumption is included to limit the scope of the present analysis, not as a defining characteristic of an ecosystem or informal organization.

g. Since by definition an uncontrolled system does not engage in any behavior as a unit, two such systems cannot interact as units in the sense that two persons or corporations interact or a mongoose fights a cobra. Neither does it have any inputs or outputs as a unit, as distinct from the sum of inputs and outputs of its components. Components of one system, however, can interact with components of another. If we view the flora of an area as one system and the fauna as another, components of one interact with components of another when the deer eats saplings and his excrement fertilizes them. And although all land-based fauna interact with sea-based phytoplankton via the oxygen cycle, such actions represent the uncoordinated behaviors of individual specimens, and I see no fruitful way of viewing the situation as two large systems interacting as units.

Similarly, in a competitive market system one buyer can interact with one seller, but the totality of buyers does not interact as a unit with the totality of sellers. Or the individual rich can interact with the individual poor, but the rich do not interact as a unit with the poor as a unit—in the pure model.

h. There is no major bargain specifying the terms under which a component enters, remains in, or leaves the system. Nor is there any exercise of authority on behalf of the system. This assumption does not preclude such bargains over membership within a subsystem.

i. As elsewhere, the boundaries of a system are set by the observer. An ecological system under study might be all the life among the billion or so bacteria that inhabit one gram of soil (Williams, 1964, p. 1), all the flora of a pasture, or all the flora and fauna, including humans, in a river basin. Among human systems it may be five workers in an office or all people on earth.

When this model is compared to that of pure formal organization, several additional characteristics of informal organization can be deduced without making further assumptions. To a large extent they apply also to ecosystems. First, the informal organization has no sponsor, staff, or recipient roles; those roles apply only to organizations that in some respects act as a unit. By corollary, since organizational types are here based on the relationships among those roles, informal organization cannot be classed as cooperative, profit, service, or pressure. (Subsystems can take any of these forms.) Relations of the sort the biologist calls *symbiotic* are seen here as mutually advantageous transactions, communications, or possibly simple organizations among subsystems. Second, since subsystems have no bargain of affiliation with the main system and no authority, there can be no question of legitimacy. Nor can there be induction ceremonies signaling entrance into membership in the main system, though there can be in the subsystems. Objections to these first two points should be withheld until the reader is familiar with the semiformal model. Third, different subsystems or their role occupants can have differential power within the whole system since they have differential access to or control over flows of things wanted by other subsystems. And fourth, the informal organization can be simple or complex in the everyday sense of the term but not in the technical sense of 4.11.0 and 4.12.0 since that distinction is related to the content of the major bargain.

4.21.2   *Structure*

The concept of structure was introduced in connection with complex formal organization because it is easy to conceptualize in that context. Nothing in this model, however, prevents pure informal organization from having a structure or, as we will see in the next section, from having a structure that is remarkably persistent and self-preserving. The concepts pertaining to structure as developed in connection with formal organization seem applicable without change to informal organizations and ecosystems. If we continue the definition of structure as the "pattern of the main system described in terms of its subsystems and their roles," the ecologist might describe the structure of a system by identifying the various species present and specifying the actions and interactions of each that affect other species and the balance of the whole system. However, it is irrelevant to the definition whether he chooses to view the individual, species, phylum, or order as the lowest level of subsystem; whether he puts all plants into one subsystem and all animals into another; whether he distinguishes subsystems by some general trait like carnivorous and herbivorous; or whether he separates them by size. His choice will affect the description of the structure but not the concept of structure. Similarly the social scientist might describe the structure of a society in terms of such subsystems as families, churches, firms, and governments. However, he might focus instead on class levels, ethnic groupings, or occupations as his relevant subsystems. The political scientist might choose interest groups, government levels, government branches, or departments as his relevant subsystems. Whereas the structure of a formal organization is perhaps most likely to be described in terms of the same subsystems that the management has planned and specified, the absence of such conscious specifications leaves more leeway for the student of informal organization to choose units for his purposes. However, a sociologist or anthropologist studying a corporation will probably slice it up into a quite different structure than does the president.

The definition of role can also remain the same for informal as for formal organization: "the set of system states and actions of a subsystem of an organization including its interactions with other systems or nonsystem elements, in or out of the organization, at the same or different levels." In seeking parallels between the two we can also substitute ecosystem for organization at both points in that definition. Because we have seen that an uncontrolled system cannot interact as a unit with any other system, we must apparently add that only a controlled system can have a role. The concept of role is independent of the number of occupants. A thousand vultures eating carrion perform the same role as one, even if they eat more. And it is the vulture, strictly speaking, that is the acting subsystem, even if for convenience we refer to the collective category of vultures. We thus use language parallel to that of considering a machinist a subsystem of a manufacturing plant, not the collectivity of machinists who occupy identical roles.

A role in informal organization is the equivalent of an ecological niche; an unfilled role is a niche with no inhabitants. In a market economy an industry is one type of niche; demand for some unproduced product is an empty niche. (One is tempted to define a niche as that which, if there were nothing in it, would be empty.) The equivalent for a formal organization is an identified role without occupants.

In contrast to formal organization, in which the structure is consciously prescribed by the sponsors (4.12.3c), the structure of informal organization and the ecosystem is solely that which evolves from the interaction of the subsystems. That is, the assumption that the sponsors totally determine the structure, partly relaxed in 4.12.52, is here carried to the extreme where there is no sponsor control. For informal organization the distinction is a partial parallel of that between *ascribed* and *achieved* roles (Linton, 1936, p. 115) but here applied to the origin of the role itself rather than to its occupancy. There is, of course, an important distinction between roles in informal organizations and those in ecosystems. Despite some flexibility, the roles of organisms in ecosystems are for the most part genetically determined: the squirrel cannot change roles with the salmon or the eagle. Hence the problem there is mainly one of numbers of occupants in each role, not one of role content. In fact, under the assumption of a closed system if the number of occupants in a genetically determined role ever reaches zero the role will remain permanently unfilled thereafter. By contrast the flexibility of human learning allows role modifications and shifting of occupants into vacant roles. The obvious exception is a role that requires esoteric skills which disappear with its possessors. Thus there is vastly greater potential for role modification and individual transfers across roles in informal organization than in ecosystems, even if the constraints of cultural transmission sometimes allow little more flexibility in fact. Where the roles come from in the first place is a question we will examine with social change in Chapter Sixteen. Why roles tend to be self-stabilizing under many circumstances is discussed below in connection with equilibrium.

A sharp distinction between the pure formal and informal models, the latter supplemented by the ecological prototype, helps clarify an aspect of so-called functionalist analysis that is sometimes distressingly fuzzy. In the pure model of formal organization it is wholly legitimate to state that a particular role exists *in order to* perform a certain function since that is a precise explanation of why the sponsors created it. It is a consciously instrumental means of accomplishing an end. In the pure informal model, by contrast, it is no more allowable to say that a role exists in order to perform a certain function, or because that role is needed to make the system work, than to suggest that it rains on the forests but not on the desert because the forest is where the plants are that need the rain. If a subsystem role had not existed a different overall system would have developed that did not need that role. Because role changes by subsystems of an ecosystem are narrowly circumscribed, this concept is perhaps clearest in that case. To display an economist's bias, I would add that in informal organization a certain role may be filled because other subsystems want its services and reward those who provide them sufficiently to induce them to enter the role and provide the service. I find this language somewhat less mystical than merely to assert that the role is filled because it supplies a needed function. Intermediate cases are examined here in connection with semiformal organization, and I suspect that some ambiguities of functionalist analysis have arisen because it has traditionally been used most often in those relatively murky nonpolar cases.

The concept of integrative processes (4.12.5), like those of role and structure, can be carried from formal to informal organization with no change in basic meaning. The important difference is that in the pure formal organization the details are all specified by the sponsors (4.12.3c) while in the pure informal case none are thus specified. Except for that difference, all that was said in 4.12.51 about coordination, integration, cohesion, and attraction applies to the informal organization as well.

4.21.3   *Ecological Equilibria*

Before we discuss equilibrium tendencies we must identify the variables that must be reasonably constant if we are to say that an equilibrium exists. For reasons indicated above, in the biologist's ecosystem these variables are presumably numbers of specimens, and we say that the system is in equilibrium if the relative population sizes of species do not change much over some extended period or if certain cyclical changes are of relatively constant duration and amplitude. For human ecology the content of the roles as well as the number of occupants of each would have to be stable or reliably cyclical to constitute an equilibrium. How long may a cycle be and still be considered cyclical rather than permanent change? We leave the answer to the investigator in the particular case. Because the problem of equilibrium is simpler if we attend to number of role occupants while leaving role content constant, we focus heavily on bioecology.

⌣   For an ecological system with numerous interdependent controlled subsystems there does not seem to be any basis to conclude with confidence either that the system definitely will reach a reasonably stable equilibrium or that it need not do so. For bioecology, for example, Poore (1964, p. 220) states that little or nothing is known about the nature of the community under uniform conditions and that no extensive quantitative studies have been made of a single community. Boulding (1970, pp. 24f) similarly states that in an ecological system, where "everything depends on everything else," an equilibrium is possible and often occurs in fact. But he does not say that one will necessarily occur. Nevertheless a theoretically closed (that is, static) microeconomic system, equally uncontrolled, definitely will reach an equilibrium. The hypothesized conditions under which it will assuredly do so are narrowly circumscribed, however, and there is no assurance that a real economic system, even if closed, would in fact reach an equilibrium, or if it did that the equilibrium would fall at the point specified by the theory. An open system could presumably go indefinitely without reaching equilibrium—though it might reach one. The somewhat mystical-sounding "tendency to equilibrium" may mean that a set of nonequilibrating components will continue to undergo change in their relationship until they stumble into some condition that happens to be equilibrating. If the boulder keeps rolling downhill until it gets stuck or reaches bottom, there is no magic in the fact that on the average the observer is more likely to see it still than rolling. For uncontrolled systems we should be careful not to read anything more than that into anyone's assertion of a "tendency to equilibrium."

⌡   Even more basic is the question of whether an ecosystem should be considered uncontrolled—a question that biologists have long debated, though in somewhat different language. Following Hawley (1950, p. 180) in using *community* as a close parallel to the present concept of ecological system,[1] Clements (1916) early regarded the biological community as an organism, a viewpoint that would make it a controlled system in present terminology. By contrast Gleason (1939) regarded it as a "fortuitous juxtaposition of plant individuals." Assuming interaction among the juxtaposed individuals, however, Gleason's

---

[1] Poore (1964, p. 213) defines *community* as "any collection of plants and/or animals occurring together which has as a whole a certain unity." Integration is definitely part of his concept, and an "unorganized" group is not. In connection with this chapter the reader should certainly see Boulding (1970, chap. 2) on economics as an ecological science.

concept is that of an uncontrolled system. After describing the contrast, A. S. Watt (1964, p. 204) indicates that the organism view is nowhere accepted now. That is, the ecological system is not presently regarded as controlled.

By defining an ecological system as an uncontrolled system of controlled subsystems this model takes an unequivocal position on that question, which is either left obscure or ignored in many places. I state here (1.20s and 1.4) that empirically there are only two types of controlled systems: organisms and certain man-made systems, the latter consisting of formal organizations and servomechanisms. Chapter Fifteen explicitly posits a continuum between wholly controlled and wholly uncontrolled systems at the level of human organizations. But I see no such continuum with regard to other types of system (1.20s). In particular I view the biological ecosystem of forest or pond as unequivocally uncontrolled, except to the extent that humans may impose conscious controls on it. There nevertheless *is* a question, which is identified and handled as follows.

By definition a controlled system has mechanisms for maintaining a particular equilibrium and returning to it following disturbance whereas an uncontrolled system takes whatever equilibrium it happens to reach. The question is whether we should firmly classify ecosystems as uncontrolled despite the fact that even after severe disturbance they often return to the *same* equilibrium as before, not merely *an* equilibrium. The climax forest indigenous to a given ambient environment is a clear example, as are tundras and many aquatic systems.

There are two main dimensions to this problem. The first is the reaching of a particular equilibrium through the operation of an explicit control mechanism as contrasted to a situation in which there is only one condition that will equilibrate all the forces at work. By simple analogy, the fact that a pendulum reliably comes to rest at the same position does not make it a controlled system. Here it is again important to distinguish constraints, which limit possibilities, from controls, which detect and correct error. The fact that an ecosystem contains a vastly larger number of variables does not change the principle. Hence I assume that for an ecosystem reliably to return to the same climax state in a given environment means merely that the climax is the only condition which durably equilibrates all forces, not that the system contains some control mechanism as a whole unit. As noted (1.20s), in case of doubt "if the investigator can isolate subsystem behaviors that seem usefully construed as detector, selector, and effector functions he will presumably understand the system better if he treats it as controlled. Otherwise not." Thus far I have seen nothing that can be construed as DSE functions for an ecosystem as a whole. Incidentally, if there is any tendency for an ecosystem to stabilize at a point that maximizes its biomass or information, as suggested by Colinvaux (1973, pp. 564f), the conclusion would apparently have to come from empirical investigation or specialized theory, as I see nothing about the general theory of uncontrolled systems to support it. In fact I expect that little more could be said from even intensive study than that biomass and information tend to reach the maximum value that they can reach under the circumstances—which statement may not be very informative.

Regarding the second dimension of the controversy, an ecosystem reliably returns to the *same* climax condition only if the environment has not changed much. To illustrate, if the average outdoor temperature drops from sixty to forty degrees, a controlled heating system can maintain the same equilibrium indoor temperature. But a comparable drop in average temperature may well shift the climax position itself in an ecosystem, as from maple-beech to pine-birch dominance, and other changes of equilibrium position would result from changes in rainfall or soil nutrients.

Let us now return to the problem of changes in role content in informal organization. Neither the model nor other evidence or reasoning seems to prescribe that successive mutual role modifications (4.12.52) will either reach an equilibrium or continue through a series of indefinite changes. A repetitive cycle can be construed as an equilibrium subject to oscillation. It can be concluded that any system empirically observed to be relatively stable *has* reached an equilibrium, in which case each role has shaken down to a settled position. We may hypothesize that the equilibrium position is reached initially by a positive-feedback relation in which the images held by each role occupant move successively closer to the actual behavior of other role occupants while the actual behaviors move successively closer to the images with an asymptotic limit at identity. The process is a perceptual parallel of the communicational consensus discussed in 3.18.4 and .5. Once a new role has been carved out and shaken down in the expectations of its occupants and the occupants of related roles, the new occupant will find his interactions with other subsystems easier initially, *cet. par.*, if he conforms to the existing expectations. That is, it is easier to move into an existing role than to create one in the first place, an important exception being a role so peculiarly a product of its originator that no one else can fill it. Several points already cited support this generalization. Among others, role violation creates dissonance, it forfeits the communicational and transactional efficiencies of an established relationship (3.18.4), it forces modification of the performance of other roles and perhaps the roles themselves (4.12.513 and .52), and it questions or upsets established statuses. Other effects have been indicated in Chapter Eleven. If a role is eventually changed by a new occupant and all related adjustments are made, there is no apparent a priori basis for assuming that the new role will be better or worse than the previous one for the whole system or any of its subsystems.

Several generalizations relevant to informal organization can be approached through the biological level, to which we now return. In an uncontrolled system with many types of controlled subsystems it seems inescapable that equilibrium positions will occur in certain subsystem interactions, particularly when oppositely paired interactions occur. The predator-prey interaction is such a situation: the number of predators varies directly with the number of prey and the number of prey varies inversely with the number of predators. A similar relation exists between parasite and host if the parasite is damaging to the host. The period and amplitude of oscillation around the equilibrium position in such cases depend on such factors as length of reproductive cycle, typical brood size, and ability to hide or escape. Relatedly, the greater the number and diversity of species in an ecological niche or role (for example, the larger the number of different species that prey on a species and the larger the number of species of prey for a species of predator), the greater the apparent stability of the equilibrium and the smaller its oscillation. "Greater number and diversity" can also be described as a larger variety of substitute performers in a role and perhaps as greater redundancy. This notion is an empirically based hypothesis taken from Pimentel (1966, p. 25) that reportedly dates back at least to Herbert Spencer. I suspect that a simple logic supports it, and I would expect the conclusion to apply equally to informal organizations.

According to 1.31.4 a system in stable equilibrium is less subject to entropic change than one not in equilibrium—that is, less subject to loss of diversity. In this connection we note that in a stable ecological equilibrium each species plays a relatively unchanging role relative to the other species. In pragmatic terms this proposition may represent little more than the truism that a species whose food supply reliably continues at a level already demonstrated to be adequate, and the threats to whose existence continue at a level already proved tolerable, is less likely to become extinct or depleted than a species for which those two conditions do not prevail. Less tritely, we can designate a species or other biological subsystem as s and the totality of all other biological systems as S. Then once a relationship among numerous subsystems has reached an equilibrium, a condition exists in which every *s* has a good, or at least viable, other-system environment *S*; and any *s* for which this is not so will already have disappeared. Hence the greater the importance of environmental factor *S* relative to non-*S* factors (temperature, rainfall, and so forth), the more stable the whole system will be in the face of changes in non-*S* factors. The role in informal organization is apparently analogous to the species in the ecosystem; the role occupant is analogous to the individual specimen. There is the difference, of course, that the role of the individual human may not be determined until he is twenty years old or more whereas the role of each species and specimen in the ecosystem is determined by its genes. That distinction is not crucial, however, within the level of generality in this chapter, although it would presumably matter for more detailed development.

For reasons related to the first point in the preceding paragraph, a system in stable equilibrium resists the effects of environmental changes better than one not in equilibrium. Here I am presuming that Poore (1964, p. 217) uses "integrated structure" with about the same meaning as "stable equilibrium" here when he says: "Communities with an integrated structure are more resistant to environmental change in time or space than are nonintegrated communities." He notes the "persistence of relict whole communities in areas which have become climatically more suited to other vegetation." We may hypothesize that this statement about the biological can be carried directly over to the human as well. As is the case with controlled systems it is probably not possible except by accident to optimize both the whole system and all its subsystems simultaneously (1.35).

Within an environment whose variations do not exceed the adaptive capacity of the system, once a system has reached a stable equilibrium it is sufficient for each of its subsystems merely to continue its behavior unchanged in order to maintain the system equilibrium indefinitely. This proposition is also valid for formal organization (Section 4.12.3) and seems to hold an important theoretical point— namely, that once a balanced, coordinated structure has been developed, whether by accident or design, the continued effective performance of either a controlled or uncontrolled system requires nothing more than that each controlled subsystem meet its goals or merely continue its behavior unchanged. The stable equilibrium is thus an important connecting link between the theories of controlled and uncontrolled systems. By the same token the concept of equilibrium has large potential for analyzing the organizations between the pure formal and pure informal—a matter for the next chapter.

4.21.4   *Summary*

By the nature of the models, a tighter and presumably more detailed theory can be built about formal organizations and other controlled systems than about informal organizations and other uncontrolled systems. If the available information, the goal, and the internal and external parametric constraints of formal organization are known, it is reasonably easy to imagine that a determinate theory could be developed about its external behavior and the way it would tailor its internal structure toward its goals. Even though it is arguable how complete such a theory could be, the point is that in the present framework the analysis would hinge around the within-system concepts of detector, selector, and effector functions.

By contrast, since the pure informal organization, like the ecological system, does not prefer one state or condition to another, has no homeostatic processes for maintaining any such position, and does not specify its own structure, any theory about the states it will reach must presumably be based on the intersystem analysis of the relations among its subsystems. Intermediate situations among humans include the informal aspects of formal organizations and the formal aspects of informal ones. Since these cases are mixed, so must be the analysis as in the next chapter. And since a model of the mixture involves more assumptions than the model of either pure case taken separately, its analysis is more complex. For reasons parallel to those of 3.47 on nomothetic and idiographic approaches, there is no reason to assume that additional basic concepts are required.

When an organization is so complex that a tight analytic model becomes too unwieldy to permit deduced conclusions, we have two main alternatives. The first is to abandon the general-purpose model in favor of the special-purpose, or simulation, model discussed in Chapter Sixteen. All simulation models (in this framework) are nevertheless constructed from the same analytic building blocks. The second alternative is to use statistical generalizations based on empirical observations of particular types of organizations and their behaviors. Here the units of observation, or "protocol data," should be selected to match the concepts defined in the present framework. As indicated in Chapter Eleven, many of those concepts should be relatively easy to operationalize.

It is obvious that a given type of uncontrolled system can have a theoretically determinate solution, the conspicuous case being the neoclassical, microeconomic system. Whether other determinate models can be formulated satisfactorily future investigation alone can answer.

4.22   **MARKET ECONOMY:**
              A SPECIAL-PURPOSE INFORMAL ORGANIZATION

Chapter Twelve developed the general model of formal organization; Chapter Thirteen used that model as the base for an explicit model—that of government. The technique was to start with the general model and then modify it by changing certain assumptions —notably by making the transaction of membership involuntary for both organization and member and by giving the government a monopoly on the use or authorization of legitimate force. This section repeats the procedure, but with different content. Having now stated the general model of informal organization we construct from it a model of a market economy. I will be considerably less explicit about details of this model, precisely since economists have themselves already constructed an inordinately explicit model of their system and it would require an encyclopedia merely to abstract the applications that have already been made to particular situations. Conspicuous are the model's exclusive focus on transactions in goods, not bads, and the government's coalitional function of joining against the violators of contracts to enforce the fulfillment of promises. Other details are mentioned below without any attempt to be comprehensive.

Real economies occupy a spectrum from highly controlled to mainly uncontrolled. As with other analytic tools we deal first with the theory of the pure cases at the ends of the spectrum. As a controlled system the controlled economy was discussed briefly in connection with formal organizations in Chapter Thirteen. The uncontrolled system belongs here. Explicit governmental controls in mixed economies are formal intrusions into otherwise informal organizations. They are not semiformal in the sense that term is used here—though many semiformal behaviors do occur because individuals have many ideas about how the whole system ought to behave and they often try to implement them.

The theory of the microeconomic aspects (firms, product markets, and factor markets) of a pure competitive economy is conceptually parallel in main contours with that of an eological system even though ecology deals with the allocations of the controlled subsystems themselves and economics deals with the allocations of their artifacts. Each component subsystem within the market economy behaves solely on the basis of its own inner logic—its information (detector), its goals and drives (selector), and its performance capacities (effector). The precise reason for the conspicuous success of theoretical economics is that it is explicit about the system states of its main subsystems. The assumed detector state of every firm in pure static theory is perfect information; the assumed selector state is unalloyed profit maximization. Although for purposes of welfare economics the additional assumption of utility maximization for consumers and factor suppliers is also necessary, the market equilibrium is nevertheless theoretically determinate if there are factor supply and product demand curves, whether or not they represent utility-maximizing positions for the suppliers and consumers respectively. The main system has no goal, does not specify its own structure, and issues no instructions to its subsystems.

Equilibria abound because oppositely paired relationships (negative feedback) occur in each significant economic subsystem. In each industry, for example, the quantity demanded varies inversely and the quantity offered varies directly with the price in downward-sloping demand curves and upward-sloping supply curves. We can also express this relationship by saying that for each successive unit of a given good the price offered for it varies inversely, and the price asked varies directly, with the quantity. In a special case, the long-run supply curve in a decreasing-cost industry is normally assumed to slope downward. To amplify 2.56.3, this conclusion is tenable only so long as the whole industry is an insignificant fraction of the whole economy. If an industry expanded until it could purchase additional inputs only by bidding successively more of them away from other industries, its (opportunity) costs would eventually rise. Hence we assume that all cost curves rise in due time, thereby making the upward slope of supply unexceptionable.

The firm represents a second equilibrium. On the input side (factor market) the benefit to the firm of each successive unit of a given factor eventually falls and the cost of each successive unit eventually rises. On the output side (product market) the benefit to the firm of each successive unit produced, in the form of marginal revenue, eventually goes down and the cost of each successive unit goes up. Within each consuming unit, the benefit of successive units of a good goes down (due to diminishing utility) and its cost (in the form of other goods sacrificed) goes up. At the macro level, savings and hoarding vary directly while borrowing and dishoarding vary inversely with income.

Because of these oppositely paired relationships an equilibrium is theoretically possible in all the major subsystems of the uncontrolled economy: the market for each product, the market for each factor, the firm, the household, and the whole economy. In a closed system (static analysis or steady state) numerous other subsystem equilibria also occur, as does an equilibrium position for the whole system. Under the theoretical assumption of single-minded profit or utility maximization by all subsystems these equilibria are not only possible but inescapable. A closed, competitive economic market system is thus a form of ecological relationship in which all subsystems are coordinated into a smoothly functioning net, in which both the whole system and all subsystems are in stable equilibrium. Once that static equilibrium is reached it will not change. As Boulding (1958) puts it, anyone with the power to change it lacks the will and anyone with the will lacks the power. The power and the will, incidentally, reflect the effector and selector functions.

It is no accident that much of the core of microeconomics, which is static theory, is compatible with the static model of the complex organization in 4.12.3. One aspect of that similarity needs special mention for economics, where the complex organization in question is the firm. As the static model of organization is amplified in 4.12.31 and is modified in 4.12.32 "to allow for moderate variations in quantity, quality, or value of inputs, outputs, or techniques," the organization (firm) can operate homeostatically. In a well organized firm such homeostatic operations can be carried on without the intervention of top management. To exaggerate only a little, the core of microeconomics encompasses only those decisions within a firm that could be made satisfactorily if the whole of top management were off on vacation. I continue to entertain a feeble hope that economics may eventually develop better theory (possibly by joining with other disciplines for the purpose) about the nonmarginal, nonhomeostatic decisions that occupy most of the time of real managers. Those decisions include those of 4.12.33 and many others.

To shift attention back to the equilibria of the whole economy, no real economy can remain closed. People age and die, and their replacements are never identical. The total population changes, fields erode, forests burn, mineral deposits are depleted. Even stable cultures evolve new techniques, and most cultures adopt techniques from other societies. Despite these changes, the static analysis has clarified many aspects of real economies.

Even in the course of change, however, at any given moment there exists a theoretical equilibrium at which the economy would rest if all forces acting on it were frozen—the logical equivalent of a closed system. That point is generally presumed by economists to be an optimum, any change from which malallocates resources. Clearly *optimum* and *mal*allocation represent value judgments that depart from science to ethics. Even within the latter realm they are open to grave question, as the literature of welfare economics abundantly attests. But our problem here is one of science. There are two important conclusions: that an equilibrium for the whole system can be reached while numerous subsystems are simultaneously in equilibrium; and that although the equilibrium position is theoretically determinate, the system does not "prefer" or maintain a particular equilibrium. If any parameter of the system changes, the system will move to a new equilibrium; it will not return to the old. To those who add the value judgment, any equilibrium reached by a competitive market is optimum. As economist Frank Knight reputedly put it, equilibrium is equilibrium is equilibrium.

To use a loose parallel in the language of ecology, the dominant species in the economic forest are firms, households, banks, and governments. There may be subspecies of each: firms fall into the subspecies known as industries, banks into commercial and savings subspecies, and governments into those subspecies that create money and those that do not. Each species and subspecies fills an ecological niche; and each niche is a subsystem that performs a particular role in the system. In the informal organization each species and each individual finds or carves out its own niche since the roles and their occupants are not specified by the main system (the main exception being in the creation of money). Each firm, household, bank, and government is a controlled subsystem; their interactions are not controlled. The noncollusive industry is an uncontrolled subsystem. To pursue the ecological language further, the main interactions are interspecific; intraspecific interactions are minimal or nonexistent. That is, within the economic realm households deal primarily with banks, firms, and governments, not with other households. Firms engage in many exchanges with other firms, but primarily with those in other industries. The logical basis for this arrangement is that transactions are based on differences in the relative valuations of things. Systems that occupy the same role tend to have similar valuations whereas occupants of different roles tend to have disparate ones.

In an industrial economy an equilibrium position shows considerable self-preserving qualities. It is not merely the current rate of output of steel, textiles, wheat, and so forth that constitutes the equilibrium; it is also the large amounts of capital equipment these industries possess. To a new party appearing on the scene (person, household, firm, bank) the existing system is the environment into which it must fit. If the system is closed and in stable equilibrium there are no open ecological niches. There is no room for an existing industry to expand (the will might exist but not the power); and for an existing industry to contract would mean to forgo potential profits (the power might exist but not the will). Hence theré is no place for new members of the community to go except to fill existing roles as they become vacant. This is at least one kind of logic by which the equilibrium position tends to perpetuate itself. A system out of equilibrium has open, or partially open, roles, compared to which other roles are relatively overfilled. Such a nonequilibrium position may induce changes in some parameters of the system, such as preferences or technology, in which case the equilibrium position itself may change. In short, stability at an equilibrium seems to show positive feedback: the more stable it is, the more stable it tends to become; the less stable it is, the less stable it tends to become.

We can apparently also apply to the economy the proposition that the larger the number of substitute performers in a role the less subject the system is to fluctuation. If bricklayers can promptly become carpenters and vice versa, if an industry can readily shift back and forth from copper to aluminum, if natural and synthetic rubber are ready substitutes for one another, and if consumers can promptly substitute chicken for pork, shortage or surplus of an item is less likely to disrupt the overall balance than if no substitutes are possible. By extension, perfect competition is more stable than oligopoly and probably less innovative.

The discussion thus far may have left the impression that the analysis of a market system as informal organization depends on the assumptions of economic theory. That is not the case. The profit-maximizing assumption is necessary for a theory of *determinate* equilibrium but not for the general notion of equilibria in uncontrolled systems. The equilibrium position under profit maximization differs from that under "satisfactory" or "comfortable" goals, but the equilibrium under such looser goals is no less an equilibrium and not necessarily less stable or less desirable. The only conceptual requirement of the uncontrolled economic system is that each firm be guided by its own DSE processes, with none specified by the main system. Similarly, conventional economic theory assumes perfect information by each subsystem; the general theory of informal organization does not.

4.22.1 *Intrasystem Interactions*

As in the formal organization, the internal interactions of the subsystems are communications, transactions, and suborganizations.

In the pure formal organization these interactions are directed by the main system toward its goals. In the informal organization they are conducted solely by the subsystems in their own interest.

The initial model of the transaction assumes that all necessary communications occur (3.21o); the initial model of economic theory does the same for its market exchanges. Economics then elaborates the social structures and processes built on transactions and the decisions behind them. Economics, however, has not gone far into the communicational aspects of an economy, much less the interactions of its communications and transactions along the lines of Chapter Eleven —except to note that prices are potent communications. I consider this a serious omission. I think we must pay more attention, for example, to the information and decision processes within firms and to the difference it might make if advertising were to delineate the shortcomings of products as clearly and as regularly as their virtues (that is, if it were honest).

We have tried to show how an important aspect of political science can be viewed as dealing with a particular kind of social configuration, the formal organization, whose logical and analytic underpinnings consist of the intrasystem and intersystem ingredients described in earlier chapters. This chapter performs a parallel task for economics. The preceding section has begun this task by identifying microeconomics as informal organization and hence subject to whatever generalizations arise from that field. The next task is to view the internal workings of that informal system as a particular configuration or elaboration of the intersystem relationships discussed in earlier chapters.

Historically, economic theory developed around perfect competition. Only relatively recently has it moved into the analysis of oligopoly, monopolistic competition, and bilateral monopoly. As noted (3.23.5), this analysis reverses the order. It begins the discussion of transactions with the unique transaction, which in economic terminology is bilateral monopoly. This opposite end of the transactional spectrum is the most general case; the unique transaction applies to the family, to international politics, to union-management relations, and to a host of other relationships as well as to economics. Its outcome is theoretically indeterminate. Perfect competition, on the other hand, is a phenomenon peculiar to economics. Its outcome is theoretically determinate. The methodological question facing this model is whether the economic market can be logically encompassed within the combination of communication-transaction-organization.

A market, as a multiplicity of interacting transactions in the same good, is obviously transactional. Supply and demand curves for either products or factors can be derived by simple summation of EP's of multiple sellers and buyers (Kuhn, 1963, pp. 570–573; and 3.27.33 above).

That a single price will prevail in a perfect market is explained by simple extension of 3.27.33 from two parties to multiple parties transacting the same good. That is, in the transaction between A and B the best terms available to A from any other B set the floor on A's bargaining power and the ceiling on B's. Similarly the best terms available to B from any other A constitute the floor on B's bargaining power and the ceiling on A's. More explicitly A's EP in dealing with one B shrinks to the point represented by the best terms available from any other B. Since the generalization applies to all A's and all B's, only one set of terms can prevail for a given good among a group of A's and B's, any one of whom is free to transact with any other. "A given good" in this context means the same as the economist's "homogeneous product."

This market consensus, or consensus by transaction, occurs less dramatically and less precisely in other areas of society. The logical connection is recognized in such references as the "marriage market," with the qualification that brides and grooms normally reflect product differentiation rather than product homogeneity. In societies where bride prices and dowries are the practice, the price in a particular case does in fact reflect the prices of alternative brides and grooms—although the average price may fall at a level specified by custom. Where such prices do not prevail the question is instead that of how much one party can offer in appearance, intelligence, status, security, personality, loyalty, and the like in return for how much of these factors in the other party. Although obvious exceptions occur and the subjectivity of the evaluation makes precise comparisons impossible, marriages generally occur between persons whose total "packages" of desired traits are not conspicuously discrepant in value.

There are marked differences from society to society in the amount of relative dominance of husband and wife in family decisions. A person's bargaining power in getting more dominance or less submission from a prospective mate is limited by the best alternatives he can find in his society. His bargaining power is increased if mates become available from different cultures that offer better terms. In some circles in Japan, for example, the presence of potential American suitors, accustomed to relative equality of partners, has increased the bargaining power of Japanese women to reduce their subordinate position relative to Japanese suitors.

Blau (1964, chap. 7) presents an extended discussion of market relations among persons who work together in firms. One party gives advice and in return the other gives compliance with the advice. The terms (or "prices") in such transactions are measured in the amount of compliance the receiver of advice has to give in return for a given amount of advice. Blau then traces the effects of alternatives available to both parties in narrowing the range of "prices" in such exchanges—a clear market process. The bidding process of fraternities and sororities has similar market properties, as do the terms on which the boy can get into a game of marbles or his father into the club.

The terms of an exchange, including the polar cases of perfect economic market and unique transaction, are influenced by both cultural norms and market effects. Supply and demand determine the equilibrium price and quantity of fish and pork in an economy, for example, but the numbers and orthodoxy of the Catholic and Jewish populations strongly influence the demand. Preferences for goods in a society are molded broadly by the culture, and hence so are their approximate ratios of exchange. But individuals are never completely socialized, and individual preferences may deviate markedly from the norm. Nevertheless, if every A dealing in a given good has numerous B's with whom he can exchange and vice versa, then a market exists and the overt alternatives available to every A and B will narrow the range within which a transaction will take place. The conditions for such consensus have long been formulated in economics—many alternative A's and B's and substitutability among all units of X and Y. The more numerous the former and the more substitutable the latter, the more narrowly will the consensus converge.

Despite conceivable exceptions, cultural and market factors also affect the unique transaction. Except for the unnegotiated transactions between parent and infant, which are largely unilateral gifts, the EP's of both parties are culturally influenced. And as to market influences, though conceivably no other transactions may, through overt alternatives, affect an exchange between two lone persons whose paths cross in the wilderness, for almost any other transaction other goods are available from other persons, goods that are at least partly substitutable for the X and Y in the case. Those available alternatives contract the EP's of A and B more narrowly than otherwise would be the case and parallel the situation in economics. It is easy to define perfect competition because the notion of perfect substitutability is clear. Perfect monopoly, however, is somewhat unclear because it is hard to imagine any good for which absolutely nothing is even partly substitutable—particularly when doing without is a substitute of sorts.

Though we thus conclude that both cultural norms and market alternatives influence the terms of almost every transaction, neither transactional theory nor the theory of informal organization indicates the relative strength of these two forces in a particular case. The discussion of semiformal organization in the next chapter clarifies the context of the problem, which is essentially developmental, but provides no answer.

As noted in Chapter Ten, a person can increase his power in transaction in goods without decreasing the power of others. We have also seen (3.34.5) that A's power and bargaining power are greater when dealing with B's of high power than with those of low power. That is, as B becomes more effective in producing Y's he will be more able and more willing to give more Y in exchange for X than previously. The result is that the A with whom he conducts a transaction will have more Y and B will get more X, the balance depending on the elasticity of the desires of both parties. This conclusion coincides with the logic of Say's Law stated in barter terms: the supply of one product constitutes the demand for other products since (in barter exchange) the supply itself *is* what producers have available to exchange for other products and hence it constitutes their purchasing power. As the supply of a product rises it helps its producers because they have more purchasing power. But it also helps the producers of other products because, given a larger demand, they will be able to get more per unit for their outputs.

4.22.2   *Six Major Aspects of Market Systems*

Six major aspects of uncontrolled market systems should be mentioned before we move to the nonmarket area. First, we have noted (3.2) that a transaction reflects two mutually contingent decisions. By extension a market embodies a much larger set of mutually contingent decisions. Similarly, just as a decision reflects the relative values of two or more things in the preference system of one person and the terms of a transaction reflect the values of X and Y to both parties, so does a market price by extension reflect the values of all parties participating in the market. A general equilibrium of all factors and all products viewed simultaneously similarly reflects the mutually contingent decisions and values of all factor suppliers, all firms, and all consumers. Since in a market economy all these valuations are expressed in the common denominator of money we can say that a market is a technique, and apparently the only one, by which the value judgments of a collectivity of people regarding numerous things can be cardinally quantified, in the sense that a $2000 automobile is valued at 1000 times a $2 bushel of wheat. This comparison of values, by which we mean the relative intensity of the subjective desires for various things, is confined within the range of exchangeable goods and cannot measure the relative values of, say, parental affection and peer group status. But for marketable goods the prices do represent values in the same sense we speak of other values—the satisfactions from using goods compared with the satisfaction denied in producing them.

Second, a market economy is highly organized—a high-energy system in which relatively small changes in one part of the system have wide repercussions on other parts. A drop in the price of aluminum in the United States, for example, may lead to its partial substitution for copper. The results may be a loss of employment in Connecticut copper refining and a drop in copper imports from Chile. The latter may reduce employment in Chilean copper mines, the number of houses built near the mines, the demand for local carpenters, the interest on mortgage loans, and the income to Chilean banks.

Third, we have seen the possible relation between the second law of thermodynamics and loss of differentiation (entropy) in closed systems. Moreover we have noted that differentiation may increase (negentropy) under certain conditions. In specialization and trade, market economies display an interesting complementarity of negentropic and entropic processes. Specialization is negentropic—a clear case of increased differentiation. Specialization in turn induces discrepant valuations of things—another form of negentropy. These discrepant valuations arise because each specialist possesses a glut of the things he produces and none of things he does not produce. Trade then moves goods until the specialist has little left of the things he produces and increasing quantities of the things he does not produce, after which the relative valuations of goods are more nearly uniform for all participants. This reduction in value differentiation is entropic.

Fourth, we have observed that a decision is an interaction between preference functions (selector) and opportunity functions (as perceived by the detector). Decisions by a whole society presumably involve the same pair, which are clearly delineated in a market. In the market for factor or product the demand curve represents the preference function of buyers and the opportunity function of sellers; the supply curve represents the opportunity function of buyers and the preference function of sellers. This conclusion is a simple extension of the observation (3.2) that in a transaction A's EP represents his preference and B's opportunity whereas B's EP represents *his* preference and A's opportunity.

Fifth, the model of the transaction indicates that a gain in value can be made when each of two parties place different relative valuations on each of two things. If in place of parties we substitute places, markets, or times, the same formulation explains the law of comparative advantage in international trade, Gresham's Law, arbitrage, and speculation. Comparative advantage deals with the relative values of each of two commodities in each of two nations. Gresham's Law deals with the relative value of each of two types of money in each of two markets—commodity market and money market (or mint, depending on the case). Arbitrage deals with the relative value of each of two currencies in each of two locations. And speculation deals with the relative value of each of two commodities, one of which may be money, at two different points in time. Whenever there is a difference in two relative valuations, a gain can be made by at least one party to an exchange.

To indicate the relation of comparative advantage in international trade to the transactional analysis, we use a diagram like Figure 11 in Chapter Eight. Assume that the United States has a comparative advantage in producing wheat and England a comparative advantage in shoes such that a pair of shoes commands four bushels of wheat in the United States and three in England. As this relationship is usually stated, shoes (compared to wheat) are more costly to produce in the United States than in England, and wheat (compared to shoes) is more costly to produce in England than in the United States. To diagram this relationship (Figure 18) we express the value of shoes in bushels of wheat and begin with the transaction that would be carried through—the sale of English shoes (Y) in the United States (A) and the sale of American wheat (X) in England (B). A's EP for shoes extends to 4, the number of bushels A would be willing to give for a pair of shoes. B's EP for wheat extends to 3, the number of bushels B would accept for a pair of shoes. Given the overlap, shoes and wheat could be exchanged at a ratio of not more than 4:1 and not less than 3:1. This

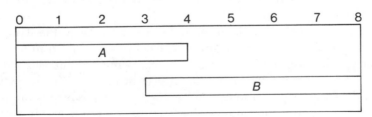

Figure 18. International transaction: exchange of American wheat for British shoes. (A = U.S. EP for shoes; B = England's EP for wheat.)

analysis shows the limits of the terms. If a unique transaction were to take place between one American and one Englishman the location of the settlement within the overlap would be determined by tactics as before. Given downward-sloping demand curves and a large number of buyers and sellers in both countries, the precise rate of exchange could be theoretically determinate.

If it is not initially obvious how to diagram the relation to show an overlap, a wrong trial demonstrates the error. Suppose we diagram, for contrast, an attempt of Americans to sell shoes to England in return for wheat (Figure 19). This time A's EP for wheat (now Y) is expressed in terms of shoes (now X). Wheat is worth one-fourth of a pair of shoes in the United States and one-third of a pair in England. Here, of course, the EP's do not overlap

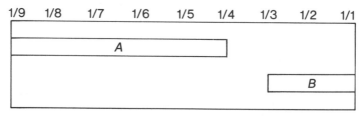

Figure 19. International transaction: nonexchange of American shoes for British wheat. (A = U.S. EP for wheat; B = England's EP for shoes.)

The same analysis can be used for less formally exchangeable goods in a small group producing a shared value. Assume that A and B would like to play Renaissance recorder music with piano accompaniment. A can play piano only modestly and the recorder not at all. B can play recorder moderately well and the piano very well. Now it is clear that there will be no duet unless A plays the piano and B the recorder. In the language of comparative advantage, A has a comparative advantage in playing the piano, even though he cannot play it nearly so well as B, for the simple reason that he cannot play the recorder at all.

To keep the logic straight we must keep the language parallel to the previous illustration. The desires of A and B to have the other perform a given instrument are represented by the difficulty of doing it for oneself—which is the inverse of one's ability to play it. This difficulty may be thought of as the cost of producing music. If one is a poor player the amount of satisfactory output per unit of effort is very low, which is another way of saying that efficiency in producing it is low and cost high. To use an arbitrary figure, let us say that for A

the relative value of having someone else play the recorder rather than the piano is 10:1; that is, *A* is only one-tenth as competent on the recorder as on the piano. To use a similar arbitrary figure, let us say that the comparable ratio for *B* is 2:1; that is, *B* is only one-half as good on the recorder as on the piano. The EP's then appear as in Figure 20.

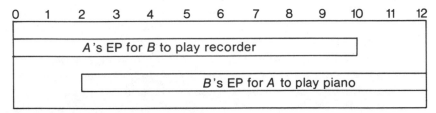

Figure 20. Musical transaction: Who plays which instrument?

The diagram assumes that there is at least enough competence to produce a desire to play and illustrates the transactional nature of the decision as to who will play which instrument.

As elsewhere, this discussion is in the language of the analyst, not the participants, and human beings have fortunately been able to do many sensible things without the conceptual set of the scientist. The scientific status of the preceding problem is a mild form of the falling leaf problem illustrated by Boulding (1970). To diagnose scientifically the path and velocity of a falling leaf is virtually beyond the capacity of the physicist-mathematician. Yet it creates no difficulty for the poet or even for the ordinary individual. The question is not whether this musical event is subject to social-scientific analysis—since it is—but merely whether it is worth the effort. Although it probably is not, it is nevertheless important to clarify that the relationship is within the competence of the analysis.

An interesting parallel to Gresham's Law and comparative advantage appears in the world of biological ecology. According to Poore (1964, p. 215), in the absence of competition all species (of certain general types of plants) do better in soil of medium moisture content than in wet or dry soils. Yet when they grow in competition some species are found exclusively in the drier and others exclusively in the wetter areas. This situation can be expressed in the language of Gresham's Law as follows. Though species *A* and *B* both do better in medium than in wet or dry soil, species *A* does relatively better in wet soil than does species *B*, and *B* does relatively better in dry soil than does *A*. Hence in competition *A* will drive *B* completely out of the wet soil and B will drive A completely out of the dry. Figuratively expressed in the language of comparative advantage, wet soil has a comparative advantage in producing species A and specializes in it whereas dry soil has a comparative advantage in producing species B and concentrates on it.

We come now to the sixth aspect of the market economy as informal organization. A firm is a special-purpose organization with the relatively single-minded goal of making money. Government, by contrast, is at least potentially an all-purpose organization. Although an informal organization cannot be said to have a goal as a unit, two different informal organizations can deal with quite different things. In this sense the whole economy is a relatively special-purpose organization (or special-function, if "purpose" seems to imply a goal for the unit). By contrast the relatively informal organization we examine next, the social system, is essentially all-purpose (except the economic purpose). This difference has repercussions on the precision of the analysis of the two types of systems since it is easier to predict the state of a system if each subsystem has only a single uncomplicated goal than if each subsystem has many conflicting goals.

Having treated four varieties of consensus separately we can now summarize them as consensus by communication and by transaction, each at a micro and macro level. Consensus at the micro level has been dealt with in connection with group decisions in organization (Section 4.12.4). There consensus by communication was seen as persuasion whereas consensus by transaction is tradeoff. At the macro level consensus by communication is the cultural process (Section 3.17) and consensus by transaction is the market process.

As indicated in Chapter One, economics is highly developed as a deductive social science. Because we are limiting attention here to those aspects of social science that are not, the preceding discussion is not intended to summarize economics, even briefly, but merely to indicate how it can be construed as one type of superstructure, both real and theoretical, built on the same analytic base as the rest of social science.

Nevertheless there are ways in which the technical apparatus of economic theory might be restructured on the conceptual base of this volume. I have identified the possible derivation of demand and supply curves from an aggregation of EP's. I consider this approach to be less clumsy than the conventional derivation from indifference curves—despite the fact that the economist's unorthodox habit of putting the independent variable on the vertical axis and the dependent one on the horizontal requires that the transactional diagram be reversed to put the buyer on the right and the seller on the left, that the prices on the diagram read from right to left, and that the diagram then be turned clockwise ninety degrees. In addition, one of my undergraduate students, Gregory Werden, has worked out a diagram that develops utility surfaces for consumers from the tips of EP's stood on end on a plane: one axis of which is the desire for one good and the other is the desire for a composite of all alternative goods. Although I have not had the opportunity to pursue this line of inquiry, I see no reason why other details of economic analysis cannot also be reconceptualized on this unified base—though I doubt that many economists would be interested in doing so.

4.23   SOCIAL PATTERN AND MOVEMENT AS HUMAN ECOLOGY

Having viewed the market economy as one type of informal organization, we now move to several aspects of social patterns of interest to sociologists and human ecologists. Because they are mostly informal we discuss them here, though we refer occasionally to the semiformal analysis of the next chapter. At any given moment a population falls into certain patterns, or groupings. Over time these patterns may change as individuals move from one group to another. The static grouping is generally known as demographic distribution, broadly construed, and in present language it is a problem of the *stock* of persons in a given category at a given moment. The movement is a question of demographic mobility, or *flow*. All changes in stock result from a difference between inflow and outflow. Hence in a direct sense all questions of stock must be explained as questions of flow. However, stocks of different sizes may have quite different impacts on their individual members and hence influence flows. The interactions between stocks and flows are therefore mutual (1.20j), which means that the stock may indirectly affect the stock and the flow may affect the flow. Thus a crucial question is whether the feedback of each on itself is positive or negative.

We assume initially that parties (individuals, families, firms, and the like) are not constrained by such government regulation as zoning or mandatory racial segregation. That is, we assume strict informal organization, or pure ecological distribution. We deal first with general theoretical questions and then move to three illustrative groupings—economic, ethnic, and status or class.

The distinction is sometimes made between micro and macro explanations of mobility. The first deals with individual parties and their reasons for moving and is thought of as a behavioral explanation. The second is aggregative, or structural, and focuses on comparisons between conditions in the place or group moved from and the conditions moved to, the explanation presumably lying in the differences between the two. These two approaches, it seems to me, are indispensable halves of a single explanation. In this model every action by a party reflects a relation among detector, selector, and effector. In logical form: *if* soil is perceived to be better in region *B* than in region *A* (detector), and *if* farmers prefer good soil to poor (selector), *then* farmers will move, *cet. par.,* from *A* to *B* rather than from *B* to *A* (effector). If we find that people do in fact move from *A* to *B* we can explain the movement if we establish that the conditions stated in both premises do in fact exist. Certainly neither condition alone is sufficient. A high correlation between population movement and soil conditions is simply a statement of fact. It gives us confidence that we have correctly established the fact but provides no explanation. Now it may be intuitively obvious why people move from poor soil to good, but we must make the motivational assumption explicit before it becomes an explanation. Furthermore, knowledge of motives explains nothing until facts (or assumptions) have been added about external circumstances. Though the relation may not always hold, in important respects the macro approach focuses on the opportunity function and the micro on the preference function. In this model both are needed. K.E.F. Watt (1966, p. 8) agrees when he says that "population processes require detailed knowledge of individual behavior."

Another aspect of mobility is that the explanation of group movement need not differ in principle from that of individual movement. The reason why five thousand farmers move from *A* to *B* need not differ from the reason why one farmer does. And the explanation does not cease to be behavioral just because many farmers behave similarly. The consequences may differ dramatically, however. For one thing, five thousand farmers may be able to do well what five cannot do at all—a problem of critical mass. Second, increased numbers of persons in a category or place alter the ratio of people to something else and may bring increasing or diminishing returns on some variable. And third, a change in the number of people may induce a change in the social structure within a category and induce other changes in their interactions. We will return to these problems later.

4.23.1   *Income Groupings*

Let us now move to grouping by geographic location of residences of families of about the same income level. The basic explanation of such a grouping consists of two assumptions: that the quality of housing is perceived to vary directly (not necessarily linearly) with its cost (detector element) and that people generally prefer high-quality to low-quality housing (selector). The conclusion is that people purchase high-quality housing if they can (effector); and the range of quality in an area is roughly proportional to the range of income in that area. The remaining step is to explain why housing tends to cost about the same throughout an area. The market process, which is an aggregation of transactions, yields a partial explanation, given alternative uses of land with differing values in each use. This explanation can be supplemented by that of status-seeking. In that connection the conclusions of 3.35.2, .3, .33, and .34 account for living in the highest-status neighborhood the status-seeker can afford. In a given neighborhood those with incomes higher than they need to live there will voluntarily move out and those with lower incomes will be unable to move in. Hence incomes will tend to be equal in the neighborhood. Those with the will to breach the uniformity lack the ability and those with the ability lack the will. If status differentials are clearly identified by factors other than location—as between servants and their masters—persons of different income can live in close proximity. This relation does not dilute the market explanation if the property on which the low-income persons live belongs to the main landowner.

4.23.2   *Ethnic Groupings*

Ethnic groupings seem readily explainable on the basis of communications and transactions. We need examine only a few facets. If we assume that people want information from others and prefer messages in which they can place confidence (3.17.6a), they will prefer messages from those who share their own culture (3.17.81). People of the same ethnic group will normally share the same culture and hence be preferred as sources of information. They will also normally share the same main language and the same dialect. Without such similarity, words will not have the same meaning to sender and receiver (a violation of condition 3.14.2 for accuracy of communication) and may not even be identifiable (a violation of 3.41.1d). Cognitive dissonance will also be smaller within than between cultures and will reinforce avoidance of communications with persons of different cultures (3.17.89). Since by far the largest quantity of communication among humans consists of face-to-face conversation, under the additive effect of these factors (3.17.9) people will prefer to live close to others who share their culture.

Transactional factors augment the effect. Transactions are

facilitated and the range of uncertainty in bargaining substantially reduced if each party knows at least the approximate magnitude of the other's EP. This condition occurs more often between parties of the same culture since both have acquired approximately the same set of preferences. Furthermore, since he knows the relevant preferences, one will be less likely to make errors, and hence suffer loss of power, if he produces goods to be exchanged within his own cultural group than if he produces goods for outside the group. The communicational and transactional factors are interrelated. Dissonance is less within the same culture, as we have seen. Smaller dissonance leads to greater liking of one another (3.41.41B2), which leads in turn to greater power and bargaining power (3.26.11 and .12 and Sections 3.26.2 and 3.28.42).

Ethnic groupings are thus merely a special case of the general tendency of people to associate with others of convergent views and to avoid those of divergent views because there is greater advantage in doing so. Status may also be involved. If a majority group is perceived as having higher status than some minority, then, since status is both a relative and a perceptual phenomenon (3.35.0 and 3.35.67), a minority member will have higher status in his own group than in the majority one. And since higher status is preferable to lower, it is more advantageous on this score for the minority member to associate with others of the same minority. Contrary forces also operate. If he can manage it, the minority member may get still higher status if he can break into the majority and share its glamour. Hence the longer-run attraction is to break out of the minority—though one is likely to do so only after he has become familiar enough with the language, concepts, values, and roles of the majority to avoid losing the other advantages of staying with the minority. Though these two movements are opposite, the basic analysis is the same (the same force of gravity can float the ship or sink it, depending on the circumstances).

Other types of interaction would help round out the explanation. But since the purpose here is only to illustrate the approach, we need not go into them. We should note, however, that this approach in no way suggests that an ethnic settlement is "nothing but" the preceding ingredients. It is also an informal or semiformal organization.

4.23.3   Class and Status Groupings

Class and status groupings are a fuller version of the income distribution of neighborhoods discussed above. The advantages of power and status are great enough (Section 3.3) to justify a motivational assumption that a substantial fraction of the population will prefer higher to lower levels of them. This assumption does not specify the type of power or status sought (3.35.0 and 3.32.5). One may sneer at social climbing while he develops ulcers for fear his paper will not be well received by his professional association. The quest for status must be studied separately in each area.

The general sources of power and status have already been spelled out (3.3). Classes or levels arise because everyone who seeks a particular type of status will try to rise as high as he can; a class consists of all those who have managed to rise to a particular level (3.35.0). Since status is relative, one's position can be furthered by keeping others down; hence the jealousy with which status levels are often guarded. And since status is affected by that of the group one is perceived to belong to (3.35.34, 3.45.2, .21, .5, and .6), there is mutual benefit in maintaining the status of those already in one's own class. Semiformal organization (Chapter Fifteen) is an obvious technique for doing this for a class.

When types of power are widely discrepant, as between wrestling champion and concert pianist, no question of cross-comparison arises—except, perhaps, when they meet on the common ground of money. When they are more nearly similar, as between old-line once-wealthy families and new-line now-richer ones, cross-comparisons are important, and one's own position is enhanced if he can successfully assert the superiority of his own type of power and the inferiority of the other. The less obvious one's status, the greater the need to display its symbols (3.36.3). If those within a class feel that their status is challenged or questionable they can use the techniques of semiformal organization to make sure all members display the appropriate symbols. If the class members act uniformly enough to constitute a dominant coalition or subcoalition in the decisions of some larger entity such as government, they may attempt to control the size of their group in accordance with the principles of 4.12.414 and 4.13.5143.

With this background, we return to the question of mobility and the mutual interaction between stock and flow. Certain kinds of mobility, such as geographic migration, can be almost totally an individual decision, and our assumption that no regulation is present makes it so. Yet a change in stock resulting from prior movement may change the ingredients of the decision. Movement of people from region A to B, for example, leaves more resources per capita in A and fewer in B than before the movement and reduces the desirability of moving. Movement from lower-status group B to higher-status group A may similarly dilute the amount of status to be acquired from A and reduce the incentive to move. If membership in a status group can be denied by those already in it, as by refusing to extend or accept invitations or otherwise boycotting expectant entrants, this semiformal action may bar movement as effectively as government proscription or more so. If those already in a category can control the number of new entrants, they will presumably aim toward that total number which would maximize average benefits of the relevant sort, according to the logic of 4.11.4. If some reasonably identifiable sponsor subgroup within the category receives the benefits, they will aim toward a total at which marginal benefits of additional members equal marginal costs, as in 4.11.5. If no one exercises control (pure informal organization), the additional members will join or refrain from joining the category solely on the basis of *their* benefits and costs. As a result new members may raise their own position while lowering that of the rest of the group.

Other interactions inhibit or facilitate movement. Those who have already moved may communicate to those who have not moved about how good or bad their new position is. If it is good they may, by generous transactions, assist friends or relatives to rise to their level. Factors like these presumably determine whether movement leads to still more movement (positive feedback) or to reduced movement (negative feedback and equilibrium). Detailed models could specify the conditions for each but need not be constructed here. For those who like the analogy, in connection with decisions we have already seen that benefits can be regarded as pressure and costs as resistance, with a decision made in a certain direction when the pressure exceeds the resistance (2.99). Since we have already identified mobility, and hence flow, with individual decisions, the analogy may be more useful for this purpose than for others. A particular movement, like a particular decision, must be explained by the magnitudes of the costs and benefits it displays; it cannot be accounted for in the general model. Flows in and out of a group that depend solely on births and deaths, as with age and sex categories of a population, must presumably be viewed as almost pure informal organization.

4.24   CONCLUSION

Because informal organization is an uncontrolled system it seems likely that a science of it will be less amenable to precise formulation than that of formal organization and that it will deal more with particular cases and less with broad generalizations. It is hard to devise a precise science of a relationship whose outcome "just happens." But because its relevant subsystems are themselves controlled systems, things "just happen" in a different sense than in, say, weather systems or volcanic systems. Whereas traditional natural science may be an appropriate vehicle for studying weather and volcanos, even if their complexity leaves outcomes uncertain, informal organization and ecology must attend to the goals and information processes of the subsystems and to the science of their interactions. In the cybernetic view the goal state of a system is just as much a *cause* of its behavior as a low-pressure area is a cause of certain weather or heat and pressure are causes of volcanic action.

As a science of one kind of informal organization, economic theory reaches precise conclusions not merely because it is explicit about detector and selector functions of its controlled subsystems but also because it assumes those states to be essentially identical among all firms. The basic theory also assumes identical detector states among all households (perfect-market information) and reliably similar satisfaction-maximizing goals among them, even if the ingredients that produce satisfaction differ sharply from household to household. Except for economic markets it seems doubtful whether even a first-aproximation model of any other informal organization can usefully assume such similarity of detector and selector states among its subsystems. If so, no other science of informal organization can be made as precise as economics.

Chapter 15

4.3 Semiformal Organization

4.31   SOCIAL SYSTEM: A MULTIPURPOSE SEMIFORMAL ORGANIZATION

The model of government in Chapter Thirteen examined a formal organization of the whole society, and the model of a laissez faire economy in Chapter Fourteen an informal one. Several patterns of social movement and structure were also broad enough to cover the whole society. Yet although each model in some sense covers the whole society, none of them alone encompasses the whole social organization nor do all of them together. This is said in the same sense that although the nervous system and circulatory system both extend throughout the whole body, neither *is* the whole body nor are both together. We can approach this problem as follows.

First, between pure formal and pure informal organization we can carve out any number of intermediate types, each subject to formal analysis insofar as it fills the definition of formal organization and subject to informal analysis where it does not. For the present we distinguish only two intermediate types—the informal aspects of formal organization and the formal aspects of informal organization. The former is given no name; the latter seems distinctive enough to justify the term *semiformal organization*. Its definition as the formal aspects of informal organization is amplified in the following discussion.

Second, we define the *whole society* as that system which encompasses all organization within some bounded area. For most purposes the unit can be viewed as the nation, though the tribe, clan, or other unit will be more appropriate for other purposes. Whether this unit is to be considered formal, informal, semiformal, or something else depends on the case at hand and is not specified in the definition. We define an *institution* as a first-level subsystem of the whole society. In industrial societies government is one institution, the economy is a second, and in medieval Europe the Church was a third. Beyond these institutions, most societies do not have well-defined first-level subsystems, and clarification of this point takes us into a terminological and a methodological problem.

As to the terminological problem, we have a name for the formal organization of the whole society—government or polity. We also have a name for the system that produces exchangeable goods—the economy. But if we use the term *social* to designate the total entity that encompasses polity, economy, and other organization then we have no distinctive name for that subentity which is the province of the sociologist but not of the political scientist or the economist. On the assumption that coining a new name is a worse sin than a double use of the same term (if it is clearly identified), we will call this third entity the *social system*. That is, the social system is one of several first-level subsystems of the whole society. For reasons to be seen below, this is the system to which we apply the semiformal model. It is also an institution, as defined, but subject to reservations about considering it an entity at all.

The methodological problem grows from the question of whether this third area *is* an entity; it is reflected in Smelser's observation (1963, p. 22) that "in practice, sociologists study a sort of grab bag of leftovers." In terms of systems studied, Smelser's point is valid: sociology can be said to cover all those many systems not studied by economists or political scientists. But in terms of analysis—and to mix present terminology with the conventional—the point is not valid: sociology encompasses, even if muddily at points, all the things viewed here as basic interactions—communication, transaction, and organization. In that sense the social analysis underlies the economic and political rather than standing beside it—or should do so.

Part of the methodological problem, as I see it, can be clarified by comparing the present definition of institution with the conventional usage. Both definitions agree that the polity and the economy are institutions. But in conventional usage courtship, marriage, religion, and political speeches are also institutions whereas in present language they are not. The present definition of institution encompasses only acting systems. Government and the economy qualify.

But courtship and religion, although they are behaviors or ideas of
acting systems, do not themselves act. My inclination is not to call
them systems at all; but for those who wish to do so they are definitely
pattern, not acting, systems. Depending on how the terms are used,
they can also be viewed as analytic constructs. As indicated earlier, a
number of generalizations apply to all acting systems, particularly
controlled ones. But I know of no such generalizations (except some
related to hierarchy) that also apply to pattern systems; in fact I suspect
sheer accident if a generalization applies to more than one kind of
pattern system. If highly discrepant entities are included within the
conventional concept of institution we thereby preclude much usefulness for it as an analytic construct.

Some intermediate cases are not immediately clear. If by
"the Church" we mean the Church of England or of Rome, it is an
acting system and hence an institution as defined here. Each is also a
formal organization. The First Baptist Church of Oshkosh is also an
acting system and a formal organization, but it is not an institution
because it is too small. On the other hand, if by "the church" we mean
the simple sum of all churches, or approximately the same thing as
religion, this is neither an acting system nor an institution. Similarly *a*
family is an acting system; but *the* family, meaning an established
pattern of organization or behavior, is not.

The remaining aspect of the methodological problem is that
the social system, as here defined, is a heterogeneous collection of discrepant types of organizations—formal, informal, and intermediate.
Nevertheless there is a kind of analysis that can be applied to the whole
collection as a unit and that justifies calling it a social *system*. That
analysis is the basis for calling the unit a semiformal organization as
contrasted to the pure formal government and the pure informal
laissez faire market. We now move to that problem.

4.32   *The Model of Semiformality:*
Formal Aspects of Informal Organization

Semiformal organization resembles informal in that it has
no sponsors, no goals, and no decisions for the organization as a unit.
Nor does it interact as a unit with any other organization or person. It
does not initially create its own structure or explicitly modify it over
time. In an overall sense the form and process are uncontrolled.

But the semiformal organization resembles the formal in
that some of its component persons and organizations have some
significant, even if amorphous, sense that there *is* a social system, that

it has a structure, that many of its roles must be filled, and that certain beliefs or behavior promote the system while others injure it and its members. In consequence some subsystems themselves take on a degree of sponsor function. This means that they have some concept of what the whole society is and ought to be like, of appropriate and inappropriate roles for making it that way, and of proper and improper behavior and system states for the occupants of those roles. It also means that they modify their own behavior to some extent, often at considerable cost to themselves, in ways they consider best for the society, while they also try to modify behavior of others toward the same goals. Detailed analysis of semiformal organization must distinguish activities that are rationally instrumental from those that reflect vague internalized feelings of "right" and "wrong." We need not make that distinction for the present but may attend solely to the performance of sponsor role by subsystems. The techniques for doing so are communications, transactions, and organizations along with the coalition for reaching a decision when communication and transaction fail.

Individuals, of course, do these things. Churches and schools are perhaps the most obvious suborganizations that take on this sponsor concern for the whole society, especially those that reflect high social consciousness. A camera club may reflect none of this concern whereas a Rotary Club may combine some concern for the whole society (semiformal aspects) with its efforts to foster the welfare of its own members (informal aspects). For the present we need not be concerned with the scope of sponsor interest in a particular subsystem; the scope may range from that of some all-encompassing reformist organization to a committee to clean up local back yards.

In the small, primitive society the sponsor interest of members may be so intense and so nearly universal that the social system closely approximates formal organization. And unlike the advanced, complex society, the social system may encompass the political and economic systems, which are not sufficiently differentiated to justify separate analysis. Given the small size and relative simplicity of the primitive society, virtually every individual can be somewhat familiar with every role in the society. This would also be the case in a family-owned firm in a small town, one with perhaps two or three hundred employees, very little turnover, and many second- and third-generation employees. Virtually everyone would know all the departments and offices (roles), most of the incumbents, how they think, what they do, and how they interact. In fact such companies often "just grow" without conscious thought about their best overall structure. In this firm, keeping an old hand sober enough to justify his wages until he is pensioned may seem quite as legitimate a goal as producing efficiently or making a profit.

The interesting condition of the small tribal society is that formal and informal aspects merge, and it may not be sensible to classify such a society as one or the other because it is both. Each individual performs a role for the self-oriented "informal" reason that the performance is rewarded. But most members can also see the whole society as a functioning unit. For reasons pragmatic or mystical they sense that failure or vacancy of some role will injure the whole society, and they feel personally impelled to see that all roles are properly fulfilled. In this sense the organization is formal (and probably cooperative), with every adult member a sponsor. Most significant roles are ascribed, as in formal organization, even if by custom rather than executive design. Induction procedures train new role occupants and induction ceremonies communicate to everyone the nature and importance of the role and appropriate behavior by and toward its occupants. The myths, folklore, and anecdotes of the society communicate many concepts and values that new members are expected to possess, and they are part of the induction process. The instructions of the organization consist of norms and the expectations of others; the authority of the organization is exercised through social pressures.

What kind of model is required to analyze a particular semiformal organization? The only way to deal with a phenomenon with so many possible configurations seems to be the special-purpose, or simulation, model discussed in Chapters One and Sixteen. Such a model could be constructed for each main type of social system or possibly for each separate system to be analyzed. Selected portions of the general models already described could be assembled to see which combination provided the best fit. As a guess, for a particular tribal society we might begin with the model of a simple organization that produces some shared (public) and some divisible (private) goods. We could then expand this model to the model of a consumer cooperative in which sponsor-recipients also serve as staff. And then we could add to it ingredients from government to accommodate its multipurpose aspects, involuntary membership, and possible delegation of legitimate force to specified roles.

By contrast I suspect that to design the simulation model for the social system of a complex, market-oriented, industrial society we should begin with the model of pure informal organization and then modify it with semiformal and formal components. The semiformal modifications would specify the major informal subsystems that display a sponsor attitude and would indicate the type and magnitude of power they wield in implementing it. Modifications from the formal model would be of two sorts: first, the types of control exercised by government over components of the social system; second, the activities omitted by government on the assumption that the informal organization will perform them.

These simulation models are only suggestive; only actual trials can determine how good a description of a real situation can be formulated by this method or how good a set of predictions can be extracted from it even if it is a good description. And as in the typical relation between science and engineering, if the technique works it can presumably be used to design as well as to explain or predict.

Perhaps the best quick description of the model of semi-formal organization is this: "Like formal organization, only looser." The major elements of looseness are that the major bargain of affiliation is not explicit and that authority, rather than being delegated from designated sponsors, is exercised by the persons or subsystems that care to take on the task. The instructions about how to perform certain roles in the semiformal organization may also be implicit in the form of expectations of others, often to be deduced by the "subordinate" from the rewards that are granted or withheld and from the attitudes expressed about behaviors he has or has not performed. *Social pressure* is the name for the semiformal organization's way of exercising authority; social pressure consists mainly of a hostile shortening of EP's. By definition, social pressures in this sense do not exist within pure formal or pure informal organization. When they are indirect they may resemble the pressure transaction, but they differ in purpose in that the pressure transaction is used for the benefit of A while the social pressure is used for some broader "social good."

Leadership in this context is similarly seen as the looser equivalent in semiformal organization, and possibly in relaxed versions of simple formal organization, of the sponsor role in complex formal organization, along with the concomitant function of management. (There could be no sponsor function at all, and hence no leadership, in pure informal organization.) As in complex organization, the sponsor function here is also of two main sorts and the effectiveness of leadership depends on how well both are performed. The first involves the intrasystem aspects of the group's performance as a unit and includes clarifying goals; using communication, transaction, and coalition (presumably in that order of preference) for achieving a group decision (4.12.48); understanding the transformational processes that are instrumental to the group goal; and achieving an effective organizational structure. The second involves the intersystem relationships of the group members. Here the effectiveness of the leader hinges heavily on the amount of authority he can muster to induce members to accept their designated roles and perform the instructions incorporated in their role descriptions (probably implicit) or in his periodic communications to them. In this respect leadership in semiformal and simple formal organization differs importantly from the sponsor role in complex formal organization. In the complex formal organization the rewards that provide the basis for authority are furnished by sponsors or recipients; in the semiformal or simple formal organization the rewards are all produced by or within the group itself, often by the leader himself. The possible rewards are varied and include successful achievement of the joint goal, praise, group approval, status, and congenial companionship. Effectiveness in this intersystem aspect depends directly on the amount of overlap of EP's in the network of transactional relations between group and leader and within the group.

This definition identifies the nature of leadership in the present framework but does not specify whether it is performed by one person or more than one or whether by relatively democratic or authoritarian methods. The distinction between the problems of the system as a unit and those between subsystems parallels the widely used distinction between the instrumental-effective and the interpersonal-affective aspects of leadership. Some empirical research suggests that leadership may be more effective if these two aspects are separated into two different persons. Nothing at this stage of the model suggests or denies that conclusion, although the model does imply that leadership will be more effective if all these aspects *are* performed. Nor does the model suggest that a "straight" division of system and interpersonal functions between two leaders is more effective than a "mixed" division. Leader *A* might prescribe the strictly technical aspects of the joint effort while *B* prescribed the organizational structure, or division of labor, for doing it. Simultaneously *A* might provide rewards in the form of praise and encouragement while *B* concentrated on rewards in the form of congenial conversation and reduced tension.

To generalize about types, in the evolutionary and historical process ecological relations and informal organizations must almost certainly have preceded formal organization. But once formal organizations have evolved it seems far easier to describe the semiformal model as a relaxation of the formal than as an emergent version of the informal. This is so because such notions as major bargain, induction, authority, and role definition are essential to an understanding of semiformal organization but do not arise at all in the informal model. Hence they are available in sharp form for semiformal analysis only if we jump to the pure formal model for precise definitions and then relax into the intermediate position.

Perhaps the best way to encapsulate the concept of semi-formality is to identify the main equivalences between formal and semi-formal organization:

Formal	*Semiformal*
Sponsor goals	Beliefs, codes, ideologies
Induction	Socialization, myths, anecdotes, rituals
Managerial function of sponsors or agents	Leadership
Authority relations	
Instructions	Expectations and norms
Sanctions and discipline	Social pressures
Responsibility	Liability to social pressures
Roles	Roles
Structure	Structure

Although we did not mention the fact explicitly, sanctions can be applied in formal organization to motivate the new member to complete that phase of his learning which constitutes induction, and the same is true for semiformal organization, perhaps even more conspicuously.

4.33   *Informal Aspects of Formal Organization*

Formal, informal, and semiformal are the only models of organization used in this book. There is no limit, however, to their permutations and mixtures. In the same spirit that Chapter Eleven dealt with interrelations of communications and transactions, this section deals with interrelations of formal and informal organization. The method of inquiry is by illustration and example, however, not through a formal logic like that of Chapter Eleven, and we deal here with only one case—some informal elements within formal organization. As before, the method is to diagnose a phenomenon by combining concepts already developed.

Between the pure formal and pure informal models lie innumerable intermediate or combined states. In terms of simple word patterns the two most obvious states are the formal aspects of informal organizations and the informal aspects of formal ones—as the headings in this section indicate. Although it might seem reasonable to develop a model of each—a neat parallel—we treat the former as a model in its own right and the latter simply as one of many mixed configurations of the explicit models. The reason is implicit in the preceding section. Formal organization utilizes explicit concepts of management, sponsor, induction, authority, and so on. Informal organization does not. When we deal with the formal aspects of informal organization we therefore have concepts of the formal model that we can then identify in their looser, intermediate form. But there are no comparable concepts on the informal side that we can take hold of from the formal side to pull ourselves into a middle position—though further thought might reveal a more sensible way of handling the situation.

The model of formal organization has all actions of all subsystems directed toward the goals of the sponsors; motivation of subsystems is thus directed by the application of authority through a hierarchical system. To clarify the informal aspects of formal organization we must recall that we distinguished organizations from other systems in that all lowest-level subsystems of organization are human beings. The human differs notably from other subsystems in a formal organization. The purchasing department of Mishmash Products, for example, makes sense only as a subsystem of the firm. For such a department to exist independently is an absurdity, just as a kidney is an absurdity except as a subsystem of an animal. The individual human, by contrast, does make sense as a separate entity in at least two ways. First, although a biological human is not really human except as the product of a culture, once he has acquired that culture he can nevertheless exist and make sense for extended periods alone, Robinson Crusoe style. Second, in even the simplest human societies the adult human is simultaneously a member of multiple organizations—the nuclear family, the extended family, the tribal dance team, and the outrigger crew. Even within the same organization he may occupy multiple roles, as when the father is alternately his son's teacher, disciplinarian, and playmate. Both biologically and psychologically, what happens to him in one role affects his behavior in another, particularly what he learns about human interaction and his likes and dislikes for other people. These multiple roles may complement or be inconsistent with one another, the latter causing conflict. Complementary roles create no problem. Conflicts have intrasystem or intersystem aspects or both. To illustrate intrasystem conflicts, membership in the peer group may require behavior that is "wrong" under standards learned in the family and may thus leave a sense of guilt. Most intersystem aspects can be analyzed as interrelated transactions, as when a father's toughness in authority transactions makes him inept or his son timid when the father shifts to a companion role. These are interesting practical problems. They are all special cases of interrelated transactions, and there is no point in spelling out their details here, except perhaps to recall that they are heavily modified by the expectations of the parties involved (3.28.45).

Informal organization within the formal is a special case of multiple roles. To say that the human in the organization is an entity with an existence independent of his subsystem role is another way of saying that he has goals of his own which do not cease when he becomes a cog in the organizational machine. Some of his goals may assist the organization—such as the intrinsic pleasure of doing a job well or of giving one's all. Others may directly thwart the organization—such as an antipathy to work or a desire to subvert the organization's efforts because he disapproves of them. And others may be essentially irrelevant—as when he plays blackjack at lunch or goes bowling at night. When members pursue their own goals within the context of the organizational relation, as distinct from the organization's goals, they constitute informal organization within the formal. Obviously it is to the formal organization's interest to see that the informal organization assists, or at worst is indifferent to, the formal rather than working against it. Given the nature of the human subsystem it is presumably impossible to eliminate these informal behaviors.

We have already noted how individuals may alter their own and other roles in the course of doing their work for the organization (4.12.53). To the often considerable extent that such changes are motivated by personal gain, these too are informal modifications of the formal structure.

4.34   **SCALE-MODEL SOCIETIES**

We now move to a different problem, again seeking merely to locate it within the present framework and leaving the details to specialists. Formal organization, and particularly the large one within a complex society, offers a special opportunity to study the informal and semiformal organizations within it in speeded-up, scale-model conditions. Moore (1963, p. 55) has made this suggestion explicit and the idea is analogous to the table-top models used by geologists, who by appropriately scaling densities, forces, sizes, and tensile strengths (gelatin instead of rock and milligrams instead of megatons) can duplicate within minutes or days processes that have taken eons of geologic time.

Because the formal organization brings together people on

the basis of their skills rather than their proclivity to associate, it often throws disparate types into continued, close association. By focusing on such groups the analyst should be able to discover in this small scale of time and numbers something about the processes of birth, growth, and differentiation of roles; development of common culture; and increasing formalization. His findings should clarify the same processes in large, slow-moving societies. It would be particularly fruitful to observe the development of moral and ethical codes. At this level these codes would presumably take such forms as agreements not to "squeal" in prisoners' dilemma situations (see Rapoport cited in Boulding, 1970, p. 70), or the conclusions of Siegal and Fouraker's (1960) subjects that prompt and equal division of the joint payoff of a game is preferable to extended bargaining, or the learning that honesty is the best policy by persons who gradually reduce tactical stalling and haggling as they conduct more and more transactions with one another. On the communicational axis observations could also be made about the increase in efficiency of communication as each group member comes to understand how others are using words, the congruence rate of pronunciation among persons of initially disparate accents or dialects, or the shortening of ideas from paragraphs to sentences, to phrases, or to single words. The processes of cultural fusion and diffusion could also presumably be observed in these scale-model societies. And analysts could study the question of whether fusion of markedly different cultures (after initial confusion) produces greater creativity or innovation than fusion of relatively similar cultures. In this context each person in the group can be regarded as representing a different "culture." The conditions under which equilibrium is developed should be informative, along with the general question of whether the "society" at equilibrium is really more resistant to change than one not in equilibrium, as ecological theory tentatively suggests.

The way in which Lennard and Bernstein (1960) have studied the patient-therapist relation in psychotherapy as a two-person evolving social system is a case in point. That relationship, incidentally, would apparently qualify as formal organization within present definitions since it has the specific joint goal of improving the patient. And although Lennard and Bernstein clarify that there is a *mutual* role adjustment during the course of therapy, it is equally clear that the general nature of patient and therapist roles has been planned in advance by the therapist and that his half of the "adjustment" is really a part of the induction process through which the patient learns *his* role. With a clear joint goal and predetermined roles this organization can hardly be considered informal, or even semiformal, and its evolving structure should not be confused with an informal one.

In such analysis we are, of course, looking forward from the cross-sectional to the developmental, which is discussed directly in Chapter Sixteen. The important point is that all these processes could be viewed as special cases of the central set of intrasystem-intersystem behaviors, not as a collection of largely unrelated findings as they currently appear in the literature of group dynamics and small-group analysis.

4.35   **MORAL CODES**

Moral codes are ubiquitous in human societies—including subsocieties of thieves and outlaws. The purpose of this section is not to explain their origin but to point out that many components of such codes can be described in the language of previous chapters and may represent nothing more than generalized statements of behavior that has been found workable.

From the realm of communications one could conclude that accurate information is more useful than inaccurate information and that if people want accurate information from others, everybody must tell the truth. And if you want people to believe you, you must make an investment in telling the truth even when it hurts.

From the realm of transactions one could conclude that if you want people to give you things on the basis of a promise you must display a record of keeping promises. If you want people to be generous to you, you must get them to like you—by being generous yourself, by not using bads against them, by not bargaining too hard in selfish transactions, by not acting as if you have higher status than they have, by telling them what they like to hear. Relatedly, many acts of "common courtesy" symbolize equal status. "Would you please," for example, implies the right to refuse whereas a straight order implies a subordinate position. Transactions in bads are highly destructive of social relationships (3.25.4) and are therefore generally prohibited within societies. One trouble with transactions in bads is that they are so confoundedly one-sided. So is stealing, which is also generally outlawed.

As to organizations, unless one fills his role properly and accepts orders from authority, the society will suffer. In other cases, simple uniformity is itself much more efficient than diversity and provides a basis for rule-following. Pragmatics require that children follow the rules long before they learn the payoffs from their experience, and some may never learn them. Hence it is helpful if rules seem to emanate from an authority with whom one cannot or dare not argue—as from gods or elders. It is also useful if punishment comes from the gods since this arrangement provides sanctions without setting up conflicts between persons. A disadvantage of attributing rules to deity is that stupid rules cannot be readily appealed.

Moral codes are perhaps the most conspicuous and detailed manifestation of semiformal organization, whether or not they are also formalized into law. Almost anyone can assist in enforcing them by generous extensions of EP's toward those who conform and hostile contractions toward those who do not. Such social pressures are reinforced by informal responses. If you have benefited or suffered at the hands of another, your generous or hostile stance may be simple reciprocity that mirrors your feelings. Verbal and nonverbal communications of disapproval, as by colder conversation and fewer smiles, are construed here as transactional. They might be strategic efforts to change EP's to the disadvantage of the nonconformist or simply transactions in themselves, the content of which happens to be communicational.

The sharp limits of the human mind constitute a strong argument for authoritative moral codes. As noted in connection with decisions (especially 2.55.3), the outcome of much behavior is so difficult to know that many of life's decisions are perhaps more rationally made by arbitrary rule than by careful thought. Simple uniformity then greatly reduces life's information costs, perhaps without significantly reducing its rationality.

Rules of morality often apply only to the ingroup. With outgroups one may steal, lie, cheat, pillage, rape, or murder. Religion may inhibit a person if it alleges that the whole of mankind is an ingroup, but in practice one's buddies, not his religion, define the outgroup. The reason why it is all right to treat the outgroup thus is that they deserve no better since they themselves are lazy, irresponsible cheats, swindlers, thieves, and murderers. The relation, of course, is one of positive-feedback image reinforcement. They *are* that way to you because *you* are outgroup to them. And you are that way to them because they are outgroup to you. The images prevent both from treating the other as ingroup and changing the image.

The order of preference for ways of achieving conformity in the semiformal organization are probably the same as within formal organization: the communicational, or normative; the transactional, or payoff; and the dominant coalition. The more successful the high-priority technique in achieving conformity, the less the need for low-priority methods.

4.36   **FAMILY**

The initial models of formal and informal organization were followed by applications to specific systems government and market economy, respectively. No neat example of semiformal organization is available, however, for reasons that are amplified in the last section of this chapter. All things considered, the family seems the most probable candidate for discussion although it is a mixture and not really capable of being analyzed in terms of a pure model. We will deal with the nuclear family, which is relatively similar in certain basics in various societies, rather than with extended families, clans, or sibs, which differ greatly. As before, we are interested in building a model of the family out of the materials already developed.

To clarify this procedure, we recall that the A-level models of Figure 4 (Chapter One) are the most general-purpose ones and that the B-level models are still reasonably general. As noted at the outset, most of this volume deals with A- and B-level materials—though the discussions of the formal, informal, and semiformal organizations of the whole society in Chapters Thirteen through Fifteen might be considered rough sketches of C-level materials. (We should also recall that the number of levels suggested in Figure 4 is arbitrary.) To illustrate the simulation process (at the risk of misrepresenting the proper location of the model of the family within the overall continuum), we will speak of this model of the family as D-level. At that level we do not generate additional basic concepts but rather assemble a unique collection of previous concepts that seem to fit the family—or at least those aspects of the family that concern us. If the model provides understanding, fine. If not, the model must be reconstructed by modifying the "package" of concepts or even the lower-level concepts themselves. I know of no way of doing so except by search—that is, by trial and error. In this sense the logical problem of formulating the simulation model seems to coincide with that of formulating any problem for a decision—which is to have intelligent, informed intuition (2.55.3 –.4). The following model is not directed toward a particular family problem, but rather toward illustrating the process. Investigators with different interests would select a different "package" even if they wholly accepted the previous concepts.

The family fulfills the definition of complex, not simple,

organization. It has more than two persons (after children arrive), a major bargain of affiliation between husband and wife, at least one party whose values received depend on the performance of the organization as a unit, subsidiary transactions about day-to-day matters, sufficient complexity to require division of tasks, and some investment and fixed costs.

As to type, the family is basically cooperative. In its main functions of procreation, rearing children, preparing food, maintaining shelter and clothing, personal care, and the like, it creates wanted things for its members who are simultaneously sponsors, staff, and recipients. Despite some exceptions, for the most part (and in the model) these things are not sold for a profit, given as a service, or exercised as pressure. And although some may be bought, as from servants, we are dealing here solely with the things that are done by the family members for themselves.

To identify the family as complex and a cooperative implies formal organization—that at least some goals and processes are decided for the family as a unit rather than being the fortuitous outcome of the separate members seeking their separate goals. Even the most formal of real organizations nevertheless contain important informal and semiformal elements. So, too, does the family. The main elements of its structure are sufficiently constrained to bring it close to the formal model, even if the structure is not consciously planned as that model prescribes. The reasons for this conclusion appear shortly below, and a review of the traits of the family appears at the end of this section.

In addition the family shares important attributes of government. First, it is relatively all-purpose—not segmental. It obviously does not do everything, any more than government does; important functions are performed outside the family by school, church, and peer group among others. To a degree that varies much with the society, the family nevertheless makes many decisions that the functions are to be performed outside the family, or perhaps not at all, rather than in and by the family. By the same logic that applies to government, though with less force, it can also be said of the family that its inactions are as important as its actions and that its inactions plus its actions equal all possible actions (4.13.1c). In any event the family is a much more multipurpose organization than any other except government.

A second attribute of the family that parallels government is the legitimate use of force, normally by one or both parents, within the family. Whether or not force is actually used, parents may use coercion to control their children, either for physical blockage or as sanctions. So far as young children are concerned, the family shows a third trait of government—involuntary membership. Children make no decision to be born into the family, and until reasonably mature they cannot leave if its costs seem to outweigh its benefits. The parents, by contrast, initially join the organization voluntarily, and with greater or lesser difficulty can also leave it.

A fourth trait is not shared with government or perhaps with any other type of organization: the absolute dominance of parent over infant, along with the inescapably generous relation when the infant can give nothing in return. (The parents may "receive" pleasure from the infant, but the infant at first does not give it transactionally any more than a sunset transactionally gives pleasure.) Fifth, the family is unique in having procreation and a continuing sexual relation as key functions. Sixth, and related to the preceding point, certain crucial aspects of the family structure, involving both role description and role occupancy, are biologically determined. We do not wait breathlessly to learn who is to be father, mother, son, or daughter. Though drugs plus surgery may accomplish some role changes, we will ignore these as unusual cases; the parent-child role relation has a definitional irreversibility that even surgical miracles will not alter. Male-female differences in size and strength influence other role differentiation, and we have already noted the helplessness of the infant.

If we start with a general-purpose model of complex organization, note the small numbers of people in it, and then modify it by the six factors just listed we have produced a tentative simulation model of the nuclear family—that is, a "theory" of it. Within this framework the theoretical analysis of the family then consists of tracing the traits of this particular combination of ingredients. If that analysis provides an unsatisfactory approximation of reality we must modify the mix of ingredients that constitute the model, modify the ingredients themselves, or both. To illustrate the method of applying the model we examine two distinctive aspects of the family—transactions over affiliation and transactions over sex—and conclude with some reasons why the family seems a potentially rich scale model for studying social processes.

4.36.1   *Transactions over Affiliation*

Marriage is a bargain of affiliation. But since each party "gives himself" the question arises whether transactional analysis applies and whether there can be terms or bargaining power. The answers are affirmative. If terms of transactions are based on values and the things "given" in marriage are persons, then the transaction of marital affiliation will revolve around the "values" of the persons, which is another way of describing how much each is wanted. (We have already noted in 3.43.6 that all interpersonal power except that of bads lies in being or possessing something wanted.) The content of what is wanted varies with the society and the individual, and the power ingredients in a particular case are therefore relatively unique. Generally speaking, however, physical attractiveness, physical or mental capabilities, wealth, and status are the main power ingredients of mates. In fact the last two factors are themselves forms of power. From the bargaining power model we hypothesize that one can command a mate who rates high on some package of these factors only if he can offer his own package of approximately equal value, though romantic fixation often dwarfs other factors.

Where the bride price is in vogue the prospective bride must offer high levels of personal charm, prestige, money, or talent to command a groom replete with large "price." The damsel whom fortune has skimped commands only a small one and is perhaps pleased to get just the man. With a reversal of actors the same logic applies to the dowry. If some cultures ascribe status or desiredness to other factors, the content of the transaction changes but not the analysis. We are of course speaking of general tendencies, and there may be no better explanation for a marriage than "It was Sunday afternoon and we were in Peoria and it was raining." And the poor boy *may* get the rich girl because she wants to defy her domineering father.

To note the reciprocal relation between fact and theory let us suppose that the general transactional analysis is so thoroughly accepted that an observer assumes it applies to marriage transactions. If he then observes that distended female earlobes command richer husbands than do normal ones he has established, if other things are equal, that stretched ears are more "valuable." The logic is the same as when we conclude from its floating position that cork is lighter than water. Although we can hardly hope for precisely measurable analytic units in the social sciences, a set of interlocking concepts along these lines should enable us to build a reasonable set of interlocking indicators.

The status aspects of selecting marriage partners are interesting. To simplify, let us assume that the status of the family is determined solely by that of the partner whose sex is dominant in the society and that we are dealing with a society of male dominance. If he "marries downward," she gains status but he loses none. If she "marries downward," she loses status and he gains none. Other things equal, there is an overlap of EP's for marriage in the first case but not in the second; and the greater the status difference between them, the greater the net gain—if we focus on this factor alone. We would therefore predict more cases of women marrying upward and men downward than the reverse. This condition prevails in fact in the United States (Berelson and Steiner, 1964, pp. 306f) if we measure status by education and class. Before we attach great social-scientific weight to this conclusion, however, we should note (as Berelson and Steiner did not) that if men on the average get more formal education than women (as they have in the past) women on the average will necessarily have less education than their husbands. It also follows that if men on the average marry downward and women upward, the proportion of unmarried persons will be higher than average for upper-class women and lower-class men while the proportion of married will be greater than average for upper-class men and lower-class women. The facts are in accord with this expectation (Berelson and Steiner, 1964, pp. 306f).

To the extent that status rises with age, the preceding logic also applies to age differences between mates. In the arts, professions, and executive groups income and status normally rise to the age of retirement; in manual occupations income and status remain more nearly constant through life and may decline as retirement approaches. One may therefore hypothesize that among men married in their later years, whether for the first or second time, wives on the average will be younger than the husbands and the age discrepancy will be greater among artists, professionals, and executives than among manual workers. This age discrepancy would also be greater among high-income groups than among those with low income. The tendency is greatly strengthened if the female is desired for her physical attractiveness rather than her productive skills. Berelson and Steiner report no empirical data on this hypothesized interaction of age and status.

4.36.2   *Transactions over Sex*

To explore the question of transactions over sex, one must begin with the relative sex urge of the male and the female. There seems no doubt of the greater urge in the male in virtually all mammals other than man, though the evidence is less compelling among the anthropoids. Basic logistics, however, suggests that a stronger male urge is probable. First, as related to the evolutionary criterion of procreation there is no biological point to intercourse when the female is not fertile, and in many species it occurs only when she is. Barring miraculous timing there is only one way to ensure that the male is ready when the female is: to have the male ready or readiable all the time—a not too unrealistic description of the uninhibited vigorous human male. Moreover, at least among humans fertilization can occur whether the female is aroused or not, but not if the male is unaroused. Hence male arousal is imperative to fertilization but female arousal is not. And as for the pleasure imperative—assuming the latter to be an evolutionary derivative of the former—the female can give the male basic sexual satisfaction in a normal relation whether she is aroused or not, but the reverse is not true. This pleasure aspect affects transactions over sex, whether or not the male has the stronger urge, since the female can sell her services to a male who does not arouse her but the reverse is not possible.

Since sexual relations are often intensely desired they are a potential source of high power and bargaining power.[1] When this fact is taken in conjunction with the preceding observations, the tactically competent, sexually attractive female who is willing to trade can command large returns for her favors. Societies may thus frown on such trading partly for the same reason they frown on transactions in bads —they can produce such abominably one-sided bargains.

There is also transactional logic in a sexual monopoly in marriage. For the husband it decreases the likelihood he will fritter away his wealth on evanescent pleasures. For the wife, and particularly where the man provides the main income, it prevents other women from bargaining his earnings away from her and the children. By removing his alternatives it raises her bargaining power. Whatever its other components, an important element in sexual jealousy must certainly be the sensed loss of bargaining power produced by a rival. If the wife's desires are the stronger or she provides the income, these bargaining relations are reversed.

The preceding discussion does not imply either that those bargaining power factors always lead to monopoly, since they obviously do not, or that they are the sole or even the chief benefits of it. The point is merely that there are some transactional supports for restrictions on adultery. And to the devout humanist who bridles at the thought of rationally analyzing such matters we suggest that the

---

[1] While discussing sexual relations shortly after dealing with market transactions it struck me that we need a name for an important concept here. According to the theory of markets, by pursuing one's unalloyed self-interest and ignoring the welfare of others he ends by doing what is in the best interest of others and of the economy. By contrast, by concentrating on the pleasure one gives to his partner in sex and ignoring his own he often ends by greatly enhancing his own pleasure. Since the first concept is known as Smith's invisible hand and this is an opposite principle, I take the liberty of calling it Kuhn's visible hand.

loveliest aspects of the tenderest relationships may rest on absolute confidence that no rival can touch the bond. If the humanist also cringes at language which says that each spouse has no transactional alternatives, so do I. That *is* one of the costs of science. Perhaps the transactionally deduced conclusions that nearly everyone wants to feel *liked* (3.28.42) and *wanted* (3.43.6) provide a sensible meeting ground.

A conspicuous aspect of sex is the nearly universal incest taboo. Important aspects of this taboo seem transactionally analyzable, and we will begin by citing explanations from Berelson and Steiner (1964, p. 317):

(a) The belief, and in certain circumstances the fact, that intra-family relations produce deteriorated offspring—the genetic reason.

(b) The need to defend the nuclear family, and to a lesser extent the extended family, from the intense strain that would arise from sexual rivalries, passions, and jealousies—the maintenance-of-the-family reason.

(c) The original need to develop a social group larger than the family for purposes of protection and sustenance, and the consequent desirability of pushing family members outward in order to broaden alliances—the mutual-assistance reason. (One of the major consequences of the prohibition is the discontinuity of the nuclear family, which typically contains only two generations.)

(d) The need to preserve the parents' authority over the rearing of the children by clearly separating the parental and the sexual roles, and the need to encourage adjustment to the larger society outside the family—the socialization and social-integration reasons.

(e) To account for intensity of maintenance, the fact that the taboo is taught early and severely.

To deal with a crucial point before discussing Berelson and Steiner, at least two reasons can be cited against an inborn aversion to incest. First, society would hardly need powerful controls to keep people from doing what they did not want to do anyway. Second, it is difficult to imagine a psychological mechanism that could transmit an inborn valence toward a learned category (2.32). Mother, father, sister, brother, and cousin are all learned concepts, and it is hard to imagine how the human could inherit a desire for sexual relations with non-family and an aversion for family if he must depend on someone else to tell him which is which. A presumption that the taboo is either inborn or genetically oriented is also weakened by the fact that it applies to adopted children and stepchildren and stepparents about as strongly as to blood relations. Since the taboos long antedate reliable contraception it might be expected that intercourse presumes possible pregnancy. That possibility is relevant to some of the reasons, but not all.

The genetic reason (a) is not transactional, at least not directly. If valid, it is rather the straightforward pursuit of advantage for the family by avoiding the burden of inferior offspring.

Reasons (b), (c), and (d) are definitely transactional, though (b) and (d) need amplification. We assume that sexual union is rewarding to both partners and begin with a possible relation between mother and son. First, the father's bargaining power with his wife declines because she has an alternative source of satisfaction. Second, the father's authority over his son declines because the son's increased bargaining power with his mother gives him a better chance of getting other things he wants. Third, the son's increased bargaining power with his mother may divert her nonsexual favors as well from father to son. And fourth, other things equal, the father will now have sex less frequently and the decline may be dramatic if the mother finds the son more competent. For all these reasons, at least intuitively understood, the father may be expected to disapprove of relations between his wife and his son—and the strength of the logic rises rapidly with the number of sons. With appropriate substitution of characters the mother's objections to relations between father and daughter are parallel.

Some, but not all, of the difficulties disappear if both parent-child relations are allowed. Assuming equal sex urge and an equal number of nubile sons and daughters, the direct loss of bargaining power between husband and wife is reciprocal and cancels out. The question of unequal frequency also disappears, and total frequency might well rise. The other two objections, however, are additive, not canceling. The father's loss of bargaining power (authority) over his son, whom he must train, is not offset by the mother's loss of authority over her daughter and the complementary conclusion is parallel. Further, channeling of family resources from father to daughter and mother to son leaves less for the parents. In addition, in a particular nuclear family the number of male and female children may be very unequal, in which event one parent would be relatively disadvantaged. In any case there would also be marked shifts in power as successive children matured.

Several factors suggest that the mother would be the more disadvantaged on the average, by unrestricted relations, and would be the more eager for restrictions. Some factors are definite and others conjectural; the latter are introduced only to indicate the bargaining power effects of particular conditions. First, pregnancy periodically removes the mother from competition and virtually guarantees that father-daughter relations would be more frequent than mother-son. Second, females mature earlier; thus daughters would be available to the father sooner than sons would be available to the mother. Third, the young and inexperienced female may be more satisfactory for the older male than the young and inexperienced male is for the older female. If we translate this relative to an absolute preference to sharpen the point, the father would then prefer his daughter to his wife while the mother would prefer her husband to her son, and the wife's loss of bargaining power could be catastrophic. Fourth, if the male sex urge is the stronger on the average, as is often suggested, then hers would be the greater loss of bargaining power from unrestricted relations. These four factors suggest that for the relation between parents the mother would support the parent-child taboo more strongly than would the father. But insofar as it affects the father-son and mother-daughter relations in particular, and parent-child relations in general, the support would be more nearly equal.

Except in the apparently nonexistent condition in which parents would want a monopoly on sexual relations with their offspring for themselves, none of the preceding transactional logic proscribes relations between siblings. Brother-sister relations would not affect bargaining power between parents, their authority over children, or the amount of family resources channeled to children. Here explanation (c) alone seems sufficient, along with the latter part of (b), which is a different aspect of the same point. That is, to prevent relations of children with one another *or* their parents forces them outside and enlarges the social group. This enlargement has two conspicuous advantages for self-defense: first, by enlarging the size of the friendly coalition; second, by reducing the number of enemies and increasing the average distance to the nearest ones. The larger group might also be able to exploit natural resources more effectively. Furthermore, transactions in general are facilitated and the likelihood of satisfactory exchanges increased as the number of available transactors increases. The incest taboo may thus by several millenia antedate the antitrust laws.

To the transactional difficulties are added the organizational. Without incest the roles are clear: mother, father, son, daughter. With incest they become unbearably confused. Suppose that father and daughter produce an infant. Does the original wife now treat her daughter as daughter or as co-wife? Does the infant treat its father as father or as grandfather? Does a son of the original pair treat his father as father or as brother-in-law. And does he treat his sister as sister or as stepmother? Does the child treat its mother as mother or as half-sister? These are only a few of the possible role confusions. If one incestuous relation could create this much confusion, several more would require a computer to keep score, and primitive man did not have one. This does not mean that computers will end the taboo. The transactional grounds remain.

There is no easy way to test these explanations. The origins of the taboo are long lost, and it is so thoroughly ingrained in each new generation that few question it. Whether or not these interpretations are correct, they clarify that there *are* transactional explanations which both incorporate and make more explicit important contributions from other sources.

4.36.3   *Family as Scale-Model Society*

Somehow it seems that the only decent reports of what really goes on in families are written as novels or biographies and rarely by social scientists—though the study of family therapy is rapidly filling the gap. It is good for the quality of life but is bad for science that we do not manipulate and minutely observe real families the way we do T-groups or other experimental assemblages. In any event the family seems the group *par excellence* in which to study social processes on a small scale. It has elements of formal, informal, and semiformal organization. It has both voluntary and involuntary membership. Transactions appear in multitudes: selfish-indifferent, generous, hostile, major-subsidiary, other interrelations, through agents, with goods, with bads, with and without competition, with coalitions and collective bargains, even with hints of market processes.

Decisions are made by individuals, by an "executive" on behalf of the family unit, and by the group processes of communication, transaction, and dominant coalition. Different members of the family hold different aggregate power, and there is definite stratification, at least between parents and children.

There are communications galore, both in the form of messages and in the transmission of the culture. Each family develops a private language that changes with the age of the youngsters.

As a bonus, the family inescapably undergoes developmental change as new children are added, everybody learns and gets older, some leave, and some die. Authority relations necessarily change as children mature, and as the age when children leave home approaches there is a gradual shift from the authority relations typical of government to those of voluntary organization. Because each person changes as an individual there are also inescapable shifts in roles. While incorporating on a small scale most of the social components of whole societies the family is also a subsystem subject to coercive as well as semiformal constraints from the larger unit. Depending on the selection of marriage partners, there can even be ethnic differences within a nuclear family.

In short, the nuclear family is potentially a valuable small-scale social laboratory—though in an age of galloping Big Brotherism one hesitates to mention it.

4.4 REVIEW OF ORGANIZATIONS

We have examined a model of formal organization and an example that requires specific modifying assumptions—government. We have also inspected a model of informal organization along with an instance with specific modifying or supplementing assumptions—the market. In each case it was easy to find real cases which, though not precisely like the model, are undeniable exemplars of it with no significant problems of identification. And in this chapter we examined the model of semiformal organization.

It is possible to make reasonably explicit statements about government because the model of the whole organization is explicit. The rationale of market theory differs somewhat, in that the nature of the whole system is deduced rather than assumed. Sharp deductions are possible, however, because clear assumptions are made about each category of subsystem and about the nature of the interactions. Each subsystem makes rational decisions directed toward maximizing some entity, satisfaction or profit, and interacts solely through selfish-indifferent transactions in goods but not bads. Communications are not denied but are conveniently assumed to be perfect, requiring in the pure, pure case knowledge of nothing more than that something is there and at a certain price.

But the model of semiformal organization, unlike the other two, presents no neat, obvious exemplar in the real world. Although the group comes readily to mind, it is such an amorphous concept with such multifarious examples that one would almost have to choose between a case study and an artificially constructed example to illustrate it. Thus we focused on the family, which is at least a real exemplar with substantial case-to-case uniformity.

Whether it will ever be possible to make precise statements about informal organizations other than market economies is problematic, though recent modelings of ecosystems (MacArthur, 1972) seem promising and certainly deserving of the attention of social scientists (see Boorman, 1972, on MacArthur). Much the same question arises in connection with semiformal organization, and perhaps should be reexamined by specialists following some of the materials in the remaining two chapters.

Two main aspects of the analytic apparatus of this volume remain to be discussed—social change and social idiography, along with the latter's relation to simulations. In fact, it is improper in one sense to have tried to deal with examples before the whole analytic structure has been stated since proper analysis of reality requires all of it. The congruences nevertheless seemed close enough, and there are disadvantages to holding all examples to the end. We now move to those remaining aspects of the overall model and to several additional applications of it.

Chapter 16

Social Change and the Idiographic

Most of this volume deals with cross-sectional analysis, or statics, not with the developmental analysis of dynamics and change. Most of it deals also with the nomothetic analysis of principles abstracted from time and place, not with the idiography of case studies and complex real situations. This chapter concerns both subjects— social change and idiographic analysis—and thereby rounds out the analytic structure of the volume. First we discuss the nature and theory of change and then the techniques by which we can apply this general social science to real situations that are both complex and specific.

As it turns out, the two subjects are intimately related. Hence we discuss part of the question of change under that heading. But other questions of change merge with discussion of the complex, unique situation—the idiographic. Before we examine why people act in ways that produce social change we must find out why they act at all —as the social scientist views the question. For that purpose we begin the discussion of social change with a short discourse on the role of norms and decisions.

5   SOCIAL CHANGE

5.1   *Norms and Decisions*

A possible explanation of social change is that it occurs because people want things to be done differently than they were done before. Hence we may ask about the degree to which people in societies make conscious decisions in their social relationships. We have already raised the question of the relative roles of rationally decided as contrasted to intuitive, impulsive, and norm-determined behavior (as in Chapter One, the introduction to 2.5, 2.56.7, and 2.84). We will return to aspects of the impulsive and intuitive later; in this section we deal with a distinction solidly related to two disciplines: economics focuses on behavior that rationally calculates advantage while sociology focuses on behavior that is normative, or culturally determined. In fact, if people follow either pattern strictly there is little room for impulse or intuition.

Within its realm economics (a) posits that people do what is advantageous (preference function or selector state), (b) delineates what is advantageous (opportunity function or detector plus effector states), and then (c) deduces that people will perform the behavior described in (b). By contrast sociology claims that people engage mainly in normative behavior. This claim is, of course, correct. But (and to overstate for the moment) until it is accompanied by additional statements of *why* they do so it remains an assertion, not an explanation. In present language the behavior of a human, as a controlled system, is not explained until we have specified his preference function and his perceived opportunity set. Economics has done this explicitly. To modify this overstatement we can say that, beyond asserting that patterns are internalized and enforced by social pressures, sociology does include many explanations of why people behave as they do. But, taken as a whole, the explanations are scattered and unsystematic, and there is certainly no central motivational assumption as in economics. Some sociologists follow the lead of psychologists in asserting that much human behavior is not "rational."

To describe behavior selected by the complex interaction of a preference and a perceived opportunity function is to describe a decision. That is, a behavioral explanation is to a large extent decisional, reflecting the payoffs in a field of alternatives. The problem of the "actor within a field" is a decisional problem, and the contours of Lewin's fields differ logically from Pareto's indifference curves only in number of dimensions. Every decision, however, is made within a set of constraints, natural or social. Social norms thus do not preclude decisions; rather they provide constraints within which decisions are made. Sometimes the constraints are so compelling that it is pointless to speak of choice—as if we chose to go down the stairs rather than jump from the third floor! At other times norms leave so wide a choice that there is scarcely a constraint at all—I can punish the youngster for breaking a window or praise him for hitting the ball so far. In a complex society, to violate a norm of one subculture may bring accolades from another. Even in relatively simple societies norms may be selectively invoked or ignored, depending on the situation and how people interpret the behavior (Buckley, 1967). And the same can be said of many laws.

There is thus no contradiction between "decided" and culturally determined behavior; there is merely the question of how tightly norms constrain the particular case. Further, a person may be subject to external cultural constraints even though he has not himself internalized them. These constraints are then part of the subject matter of his conscious decisions in the same logical sense the natural environment is. An obvious case is the politician who scrupulously follows the norms of his constituents even if he has internalized none of them. Even the person *within* the subculture may follow norms because it pays, not because he accepts them. Here rational calculation of advantage and normative behavior coincide in what Etzioni (1961, pp. 14ff) calls "utilitarian compliance."

Nadel (1953, as discussed in Buckley, 1967, p. 165) has proposed that the customary and normative are often advantageous. Furthermore, to the extent that preceding generations showed sense the normative may be instrumentally the most effective. As Buckley suggests, it may then rationally be selected whether or not society penalizes nonnormative behavior. For example, a host of satisfactions are available through the "normal" continuous associations of marriage that are simply not possible through "nonnormal" discontinuous associations. And society need not penalize driving through field and stream to keep most motorists on the highway. That the economist refers to "rational" behavior and the sociologist to "adaptive" does not mean they differ logically. On the possible coincidence of advantageous and culturally determined behavior one should also see Firth (1967, chap. 1) and Blau (1964, pp. 17–19). As Pepitone puts it (1964, p. 28), in the past decade there has been "a dissatisfaction with the alleged neglect of intellective and rationalistic factors on the part of the need-satisfaction model"—in Freudian language, too little attention to the ego.

In many situations, on the other hand, normative behavior is distinctly disadvantageous in itself while deviant behavior is highly rewarding. Many devices widely prohibited among ingroups of nearly all societies have a high payoff if successful in the unique transaction. In two minutes the successful thief acquires what it would have taken him a year to produce or earn, the murderer eliminates forever a particular human irritant, the rapist achieves a particular satisfaction while escaping its responsibilities, and the angry man achieves high satisfaction from flattening another's nose. If others are also seeking *their* advantage, however, it shortly becomes evident why behavior that pays off handsomely in the unique transaction will pay off badly over a series of transactions with the same people and why groups almost inevitably prohibit (form coalitions against) transactional devices with highly one-sided payoffs.

We might speculate whether it would be useful to construe consciously decided behavior as analytically parallel to the model of formal organization and unconscious behavior as parallel to the informal or semiformal. The nonconscious or intuitive (preconscious in Freudian language) would then be seen as tending in the same general direction as the conscious, but more loosely, less reliably, and less quickly.

An interesting methodological problem can be examined if we note the suggestion of some sociologists that deviant behavior must be "explained." In the preceding examples the explanation lies in the obvious advantage of the deviant behavior—if it remains unique. Although that interpretation is not at odds with the sociological explanation that the socialization process is not complete, there is an important difference in emphasis. Just as the economist uses the perfectly competitive market and perfectly economic man as his initial model, the sociologist less formally and perhaps less consciously uses the perfectly socialized individual. In effect the sociologist takes the perfectly functioning machine for granted and wants malfunction explained. Although this approach is unassailable, given the model, in the larger perspective it reverses matters. In the larger sense the surprising thing is not that so much deviance occurs but that so little does —it is not the deviance that requires explanation but the norm. In a universe in which order is improbable it is not the absence of the machine or its partial failure that requires explanation but the fact that it exists and works at all. To ask for an explanation of deviance takes the functioning machine for granted—it assumes that the improbable is the normal and wants the probable explained as if it were abnormal. As Moore (1963, p. 13) puts it: "On a strictly actuarial view of socialization, uniformities are somewhat more remarkable than variations." In the larger perspective chaos is the most probable state, and it is order, not chaos, that must be explained. But that is a problem of the developmental process, which we will come to shortly.

In this framework the cultural process is neutral between norm and deviance; any behavior that occurs can be culturally communicated. Similarly, *culture* does not explain why norms exist but only how they are transmitted. To say that norms are the product of culture is like saying the human race is a product of procreation—a perfectly sensible explanation if we assume there already *is* a human race. The existence of norms is also a developmental question to be examined below. Meantime we note that deviant behavior is as much a part of a society as the normative, even if less frequent. Even simple societies incorporate quite stable sets of alternatives to many of their norms (Buckley, 1967, pp. 10 and 25), and the complex society displays few norms common to all subcultures.

5.2   Aspects of Change

In this section we make no distinction among the terms change, development, and developmental change. We may as well face at the outset that change is a primitive term we can discuss but not really define. We might use the ploy of reducing change to terms of same and different, but that does not help much. In some respects no one is the same as he was five minutes ago; in other respects he is the same as when he was born. As with the question of whether something is or is not a system and where it is bounded (1.20o and r), change is not a question of reality alone but one of how the observer views it and what he is interested in. Our problem here is thus not to define social change but to find out what varieties of change and what principles are worth the attention of social scientists. In exploring this topic we begin with a suggestion that all change of interest to social scientists consists of alteration of structure—that is, change of roles and interactions of roles. Then we will examine social change that apparently does not involve structure and finally return to the starting point by indicating why these "nonstructural" changes might also be considered structural.

5.21   *Structural Change.* One conceptual distinction is clear, even if it is not always easy to match it with reality: systems come in hierarchies, and the first distinction is whether the number of hierarchical levels changes or not. If a college in which all departments report directly to the dean is divided into divisions of humanities, natural sciences, and social sciences with a subdean of each, a hierarchical level has been added. The reverse procedure would reduce the number of levels. Perhaps because we have cogent theories about them we will confine our attention here to the appearance of higher-level systems that incorporate and coordinate previously separate systems—and the reverse process in which a higher-level system disappears and leaves only the lower-level ones. If the college added divisions of arts and of business, or if psychology were transferred from the natural science to the social science division, the structural change would not change the number of system levels. The change from a four-cylinder to an eight-cylinder engine would also be structural without involving hierarchical level. So would be the hooking of two four-cylinder engines to the same drive train, unless a new level was added to coordinate the two engines or unless they somehow interacted with one another.

The distinction between structure and process seems essential here and must be related to the concept of cross-sectional analysis. Cross-sectional analysis does not imply that nothing is happening: there is clearly a great deal going on in the steady-state equilibrium of a forest, firm, economy, government, or social system. And so long as the position and volume of Niagara Falls remains the same we do not think that the falls has changed just because there is a complete turnover of water every three seconds. Whether a change is process or alteration of structure may depend on how long we look at it. If we looked at the solar system for six months we would conclude that it had changed structure since the parts would be in very different relationships. If we observed it for six centuries we would conclude that its structure had not changed but had merely gone through a repeated cyclical process. Six eons of observation would again look like structural change. The adding or subtracting, separating or combining of tasks on an auto assembly line would look like structural change from the perspective of a single model change; but it would look like process from the perspective of a score of such changes. The difference may also depend on *what* is observed. If we focused on the inputs and outputs of the engine and the repeated cycling of pistons and valves we would view it as process. But if we focused on carbon deposits and wear of cylinder walls we might view it as structural change. Perhaps more importantly the distinction depends on level: what is process for one level is generally change of structure for a lower level. Digestion is process for the whole human but structural change for cells and molecules. "Running" is process for the whole engine but structural change for its components: chemically for throughputs and positionally for solid components and some of their molecules. Marriage, births, maturation, and children leaving home are structural change for the individual family but process for the society. Thousands of liver cells are born, mature, and die: this is process for the whole liver and structural change for the cells. It is also possible, of course, to have structural change at a higher level while there is none at the lower level. The overall structure of a steel company can change drastically—by being nationalized, merged into a competing company, or spun off from a conglomerate— while the structure of its tubing division goes serenely on.

The point is to suggest that perhaps we should not think of something as having changed unless there has been a discernible alteration of structure. Conversely we find that process consists of structural alteration at some lower level or in some shorter time span.

We now turn to the change in number of hierarchical levels of a system with the emphasis indicated above. Such change can go in either of two directions. *Emergence* is the appearance of a new and higher system level; *decay* is the breakup or disappearance of a higher level of system such that only a certain level of its subsystems remains. As treated here, emergence is a broadening of the theory of biological evolution whereas decay is a broadening of the second law of thermodynamics. Because the latter helps us to understand the former, we deal first with decay.

5.21.1   *Hierarchical Change: Decay.* When a mammal dies, eventually only molecular levels of subsystems remain; but organs and cells have intermediate life spans, as do certain complex molecules. A human fed intravenously in a coma has lost the level of psychological coordinated wholeness that makes him a human and he functions only at the biological level. When a marriage or a business breaks up, the original social level of system ceases to exist but the individuals who were its subsystems go on. Such decay (or disintegration) can occur because it was consciously decided to be advantageous, in which case we utilize such previous concepts as group or individual decision-making, sponsor roles, and transactions of affiliation. Or decay may come by cataclysm as when the fly swatter descends on the fly or the black plague decimates a population. Someone might venture that a system disintegrates because it has lost its integration or its cohesion. For social systems, at least, this is merely a mystical way of saying that it has ceased to be advantageous to its participants.

We now turn to a broad variety of disintegration based on the second law of thermodynamics. According to that law (1.32.1), all closed systems are subject to increasing entropy. In layman's language, all matter within an insulated area, at least of Newtonian dimensions, will eventually have the same temperature. Only the introduction of temperature differentials from outside can prevent or reverse the process. But that event implies an open system, such as a break in the insulation, which violates the conditions of the law. The generalized version of this law, of which the second law of thermodynamics then becomes a special case, is that all closed systems are subject to loss of differentiation. A strong link, considered by many to be an identity, has already been established between entropy and loss of information—see, for example, Cherry (1957), Tribus (1961), Shannon and Weaver (1949), and Jaynes (1957). An interesting corollary of this relationship is the proposition that information can be lost, but never gained, in transmission. The proposed extension to social processes, however, must for the present be taken as merely a proposal whose validity requires much examination. An obvious example of this extension is that societies which are cut off from the rest of the world display less variety of behavior and belief, less innovation, greater homogeneity within generations, and less variation between generations than do societies in regular contact with other societies. Similarly, studies of group behavior indicate that groups whose members interact only with one another have greater similarity of attitude and values than do groups whose members have frequent interactions outside the group. The fear of inbreeding that leads many university departments not to hire their own PhD's as professors directly reflects at least intuitive knowledge of this principle.

At the biological level a significant theory of aging is that cells and organs are initially highly differentiated and that the coordinated interaction of these different subsystems is what makes a well-functioning organism. With time the differentiation introduced by the genes is damaged or randomly dissipates until the various subsystems are no longer able to perform well enough to service the remaining subsystems and the coordinated activity of the whole then ceases. The decay of a once highly structured tree into a pile of undifferentiable topsoil is another example; both the decay and the aging occur because no new differentiation is introduced from genes to replace what is lost.

We need not try here to assess how much social behavior is amenable to this analysis, but we should insist on its importance as a springboard for understanding the reverse process—emergence—to which we now turn.

5.21.2   *Hierarchical Change: Emergence.* In the simplest case of emergence, which might be thought of as the paradigm case, two or more systems combine into a larger system in which each of the previously independent systems is now a subsystem. A marriage, or even a friendship, makes a system of two previously separate parties. A detector plus a tuned circuit plus an amplifier make a radio receiver. Thirteen colonies plus a constitution and the will to make it work constitute a United States.

Independent emergence must be distinguished from replication. The first appearance of life on earth, the first cooperative government, the first flyable plane of the Wright brothers—all represent independent emergence. But when the first living thing reproduced itself, when other societies adopted the government, when others copied the airplane—all represent mere acts of replication. Replication is a communicational process, whether by language, genes, imitation, or other technique. The first appearance requires an utterly different explanation and is our central concern here. Strictly speaking the criterion is independent appearance, not first appearance. If five different people had independently invented the airplane at either the same or widely different times, or if life had spawned independently at five million different spots and moments, each would be an emergent act.

Since it is the opposite of entropy, emergence is often known as negative entropy, or negentropy. It is also referred to as increasing structure, increasing order, decreasing chaos or randomness, and creativity; the sense of each designation should become clear as we proceed. The relation between differentiation and structure is that structure itself consists of differentiation plus coordination. The human could not be made of blood alone, or bone, or nerves; nor could the automobile be all motor or all wheels. The systemness lies in the coordination of differentiated parts.

If loss of differentiation occurs inevitably in all closed systems, logic decrees that increased differentiation can occur only in open systems. Differentiation, or variation, is introduced only from the environment of the system or from interactions within the system that are related to receipts from the environment, under circumstances in which the forces are of sufficient number and variety that their influence in a particular case is essentially random. Among those variations some are more conducive to the goals of the system than others. The former will continue and the latter will not, thus moving the system, or the average of a population or species of similar systems, in the direction of the "favorable" variations. Further random variations then occur from that new position or average. If the criteria of the initial selection are stable (for example, survival), then new variations that lie still farther in the same direction will again be "selected in." Succinctly, the differentiation-increasing process is that of selectively and successively retaining it. The process is often subject to self-reinforcing positive feedback because the evolved state of affairs constitutes the environment within which all systems then live and from which successive variations may take place (Maruyama, 1963; Dunn, 1971).

If we assume that no two things in the universe are exactly alike, possibly excepting elemental particles, then *some* variation is the normal order. Increased nonrandom variation requires a selective process, which is a form of goal, and a goal exists only in a controlled system. Hence increasing differentiation can occur only in open, controlled systems. This leaves the question of where the controlled system came from in the first place. Since organisms and man-made social and mechanical systems seem to be the only known controlled systems on earth, the problem is essentially that of the origin of life. Life is a highly improbable form of order, which is itself improbable because it is based on differentiation and negentropy. The real question is how to account for the highly improbable. Here the answer seems to be initially sheer numbers, and Ashby (in Buckley, 1968, p. 116) adds that the large number must occur within a dynamic system whose laws are unchanging and single-valued. As it is sometimes put, among billions of billions of occurrences some instances of the highly improbable become virtually certain. The simple self-preserving system merely happens. And having happened, its self-preservingness is itself a selector—that is, a goal. I am not sure whether von Bertalanffy (1968, p. 152) disagrees with this. We are in agreement, however, when he says that "selection . . . already *presuppose*[*s*] the existence of self-maintaining systems; they therefore cannot be the *result* of selection." If we put the two views together perhaps we can say that the origin of life, of controlled systems, and of the ability to generate and retain further differentiation occurs when selectivity *happens*—initially a fortuitous event but one with thoroughly unfortuitous consequences. (See also 2.42.)

Emergence and differentiation are different but related. There are cases where mere aggregation of identical specimens gives some advantage to each unit over separate operation, as with colonies of cells or schools of fish. The advantage seems to accrue without division of labor or coordinated activity—though fish must coordinate their direction and speed if they are to remain a school. Enlarging the aggregation, however, does not alone make it a higher-level system.

Our model of complex formal organization includes division of labor both horizontally and vertically, and it seems tentatively sensible to generalize that both kinds of differentiation are essential to all emergent controlled systems. If the advent of an emergent system requires differentiation plus coordination of components, then differentiation is a necessary, though not sufficient, condition for emergence. (Starting from a given state and following it with differentiation and then recombination seems to be one way of describing thesis, antithesis, and synthesis.) Differentiation thus provides both the wider range from which "superior" specimens can be selected and the division of labor with which complex organizations can be built. For this reason we tie the problem of emergence to the causal conditions for increased differentiation.

Whereas the explanation of decay (decreased differentiation) is generalized from the second law of thermodynamics, it is clear that the conditions for increased differentiation are generalized from the essentially Darwinian principle of mutation and selection. Although the broad version is perhaps better called random variation and selective retention (RVSR), for the conscious purposes discussed below it seems better to speak of random *generation* and selective retention (RGSR). (To ensure that random means fortuitous rather than a carefully defined mathematical concept, Campbell (1962) uses "blind" instead.)

For simple self-preserving systems the sheer fact of survival is the only criterion of selection. To avoid teleological implications we must be careful how we state these matters. We can say that all self-preserving systems display survival-oriented behavior. If they did not they would not survive or qualify as self-preserving. When we say that survival is the selector, which we equate with a goal, we do not mean that the system *wants* to survive in some anthropomorphic sense (2.2). We mean merely that to an observer all self-preserving systems act *as if* they wanted to survive; otherwise they would not be there to observe.

We have construed the advantageous as that which contributes to achievement of the preferred; the preferred is derived by few or many steps in detector and selector from what contributes to survival. There is thus a basic sense in which the advantageous is that which contributes to survival. However, this statement provides no detail for complex systems. For them we can say only that selection is determined by whatever criteria of advantage happen to be operating in the particular case, which in organizations may mean merely what is perceived as advantageous for the subsystem or coalition that happens to be dominant at the moment. In some cases the natural environment provides the variability and the society provides the selection (Moore, 1963, p. 20). With respect to survival the language is in one sense a tautology: some trait or behavior survives because it has survival value, the measure of survival value being that it survives. There are nevertheless enough instances where survival value (or advantage) can be independently measured to make the statement useful.

Although the original Darwinian explanation has subsequently been modified, including attention to internal factors (Whyte, 1965), the basic rationale of variation and selection remains central, and that is all we need here. Because the extension of that rationale to other types of emergence is relatively new and not widely used, it is offered here only as a hypothesis. Though I have considerable confidence in it, my purpose here is to indicate the main implications of this explanation, if accepted, rather than to urge its acceptance. Further detail can be found in Pringle (1951), Campbell (1962, 1965), and the comprehensive review in Barringer, Blanksten, and Mack (1965). Environment provides "chance" variability and society provides for selective adaptation (see Moore, 1963, p. 20).

To accept this explanation has several consequences. First, the existence or appearance (as contrasted to the structure or process) of an emergent phenomenon can be explained only in retrospect; it cannot in theory be predicted—subject to modest qualification below. Although some early followers of evolutionary theory in biology construed it as the technique by which Nature produced its preordained highest creature, Man, and thus gave it a teleological and predictive quality, no tinge of such implication is intended here. What happened happened. If the circumstances had been different, something else would have happened. And despite this apparent nondescript character, the hypothesis is nevertheless useful when details are added about what survived under which circumstances and how the surviving patterns are preserved, transmitted, and multiplied through genes and culture. Furthermore, to accept this hypothesis is to reject alternative explanations. And even if the outcome of the process is in principle unpredictable, the theory nevertheless shows what process we must use if we wish to speed emergent development and increase creativity —namely, to create the conditions for exposure to numerous variations and combinations of the patterns already in existence, thereby increasing the likelihood that new patterns will be found that better meet our selective criteria. Ashby once identified genius very simply with requisite variety, and "A Case for Aimless Browsing" (Williamson, 1966) beautifully catches the spirit of the process.

The evolutionary process can operate at the matter-energy level, the information level, or both. Biological evolution is clearly the former: material individuals and species appear, adapt, compete, procreate, and the species multiplies, declines, or disappears. The evolution of ideas, or creativity, within or among persons is clearly the latter: ideas and mental patterns similarly appear, adapt, compete, produce ideological offspring, and flourish or decline. The random generation (rather than variation) that characterizes the information level is typically referred to as *scanning* and *search*. Tools and other artifacts may be viewed as a combination of, or midpoint between, the matter-energy and information levels. They follow the logic of ideas in that one may first appear as a pattern in a man's head. It is then reproduced externally in matter and is used. How satisfactorily it performs may determine whether the man will cast about in his head for a better pattern. Or a variation may occur in external matter through accident, whereupon the mental pattern will change when that variation is observed (3.18.4).

Evolution at the information level has distinct advantages. It is cheaper, faster (especially when assisted by computers), and can preserve in heads and libraries information about tools, social structures, or anything else, even when there are no real counterparts. And whereas the new patterns of biological organisms are limited to modest variations (mutations) of existing species within the narrow limits of cross-breeding, there is no theoretical limit to the variety of information patterns that can be formed quickly or to the breadth of cross-fertilization. Information patterns have the flexibility that would prevail in biology if any species could cross-breed with any other, as could all the hybrid offspring.

Whether an idea about social structure developed in someone's head can be translated into reality depends not on evolutionary logic but on communications, transactions, and the structure of existing social organization. That is, social planning, as planning, takes place by random generation and selective retention within heads or computers. But its acceptance and execution depend on social processes and conditions, including current norms and the locus of power.

In what Boulding (1966, p. 21) calls fundamental surprise, prediction of new information patterns is literally impossible since to predict an idea is to have it. To predict new matter-energy patterns is not totally impossible, though it is limited, and the prediction itself presumably must use the same process of RGSR at the information level. For example, one can scan existing patterns for possible variations or combinations, guess which are most likely to occur in fact, and from among those guess what selective forces will retain some and discard others (see Thorson, 1970).

Formal organization clearly has greater opportunity than informal to experiment with RGSR at the information level since it can at least in theory institute any structure that its mental trials select. By contrast, in informal organization the structure evolves in actual operation and hence cannot first be run through mental trials. Semiformal organization would lie somewhere between. One of the current hopes or fears, depending on one's views, is that greater understanding of this process may lead to more conscious planning, which means more formalization of the social system.

Blau (1964, p. 288) feels that the informal aspects (present definition) of a society are more conducive to change than are the formal, since the absence of overall controls provides greater opportunity for variation. This may be so, particularly where the formal organization has no yen for change. But if it *wants* change, formal organization may be able to generate, evaluate, and effectuate it faster than informal or semiformal. No generalizations seem possible on this score without knowing the nature of the change and the organizations into which it is introduced. Incidentally, although evolution can also produce simple change (the larger liver or the redder fruit) it is the only process that can produce emergent change. Even when cultural evolution proceeds entirely unconsciously, and even when its content is not consciously housed in anyone's head, it seems appropriate to refer to it as social learning (Dunn, 1971).

In the language of ecology, newly evolved systems that fill vacant niches seem to have the best chance of prospering since they have minimal competition, and this idea may justify modifying the previous statement about unpredictability. It seems safe to say that, other things equal, a variation that fits a vacant niche is more likely to survive than one that fits only an already crowded one: the blundering monopolist may earn more than the competent competitor.

A permanently vacant niche lies in the fact that there's always room at the top. That is, a new organization at a system level higher than any other organization has less direct competition than new specimens of existing types. If this is true, evolution has a built-in bias toward successively higher levels, loosely thought of as a teleological "urge" toward increasing complexity. The conglomerate corporation is an obvious example within which the traditional firm is but a subsystem. The investment trust similarly makes subsystems of the original autonomous investors. Insofar as more complex systems also have greater information capacity they are presumably also more adaptive— as contrasted to adapted (Dunn, 1971)—and have greater survival value. If so, this factor also gives evolution a bias toward successively more complex systems.

If the evolutionary process is to operate, the less well adapted specimens must be pushed out by the better adapted. Again, we need not define "better adapted" but can indicate prerequisite conditions of the process. The obvious requirement is that two or more specimens must operate simultaneously in essentially the same environment; only thus is it possible to discern, by observation or actual survival, which is superior. In economics, firms selling in competition meet this condition. But the structure of an industry may be as important to the quality of economic performance as is that of the firm. Since we cannot have both a competitive and a monopolized structure of the same industry in the same economy, and hence no direct comparative test of the structures, continuance of one does not indicate its superiority. The same is true of governments, since autocracy and democracy cannot cover the same unit at the same time. Nor can a nation simultaneously have a congressional and a parliamentary system, a two-party and a multiparty system, or a pure socialist and a pure capitalist economy. Other differences between nations are likely to be too great to provide a proper test; for such organizations the criterion of relative ability to survive in the same environment is not available. Hence long-term survival of a particular governmental structure is no proof of its merit. The moral code and many other traits of a society may be in much the same position.

Since according to our hypothesis variations arise from outside the system, the system must be open for them to get inside. To apply this logic to societies we must be explicit about the meaning of openness. A society is often considered closed if it has no communications or transactions with other societies and no migration. Yet even such a society is open in at least five important respects. First, all biological systems are necessarily open since they receive and release matter-energy and information. Hence the individuals who constitute a society are subject to changing inputs from their natural environment, which may change them directly and the society indirectly. Second, mortality guarantees turnover of personnel. Since no two sets of genes or lifetimes of experience are identical, no role occupant can be precisely like his predecessor. Third, all communication, including cultural transmission, is inexact. Even if the words of a folktale are invariant for centuries their meanings can change, and the vase, robe, or totem is never copied precisely. Fourth, however much the culture may impinge on the individual the initial randomness of his brain is never completely eliminated. Hence new combinations perpetually arise in people's heads and some of them see the light of day. And fifth, accidents happen even in the best-regulated societies and provide unanticipated new experiences. For such reasons every society is necessarily open, even if it has no interactions with other societies.

Other influences can, of course, come from outside—as from cultural fusion or changes in the natural environment. Depletion of resources by extraction is brought on by the society, yet once done it exists externally in the environment. The amount of change induced in a society by these external factors presumably varies with the amount of interaction between the society and the factor and also with the amount of change or difference in the factor—change in the case of natural environment and difference as between a given society and another with which it has contact. Between cultures the degree of interaction might vary directly or inversely with the degree of difference, and there is no ready a priori basis to predict which (4.14.5).

5.22   *Nonstructural (?) Change.* We now turn to change that is not structural—or at least may not seem to be. One form of such change was introduced as an elemental sequence of behavior in controlled systems capable of learning. In the first stage (2.07.1) the system detects the state of the environment, selects a preferred response, and effects it. Since we have seen that the first two steps can be construed as causes and the third as effect, this sequence certainly seems to qualify as cross-sectional analysis. But having acted, the system then receives a feedback of information about the consequences of its first action, evaluates it, and gives a second response (2.07.2). Either response can, of course, be inaction. Unless the first action leaves the system (or brings it back) where it was to begin with, the second action starts from a different base than the first and would have taken a different form if the first action had not occurred. If the system also learns from this experience it will thereafter display a different relation between its information inputs and its behavioral outputs—certainly a form of developmental change. Even without learning, a series of such behavior cycles can bring the system into successively different environmental circumstances that seem worthy of being considered developmental.

Let us now move to social interactions of humans. Before a transaction is completed, A has X and B has Y. After it is completed, A has Y and B has X. Under normal circumstances we can assume that

both the desires and the capacities of both parties toward future transactions between them or others will be different than if the first
transaction had not occurred. Similarly a communication will alter the
detector states of one or both parties and have at least the potential for
altering their subsequent communications. The same logic applies to
the whole gamut of transactions, communications, and combinations of
them. The point is not that an interaction necessarily moves one or both
parties to a point from which subsequent interactions start from a different base. The point is that if and to the extent that such is the case
the situation involves developmental change. We have examined the
intrasystem and intersystem developments with two persons. The logic,
however, extends to any social unit—formal, semiformal, or informal
—up to and including the whole society. Put in its most general form,
any action or interaction that produces an irreversible consequence (or
one that is not in fact reversed) provides the basis for developmental
change. Expressed in symbols and with respect to interactions, developmental change can be said to have occurred if interaction I so affects
the system states of A or B that subsequent interaction I_1 is different
than if I had not occurred. The cultural succession of ideas and artifacts
is an obvious parallel in which each development occurs only because
it is built on preceding ones. Cycled repetition that regularly brings the
system back to a prior point clearly does not qualify. This situation can
also be thought of as stochastic. Since each successive step takes the
system farther from its starting point, it can also be said to involve positive, or deviation-amplifying, feedback (Maruyama, 1963) rather than
equilibrating negative feedback.

  Let us briefly introduce the question of whether this form
of developmental change requires a theory of change, as distinct from
the kind of cross-sectional analysis that occupies most of this book. It
seems clear that we can use nomothetic cross-sectional analysis for
interaction I by tracing the effects of the system states of A and B on
the interaction. We can also use it for I_1. The missing link is the effect
of I on the system states of A and B, since without that information
we cannot know the new base from which I_1 proceeds. Having identified this gap we will defer further discussion to the next section on the
idiographic and the nomothetic.

  We began this discussion of social change by suggesting that
nonstructural changes are perhaps also really structural. Although
the point is not momentous and I would not push it very hard, I think
it worth exploring with some examples. The most obvious case is
formal organization and the informal modifications of structure that
occur within it (4.12.53 and 4.33). Since interactions are parts of roles,

if one interaction changes subsequent ones the role itself has in some degree been changed and hence the structure. And since structure is given the same definition for informal and semiformal organization as for formal the conclusion applies to them as well.

As to intrasystem change, the personality of an individual, or the total set of his less transitory DSE states, constitute a pattern system. It, too, has a structure (in ordinary language, though not as defined for organizations). That is, it has numerous parts that are related to one another in some ways but not others (which is perhaps a suggestion that the excluding aspect of selective retention may be more important than its including one). If one part is changed, then except in the unlikely event that it is unconnected to other parts the structure of the whole set has been altered.

5.23   *Another Approach to Change.* Another way to discover the causes of change is to delineate the conditions of stability and then attribute change to their absence. For reasons just indicated we are interested primarily in change in the informal organization. Here are some obvious candidates for a tentative list of conditions of stability in informal organization:

a. No subsystem develops new ideas about how it can enhance its own advantage.
b. No subsystem undergoes a change of goal or motives and hence of its EP's in dealing with other subsystems.
c. There is no change in relative bargaining skill (tactics and strategy) among the subsystems.
d. Either there is no turnover of personnel in a subsystem or the replacements have system states essentially the same as those replaced.
e. There are no changes in the natural environment that alter the exchange ratios of inputs and outputs among the subsystems.
f. There are no internal changes in a subsystem that alter the transformational ratio between its inputs and outputs of things that are exchanged.
g. Communicational errors or failures do not produce irreversible effects. Among other circumstances this condition would be met if the communication was disregarded, dealt with matters irrelevant to the situation, or was later corrected.
h. Any deviations from the preceding conditions are subject to negative feedback.

All eight conditions, and whatever others need be added for a comprehensive list, are presumably subject to liminal levels, but we need not pursue the question of how large a change must be to be significant. This technique for discerning the conditions of change is logically foolproof, albeit a mite underhanded. Its main practical difficulties are twofold: the complete list may not be formulable; and merely to state that some condition for stability is absent may not be informative. Even so, the attempt might be a useful heuristic.

5.24   *Evolution Versus Revolution.* I am not sure that the distinction between evolution and revolution is worth formalizing, but it is too widely used not to mention. *Revolution* is often used loosely and tends to mean nothing more than that a change has occurred which the speaker finds impressive, whether in speed, breadth, depth, or ferocity. The conspicuous cases of political revolution mean, in present language, that after a relatively long dominance by one coalition a different coalition with different goals has become dominant, and the change is most evident if force is the main type of power relevant to the changeover. A change in decision rules that brings a marked shift in the locus of power might also be considered revolutionary, as between the Articles of Confederation and the Constitution in 1789. So might be a reformation of the constituency of the dominant coalition, as in the alleged break-up by Nixon of the coalition first assembled in the 1930s by Roosevelt. Sometimes conditions become so bad that limited measures seem fruitless and a straightforward decisional process undertakes more spectacular ones—as in the 1930s. And occasionally, of course, highly improbable combinations of circumstances set off spectacular activities with great consequences—like the Crusades. The heroic ethic, as contrasted to the economic or incremental ethic (Boulding, 1958), does sometimes prevail.

Although the social scientist will keep his focus on the massive continuities of a culture that are often remarkably untouched by even large alterations of top-level power, under such headings as "thresholds" and "critical mass" we often observe changes that are small at the margin but throw a system into a whole new set of relationships. Furthermore, with society as with science, whenever we observe an apparently rapid change in norms we find on more careful examination that its seeds had beeen planted long before. It is also true that certain technological advances alter the parameters of social relations and bring otherwise improbable changes—as have electronic communication, the computer, the automobile and airplane, widely available contraception, and medical advances that greatly extend the average life span. The technology may have a positive-feedback relation with norms that further affect change—as in the period between World War II and the point at which fears about the environment became widespread. During those two decades lack of rapid technological change was deemed abnormal and bad. In fact, recent fears about the ecological consequences of technology and affluence have helped accelerate a move in sociointellectual circles to speed the rate of social coping, as evidenced by courses, programs, and even academic degrees that now deal with "speeding organizational change" or training people to become "change agents" in schools, cities, corporations, or governments. As Deutsch, Platt, and Senghaas (1971) have found, we have already entered the era in which the time between the introduction of an important idea and its relatively widespread use is remarkably short —though there have been quibbles about their choice of "important ideas."

Overall, I see nothing to suggest that the main social processes are different with respect to relative stability, slow change, and rapid change. Only the particulars are different, which is a matter for the next section.

6   IDIOGRAPHIC

The two dichotomies—idiographic and nomothetic, developmental and cross-sectional—are conceptually independent yet partially related in fact. To indicate their independence we can cite examples of all four possible combinations. We begin with the assumption that nomothetic analysis uses propositions extracted from time and place and that idiographic analysis deals with the rich detail of the particular case. We assume also that cross-sectional and developmental are given the meanings indicated previously. Then nomothetic, cross-sectional analysis appears in the law of gravity and in the statement that bargaining power varies inversely with one's EP. Nomothetic, developmental processes occur when a completed transaction alters the EP's of the parties or when the hard steel ball, having rolled down the inclined plane, is now in a different position than when it started; such processes require attention to a different set of variables and principles to diagnose what will happen next. Idiographic, cross-sectional analysis is illustrated in the complicated interactions of all the parts at a given moment in a biological human, a personality, or the U.S. Congress. Idiographic, developmental analysis involves tracing the changes in those interactions over time. Despite conceptual independence the two dichotomies seem related in certain ways. We will tentatively explore that relationship and note the difficulties of making sharp distinctions within each pair.

If evolution is the only source of emergent change and, because of its random component, is in principle unpredictable, then each emergent development is unique. This does not mean that a particular pattern occurs only once (life may have originated independently many times) but that its background situation involves so many variables that its likelihood of emerging is small. The probability is also high that many other patterns are possible, although the probability that a particular one of *them* will arise is also small. That is, the situation is so complex that we can predict its outcome only on a probabilistic basis based on prior observation of outcomes in similar situations, not by calculation based on measurement of important variables and their interactions. (We do not compute how long it will take a feather to fall a hundred feet; we drop some feathers and count.) Each emergence is a unique, idiographic phenomenon.

The study of complex phenomena constitutes idiographic analysis, or the clinical approach, whether the subject is a psychiatric patient, a political convention, a power struggle in a corporation, a race riot, or the total set of interactions in a family. But to identify a phenomenon as idiographic does not tell us how to analyze it. Is there, in fact, such a thing as idiographic analysis? Let us begin with a simple contrast. Nomothetic propositions are of the sort: If A, then B. Put in similar terms, idiographic problems are of the sort: If $A, B, C, \ldots,$ N, when $A, B, C, \ldots, N$ are partly conflicting and partly supplementary forces, many unidentifiable and many identifiable ones of unmeasurable magnitude—what then? And what if one cannot know whether Q is present, when Q might be the dominant causal factor?

Several points can apparently be stated. First, the outcome is not knowable until it has happened, and in complex cases it may not be knowable even then. One can know that the United States fought in Vietnam. But who can say precisely what its effect was on American internal or external politics?

Second, to say that the answer cannot be known is not to say that analysis is fruitless. Complex situations apparently cannot be analyzed except by a kind of thinking that when formalized we call the simulation model. In using it one develops out of the thousands of forces at work some conscious or intuitive image of the two, three, or half dozen forces he considers dominant. He assesses the approximate strength of each, the number being small enough so that the mind can follow their possible interactions. Multiple correlation and analysis of variance can help identify and measure these variables if the necessary data are available. If he is lucky, his prediction based on this simulation will be good. If not, he can revise his list of dominant factors, improve his measuring techniques, or make some other adjustment. A formal

simulation model of a city, the U.S. Congress, or a war, or the Club of
Rome's simulation of world population merely makes explicit what
seems the inescapable nature of thinking about such things. That
foolish things are sometimes done with simulations no more demonstrates that the technique is bad than lying with statistics impugns
statistical theory. The technique is subject, of course, to such errors as
bad models, premature closure, or faulty inference from the model.

Third, when one gets down to it much analysis of the
unique, developmental situation is actually nomothetic and cross-
sectional. As noted in Chapter One, a simulation model designed as
a special-purpose tool to deal with a particular situation does not
abandon the general-purpose analytic tools but simply groups them in
a way that seems to deal best with the problem. For example, however
complex the details of a party convention, corporate power struggle,
or race riot, to understand them seems to require such nomothetic
propositions as: candidates for lesser office prefer a presidential candidate with a "coat-tails" effect; friends on the board help one's struggle
for a promotion; and frustration tends to produce aggression. That the
analyst may leave the propositions implicit does not negate their
presence.

Or let us consider the rise in power of the merchant class
in Western Europe after the eleventh century. This event certainly
classifies as social change and hence is developmental. It is unique to a
particular time and place and hence is idiographic. Yet when we get
to details these categorizations are not at all clear.

Chapter Ten described successive favorable transactions as
a technique of accumulating power. A can accumulate money if he
sells something for more than he paid for it—clearly a cross-sectional,
nomothetic proposition. But suppose that event is repeated two, ten,
and eventually ten thousand times; by two, ten, and eventually ten
thousand persons; over a period of six centuries. At what point does
the analysis cease to be cross-sectional and become developmental? At
what point does it cease to be the detailed chronicle of an idiographic
event and become simply one more instance of a nomothetic principle?

If the distinction between the cross-sectional and developmental is to be made at all in a case like this it must apparently be
made at the very low level at which we shift from the single transaction
to an accretionary succession of interrelated ones as in 3.3. Although
the same basic transactional analysis applies to each successive transaction, what A can do in his second transaction depends on how he is
affected by the outcome of the first and so on. But his response to the
outcome of the first cannot be forecast from the nomothetic proposi-

tions. We must add numerous details about who B is, what X and Y are, the specific magnitudes of $AX, BX, AY,$ and $BY,$ the explicit alternatives available for the initial transaction and for all subsequent ones, and so on.

To shift illustrations, from nomothetic analysis we can state the kind of modification of terms that A's hostility, as contrasted to his indifference toward $B,$ will produce in a transaction. But to know whether B will react to that hostility with counterhostility, indifference, amusement, or sadness may require a detailed clinical history of B taken in light of the present and immediately preceding situation and the state of his health—and even that knowledge may not help much (Riker, 1962, p. 191).

The developmental process consists of a chain. System states affect interactions, which affect subsequent system states, which affect the next round of interactions, and so on. The first and third links are subject to nomothetic analysis, at least in principle, and so are all subsequent odd-numbered links. But all the even-numbered links are highly particular, not subject to nomothetic analysis at all, and certainly not predictable. They are idiographic phenomena. With every other step unpredictable by "science" we must conclude that the total process is unpredictable. To shift the analogy again, cross-sectional analysis can state with confidence that if you drive west from Albany on the throughway you will come successively to Syracuse, Rochester, and Buffalo. But it cannot tell whether you will, having reached Rochester, actually go on to Buffalo. To answer these questions requires additional knowledge of a very different sort.

Many general propositions about such matters appear in Chapters Seven through Eleven; the problem in the particular case is to discover which parts of the general model are applicable—that is, to create a simulation model, though a more explicit one than the general models of family and government presented here.

In short, cross-sectional analysis can be nomothetic though it can also be idiographic. Many detailed processes by which developmental change occurs are subject to nomothetic, cross-sectional analysis though it may be inhibited if the necessary analysis is not yet extant. But the direction and speed of development are subject to idiographic analysis alone since they depend on facts of the particular case.

Once all the necessary analytic tools are available a simulation model may coincide with a set of constraints. A toboggan run can be complex; yet if all the turns are appropriately banked there can be no doubt where the sled will go. The biological development from fertilized egg to adult human involves an incredible number of variables. Yet each is so constrained that the end product is predictable in thousands of details.

The rate of change of the species is constrained by the fact that many generations are required for significant alterations of the gene pool, with each step only a slight modification of the previous one. Basic mechanics provides constraints even in long-term biological development. It could be predicted, for example, that all protoplasmic animal life above a certain size living on land will have internal or external rigid parts, that openings for ingestion will be in the front and those for excretion in the rear, and that locomotion will be accomplished by reciprocating motion, not wheel-and-axle arrangements. The fact that weight is a cubic and strength a squared function of linear size, taken in conjunction with the nature of the structural materials, places limits on the absolute size of any organism that is to move on land.

The development of a business enterprise depends on its constraints. Does it sell in a market of 50,000 or 200 million? What antitrust laws are there, and how are they enforced? Does the enterprise operate in a formal or an informal economy? Comparable questions would have to be asked about developmental change in family, government, church, or any other social structure or interaction.

The logic has now come full circle. We say that accurate prediction can be made only about nomothetically analyzable things. Development is unique and generally too complex for prediction. But if the developmental situation is characterized by sufficient constraints we may nevertheless be able to make reasonably accurate predictions about it—if we know the constraints and the rules by which they operate. By the time we have specified enough constraints to make the situation predictable in the individual case, however, we have converted the analysis to nomothetic and cross-sectional.

Thus reasoned, scientific prediction is possible only within nomothetic, cross-sectional analysis. By reasoned and scientific we mean deducing an anticipated effect from a known antecedent condition on the basis of some principle of explanation. We do not mean unexplained extrapolation of the past, whether as a highly certain prediction that the sun will come up tomorrow because it always has or a probabilistic one that it may rain tomorrow because it sometimes does. The comparable reasoned predictions are based on a model of a rotating earth and one of precipitation as an outcome of certain moisture-temperature relations. This distinction, incidentally, coincides closely with Kaplan's contrast (1964, pp. 346ff) between explanation and prediction.

In short, if the preceding discussion is tenable, prediction about developmental change is possible only when it simply assumes that past trends will continue but without knowing why, or when the development is amenable to cross-sectional analysis. Furthermore, there is serious doubt whether anything predictable by mere extrapolation is really developmental in any case. Except as guesses, the emergence of higher-level systems can be predicted only if the constraints on possible variations and selector factors are so narrow that the analysis becomes cross-sectional and nomothetic.

To digress to other sciences, we have already noted that astronomers can make precise statements about the movement of planets that are based on nomothetic, static propositions. But all they can say about the size, number, and distances of the planets is: "That's the way it happened." The astronomer could no more predict how many planets would result from an explosion of the sun than the physicist can predict how many pieces a broken window will make when hit by a baseball. We accept that such matters are unknowable. The geologist can tell us much in retrospect about how the earth's surface was formed. But every single detail rests on a nomothetic principle of chemistry or physics. Continents drift because they are a less dense solid floating on a more dense liquid substratum. Rainfall runs into the ocean because of gravity. Mountains rise because cooling or other contractual pressures squeeze them up. Mountains shrink because freezing and other forces break off pieces and gravity, with or without the help of wind and water, pulls them down. By contrast, when we ask why a particular mountain is 20,000 instead of 14,000 feet high, why huckleberries grow on small bushes and acorns on large trees, or why John is six foot four while his older brother is only five foot eight, we can cite the general principles of geological squeeze, evolution, and genetic factors in growth. But to each particular case we can only answer: "That's the way it happened."

In short, social sciences are not the only ones in which developmental processes can be explained only after the fact and then only to the extent that they can be reduced to cross-sectional propositions. Beyond that, "it just happened" may not be elegant but it is probably the best we can do. If so, true developmental problems will always be a sorry area of science. In those aspects of physics where prediction reaches its greatest heights there is no developmental analysis at all—as in the paths of the planets.

If the preceding discussion is valid, what is the appropriate relationship between cross-sectional and developmental study? Clearly the cross-sectional must take priority, for without it nothing about the developmental can be explained. Moreover, it is doubtful whether development can even be *described* satisfactorily (see Chapter Seventeen on history) except in cross-sectional language. Merely to describe the formation of a mountain requires such nomothetic concepts as stress, pressure, plasticity, expansion and contraction, tensile strength, and the like; without such terms the description is primitive indeed. Merely to describe the change in structure of a corporation requires such nomothetic concepts as department, authority, communication, bargaining, agreement, decision, production, concession, and the like; and only when those terms are sharply defined can the description be precise. In present language, all the underlying principles about communication, transaction, and organization are the same whether they are producing a steady-state equilibrium or a developmental change. It is not the fundamental processes of equilibrium or development that differ but the particular circumstances in which they occur and the particular patterns they take—and these "just happen."

Whether one wishes to specialize in cross-sectional or developmental study is a personal choice. But if the preceding analysis is valid, one can understand cross-sectional analysis without ever going into the developmental although the reverse is impossible. And whereas cross-sectional generalizations can apply to a wide scope of materials, the developmental are unique—which is why this approach to the basics of social science is almost entirely cross-sectional. Even if we agree with Buckley (1967, p. 205)—and I mostly do—that "imbalance and structural elaboration, not equilibrium or homeostatic maintenance, are the characteristic, inherent features of socio-cultural systems," we have no reason on that ground alone to deny that the intrasystem functions of detector, selector, and effector, and the intersystem behaviors of communication, transaction, and higher-level organization, might be appropriate analytic tools of developmental social science. Some cross-sectional and related questions are discussed in Chapter Seventeen in connection with history and geography.

6.1   *Reductionism and Teleology*

Chapter One indicated that functionalism is the static upward look from a given system to the role it plays as a subsystem of some larger system whereas reductionism is the static downward look to that system's own subsystems. In light of the intervening chapters it seems proper to apply teleological criteria to the function of a subsystem of a formal organization. In the pure model every subsystem is there in order to carry out a function for the sponsors. In pure informal organization no controlled subsystem can be accounted for teleologically with reference to the whole organization—by definition. But its performance of whatever role it happens to have *is* teleological with reference to itself—by definition of controlled system. Semiformal organization lies between the two. In it subsystem roles are not consciously defined initially. But once they are in existence specific selective efforts are made by significant members of the society to keep them that way or possibly to modify them. In this sense biological evolution is akin to semiformal organization. The subsystems of the human organism were not consciously designed by the main system. But the fortuitous variations that gave rise to them were subject to a selective factor which retained some but not others.

Let us next look at the question of reductionism. Chapter One indicated that reductionism has two forms: one is simple concentration on parts of a system rather than on the whole unit; the other involves a "lower level" of analysis. As to the former, we can examine the operation of a whole organization and then proceed to analyze the operation of one of its (controlled) subsystems. Although the latter deals with a lower level of system, the type of analysis is no different than for the main system. Both are oriented around detector, selector, and effector. By looking downward we can also focus on the interactions of the subsystems in their communications and transactions. Although this analysis will have different content, it is in no sense "lower-level" analysis than that applied between the whole system and other systems. Nor are intersystem and intrasystem different "levels" of analysis, since with either the analyst can move upward or downward from a given level.

Psychology is typically viewed as "below" social science; and biology, chemistry, and physics are lower still. In this sense, can either system analysis or communicational and transactional analysis be "reduced to" psychology? Definitely not. In present language, psychology *is* system analysis applied to the individual, and there is nothing reductionist in moving from one level to a lower-level controlled system. Only the content changes, not the logic.

Communications and transactions both reflect and affect system states of the interacting parties. Hence psychological processes occur in the mutual interaction between system states and system behavior, but the present analysis does not care *how*. Thus psychology is not necessary in order to understand this social science. But since this approach does deal with mutual modifications between system states and system behavior, psychologists could be helpful in designing experiments to test its propositions, even if their knowledge has no direct use in formulating those propositions. They could also assist with the model of social man.

More explicitly, communications and transactions deal with the effects of certain system states on system interactions—given the system states. The nature of those effects is independent of how those system states arose or are modified. Both the communicational and the transactional interactions depend on the mutual contingency of two sets of system states. Nothing about psychology as such, which is strictly intrasystem, can tell what will happen in intersystem analysis. That some psychologists are concerned with such matters means merely that they are partly social scientists.

Here as in other situations, the society is part of the individual's environment. As such it has much to do with the kinds and quantities of his inputs and hence with the content of his system states. But it has nothing to do with the processes by which inputs are converted into system states or by which system states are converted into outputs. Basic psychological processes are genetically, not culturally, determined. To be explicit about fundamentals, nothing from psychology gives any hint about the nature of power or bargaining power or whether a communication is most usefully divided into two, five, or fifteen steps. The intersystem behavior and analysis are emergent with respect to intrasystem and cannot be reduced to it.

Within the strictly social analysis the relation is somewhat different. After the fundamental concepts have been defined and propositions about them stated they are put together into particular configurations, or special cases. Authority and markets, for example, are special cases of the transactional relation, and culture is a special case of communication. Government is a special case of organization, as are family and firm. Status is a particular combination of transactional and communicational elements, and Section 3.4 sketches many other cases. All have a reductionist element in that they can be analyzed only as instances of communications, transactions, or organization, with or without attention to intrasystem analysis. The composite, however, is always something more than the units that make it up—in the same sense that a wall and a fireplace both have patterns independent of the bricks that compose them. The operating portions of a manual calculator consist solely of gears and levers; yet the calculator is definitely something more than the sum of its parts. This, I take it, is what Blau (1964, p. xi) means when he says that "the problem is to derive complex from simpler processes without the reductionist fallacy of ignoring emergent properties." Hence all but the most fundamental concepts of the social science described here can be "reduced to" those fundamentals because the analytic structure was deliberately designed that way. But each composite (nonfundamental) has a special pattern of its own, and knowledge of the fundamentals alone no more provides an understanding of the pattern than knowledge of the basic physics of gears and levers alone provides an understanding of the calculator.

While unexceptionable, that conclusion leads to a fascinating one which Rube Goldberg would have appreciated. Mechanical aptitude tests often show a drawing of a succession of gears, pulleys, and so forth. They then show an input to the system and ask the testee to predict the output. Much more complexly, computers have been programed so that if a diagram of a radio circuit is fed into one, along with the values of all transformers, capacitors, resistors, transistors, and such, the computer will accurately describe the operating characteristics of a radio constructed to those specifications. The intriguing question is whether we an anticipate a level of knowledge in social science that would similarly predict the performance of, say, a complex organization if we fed the computer a diagram of the organizational structure and a detailed specification of each role—as structure and role are defined here. At the moment I see no reason why it is not feasible in principle. In fact I entertain a self-serving suspicion that it would be easier with the kind of social science developed here than with what is available elsewhere—though I have yet to pursue the relevant techniques and problems. Although there is no theoretical reason why the technique could not be applied to other social structures or interactions, because of its explicit assumptions and specified structure the formal organization would seem the most likely starting point. As the sociologist puts it, the question is whether we can predict macro-level phenomena from micro-level ones. As contrasted to the preceding paragraph, this one requires information about the explicit *traits* of components, not mere knowledge of their principles.

Even if such simulations can be worked out, we will not be able to predict how a real organization that replicates the simulation would work. The beauty of the radio is that one can buy parts that have the same characteristics as those specified in the model and then construct it so that there are no interactions among the parts except those shown in the diagram (though magnetic fields often set up troublesome interactions not specified in the schematic). One can also shield the whole set so there are no inputs from the environment except those that enter the energy (maintenance) and signal (information) input terminals. Merely to describe those conditions identifies the impossibility of achieving similar predictability in real organizations—though there is no reason why such simulations could not readily identify obviously unworkable structures or the relative expected performance of different structures under assumed average conditions. Nor is there any reason why computers cannot be programed for the emergent process by systematically generating different combinations of known components and testing them against specified selective criteria. But this discussion is not designed to suggest or to predict. Its aim is to clarify the nature of developmental processes, of the idiographic, of the relation between them, and the relation of prediction to both.

Chapter 17

7   Applications to Other
Areas

A major test of this framework presumably consists in attacking a significant problem or area of study and restructuring it into the present concepts. But because that task would require a book of its own, in this chapter we select three areas and merely sketch how they might be handled if this approach were applied to them in detail. In addition we identify some methodological problems. The first area deals with the major social sciences less directly touched on to this point—history, geography, and anthropology. Here we show how they would be related to this approach. Second, we discuss three related and widely used concepts—cooperation, conflict, and resolution—to show how some concepts not considered basic in this framework are nevertheless amenable to it and to show how someone using this framework would go about dealing with the kinds of problems currently encompassed under those headings. Third is the area of personality and psychiatry, intrasystem in one sense but long recognized by social psychologists, anthropologists, and sociologists as intimately tied to social relationships—and increasingly so in recent decades by psychiatrists.

7.1   HISTORY, GEOGRAPHY, ANTHROPOLOGY, AND SOME DISCIPLINARY RESTRUCTURING

It is not my purpose to tell fellow social scientists what to do, and they have enough sense to ignore me in any event. Widespread acceptance of a framework such as this would nevertheless have repercussions on the structuring of the social disciplines, and spelling out those repercussions is a useful way to clarify certain aspects of the framework itself. As noted at the outset, the book deals with the alleged underpinnings of all social science. In content (though not format) it is introductory social science—the things all social scientists would have in common. It does not move to specialist levels, though it does briefly identify how sociology, economics, and political science are related to semiformal, informal, and formal organizations, respectively. There is also a parallel rough correspondence of those three disciplines respectively with communications, transactions, and organizations. To explore the problem properly we also adopt the rosy assumption (or dismal one, depending on your view) that all social scientists know and use this framework.

Although any respectable science must span the idiographic-nomothetic spectrum, in important ways history, anthropology, and geography lie heavily on the idiographic side, with sociology, economics, and political science on the nomothetic. Even aside from a common base in this framework the three idiographic sciences would be heavily interdisciplinary and use advanced as well as elemental materials from all disciplines. The three nomothetic sciences would be considerably less interdisciplinary. If they are to remain healthy however, they would continue to forage in new areas, biting off chunks of material which might have to be handled at first at the level of description and to be examined for possible concepts or generalizations not already developed elsewhere. In that connection this framework should be viewed as a means to simplify and systematize existing understandings, not to inhibit new explorations. Along with developing generalizations needed for their own materials the idiographic disciplines could perform a useful function by perpetually uncovering new phenomena and challenging the nomothetic disciplines to explain them. They presumably should also try to form the explanations themselves with the help of nomothetic tools, all the while reporting on the uses and limitations of those tools. With this background let us look briefly at the possible role of each idiographic discipline and then take a collective look at the nomothetic ones.

7.11   *History*

First, it must be stated as flatly as possible that every "simple description" of an action or behavior uses or implies cross-sectional principles. The historian cannot avoid this imperative by using only the language of the layman: the factual statement that "Joe walked to the switch and turned on the light" inescapably incorporates assumptions about electricity, switches, incandescence, and other aspects of physics, as well as about Joe's psychomotor processes. The factual statement can be made and understood without specifying or even understanding its electrical or psychological assumptions. But it cannot be valid unless the implied propositions about electricity and psychology are also valid.

Similarly, the historian can make such allegedly factual statements as "France won the war" or "The marriage of Henry's daughter to the heir to the throne of the Netherlands helped cement relations between the two countries." But each statement contains implicit propositions about transactions, and the validity of the factual statements depends on the validity of the implied transactional ones. (If there is uncertainty about the appropriateness of transactional analysis for this purpose, then other kinds of propositions would apply. But to illustrate, we assume that the transactional is applicable.) A war is a transaction involving bads, and "winning" is the loser's agreement to do something in return for the winner's relief of stress (Section 3.25.14). A marriage between members of royalty is also a transaction whose interrelations affect the terms of other transactions. The necessary propositions about transactions may not be known by the historian or they may be sensed intuitively. But if the implied transactional propositions are false the historian's "factual" statements can no more be true than to say of a conventional electric fixture, "Joe turned toward the light and blew it out." The historian cannot mention the birth or death of a monarch without implying propositions about biology or describe Hitler's rise without implying propositions about political power. He cannot even cite a date without accepting certain assumptions about the solar system.

If the necessary social science is not available, the historian must do the best he can and hope that his statements imply no false propositions. Once a nexus of propositions about transactions is available, however, the historian potentially stands in a dual relation to them. First, if the propositions are valid they should help him to understand the things he reports and to deduce from certain observed transactions that certain other ones probably occurred—which he can then look for. Second, by having at his command a storehouse of information about EP's that no one any longer has a tactical reason to hide, he can run retrospective empirical tests of the propositions.

Beyond the above, history is the medium *par excellence* for studying the developmental process. This study can range from simple irreversible events (the treaty, once signed, may be abrogated, but the signing and subsequent abrogation are not at all the same as not signing) to the grand, if suspect, patterns of Toynbee. Some historians (Ward, 1971, p. 28) suggest that the forte of history is narrative, which may be another way of describing irreversibility. As soon as someone dissects how and why certain behaviors and interactions cannot return to the *status quo ante,* he is dealing head on with developmental change. If he approaches his materials in this light the historian might develop categories and criteria of situations that do or do not tend to be irreversible. He could also help test the hypothesis that the former are subject to positive feedback and the latter to negative feedback and equilibrium, and try to discern the conditions under which each is likely to occur. However, if the essence of narrative should turn out to be nonhierarchical structuring of knowledge, the historian might provide insights into the uses and limitations of such knowledge.

Whether or not there is a close tie between developmental and idiographic analysis (Section 6), certainly history is a field to study the unique event, the fantastic interweaving of its numberless strands, and the role of the fortuitous. And even if nomothetic analysis is helpless to deal with the total of a complex pattern, every detail of behavior is presumably subject to nomothetic analysis—though we cannot demonstrate this where the necessary principles have not yet been developed. In parallel, despite their incredible complexity and case-by-case uniqueness, we assume that every one of the billions of events that are the geological history of the earth occurred in strict accord with the nomothetic laws of chemistry and physics.

Incidentally, historians sometimes debate the relative impact of outstanding persons and underlying events—the Great Man versus Social Forces theories. In light of the theory of social change suggested in 5.21.2 it might be worth experimenting with the notion of the great man as random variation and social forces as selective retainers (Kuhn, 1971a, p. 148).

Since it would be dangerous indeed for historians to report only grist for testing social science hypotheses, they must to some extent accumulate data without regard to that purpose. But in an age when sheer records rise by hundreds of tons daily the historian has no choice but extreme selectivity. What, then, will he select? Somehow he must attend to the important and neglect the unimportant. But what is important? It is hard to imagine a criterion of importance in selecting materials other than that they help to explain something. This means, whether we like it or not, that the materials are related to some theory. Let us say that the historian reports a bad crop year in England. It is safe to expect that in the same context he will refer to such things as locusts, drought, enclosures, or poor crop rotation, and not to Ming vases, Shakespeare's last play, or Caesar's wife. If he talks of World War I he will talk about Woodrow Wilson, the Kaiser, troops, and the German economy, not the divorce rate in Kansas City or extinction of the passenger pigeon. No matter how much he hedges or denies it, he includes things that in his view are somehow causally related to the event. Having said that any "factual" statement the historian makes about events has some theory inescapably embodied in it, we now add that any decision he makes about inclusion or exclusion, unless random or arbitrary, also inescapably reflects some notion of causality. The historian can neither describe nor select without theory.

Hence instead of dividing historians into those who use theory and those who do not we divide them into those who know what they are doing and those who do not. The former could be subdivided into those who know explicitly what theory they are using and those who understand that they are inescapably using theory but do not make it explicit. One is tempted to suggest that only those in the fully knowledgeable category can write good history. For the long future, perhaps so. For the present they might write some bad history, while sophisticated intuition may produce some good history from the man who insists he wouldn't touch a theory with a thousand-page volume. Like it or not, there is theory in the way one groups his notes in his files and in the sequence and headings in his book.

This discussion suggests three main roles for history: as a laboratory for testing cross-sectional, nomothetic propositions and possibly originating new ones; as the field for describing and elucidating the idiographic in human affairs; and as the best of all possible laboratories for the developmental. That these things will not necessarily be easy will be clear from a reading of Calhoun (1969). However, the second and third assignments for historians may perhaps be best approached as simulation models. In the language of Chapter One, simulations are D-level investigations, without which I insist no historian can describe an event.

7.12   *Geography*

Any geographer who has not stopped reading long since will have noted that this volume has not even mentioned space or locale as a factor in social science and has talked of man's interaction with his natural environment in only the most general terms. Such areas must obviously be included in the whole social science. They are not included here because I see them as topics for specialists, however fundamental their impact in reality. Even though I have categorized geography as idiographic overall, it could develop a nomothetic science about the effect of space or distance on communications, transactions, and organizations. Of particular interest is the effect of distance on the optimal size, structure, and internal coordinating techniques of formal organizations. Parallel study of informal and semiformal organizations is also of interest.

The relation of man to his natural environment is crucial. Man's behavior is affected by the environment and has feedback effects on it. In the system framework the geographer might examine the conditions under which this relation is stabilizing (negative feedback) or destabilizing (positive feedback). (He might also tell us whether those are useful concepts when applied in his field.) Such investigation would presumably be part of a broad view of human ecology.

An interesting terminological question arises here. An ecological system is by definition uncontrolled: it has no goal of its own and maintains no particular equilibrium. If man alters its equilibrium, as he does every time he converts a forest to field or town, nature does not respond with negative feedback that automatically offsets man's action and restores the previous equilibrium. Instead the new equilibrium continues at least as long as the field or town does. Although agriculture is clearly "controlled ecology," as are certain other "decided" uses of land, thus far there is no question that the ecological system as a whole is uncontrolled—its overall equilibrium falls wherever the joint effects of its separate controlled subsystems, including man, put it.

Suppose, however, that man deliberately changes his own behavior to produce a *particular* subsystem equilibrium position of atmospheric content or fish species. At what point, as man increases such activities, should we construe the whole ecological system as controlled? For reasons more intuitive than explicit my inclination is to say never. Although man is inside the system, it strikes me as analytically sound to view such actions by him as constraints imposed from outside rather than as goals of the whole system taken as a unit (1.20s and t). There is an important theoretical difference between a highly constrained uncontrolled system and a controlled one, and for one species to have dominant, conscious effects on the whole seems quite different from saying that the whole system is itself controlled. (We will leave for another time the question of whether man yet knows enough to avoid producing effects the opposite of those he wants.) It remains man's goals that modify the system, not some goal of the system as a whole. However, under the model of formal organization we could view man as the dominant coalition among all species, consistently controlling all decisions for the whole because no other species can communicate or transact in the decision process (as those terms are used here) or has comparable capacities to produce conscious effects on the system. The question again is not what the situation *is*, but what model will provide maximum understanding. At the moment I doubt that the formal model will be too useful for the whole system, though it clearly could be applied to man's role in some subsystems. (Incidentally, this discussion deals with a different question of *control* than does Section 4.21.) Perhaps the model of semiformality would be more appropriate since this is a case in which one subsystem deliberately modifies its behavior to alter the whole system. The geographers might give substantial help in conceptualizing these problems.

The geographer is no more immune than the historian from having nomothetic principles implicit in his supposedly descriptive statements and is under no less obligation to identify them. My distinct impression is that geographers are more aware of this situation than are historians, and that they make more use of the other social sciences. As to their idiographic roles—and despite inescapable overlap in the two idiographic descriptions—the geographer's delineation of the spatial orientation of man contrasts markedly with the temporal orientation of the historian. The rest of social science needs the geographer's tracing of the many interdependent strands that relate man's activities both to his location in space and to one another within that space. In such study attention is focused on the pattern system produced by a society, rather than on the acting system—the society itself.

7.13   *Anthropology*

The anthropologist is the best qualified of the social scientists to deal with the whole of the pattern system that is culture. He can trace the subpatterns of beliefs, behaviors, techniques, language, social structures, and the like, as well as their interdependencies, continuities, and discrepancies. He might also identify the tolerable limits of discrepancies and discontinuities within a culture and determine how those limits vary with the nature of the society and of the discrepancy. He, of course, should identify those respects in which the whole culture is or is not more than the sum of its parts, and he might inquire whether any of the generalizations about self-preservingness in ecosystems (4.21.3), which are acting systems, also apply to the pattern system of culture. He might also report whether pattern systems change according to the same principles as acting systems—or whether that is a meaningful question.

Among many types of pattern interdependencies, that between personality and culture is one of the most interesting. To what extent is the personality a function of culture, and vice versa? To what extent are personality and culture both functions of genetic patterns, transmitted biologically, taken in conjunction with patterns in the natural environment? To what extent is the content of culture a function of the transmission process, and to what extent does the transmission process change as the content changes? To what extent are both the content and process of culture dependent on the intelligence level in a society, and is that intelligence level dependent in turn on cultural content and process? In studying these matters the anthropologist can both learn from and teach those other specialists who share particular concerns—the psychiatrist and clinical psychologist in connection with personality, the ecologist regarding relations of man to environment and of parts to wholes, the communication theorist about the science of pattern transmission, and the psychologist about intelligence. Although we have assigned anthropology an idiographic role overall, many of the above questions deal with essentially nomothetic problems; no matter how concentrated the attention on complexities of a specific time and place, the question always arises whether similar patterns appear in different times and places.

Like those of the historian and geographer, the anthropologist's descriptions also incorporate implicit nomothetic propositions, and the validity of his description depends on that of the implied propositions. Hence, although the anthropologist should never limit his investigations to areas that can be described in the language of the other social sciences, he should be alert to those situations which cannot be thus described to see whether they reflect gaps in the nomothetic sciences. Like the historian the anthropologist could distinguish communicational from transactional transactions in the societies he studies and help test nomothetic propositions about them. I would hope that a common conceptual base like the present frame-

work might lead to more mutual assistance and challenge than at
present between the nomothetic and idiographic aspects of science.

7.14   *Restructuring Other Disciplines*

The preceding three sections dealt with social science
disciplines whose subject matter this framework has not attempted to
incorporate explicitly—history, geography, and anthropology. This section discusses the three that the framework has obviously invaded—
sociology, economics, and political science. As suggested early in
Chapter Fifteen (4.31), I tend to see sociology, probably in conjunction
with social psychology, as the encompassing, basic social science, particularly if it should adopt the kind of approach presented here. It
would elucidate the basic concepts of all social science areas and also
perform the grab-bag function of providing specialized analysis of those
social concepts and systems not covered by other specialists. Perhaps we
should selectively breed a subspecies of sociologists whose function
would be to keep tabs on the gamut of social knowledge, particularly
noting both duplications and gaps in coverage as it lumbers along.
My main suggestion about their social theory is that sociologists accept
the distinction between communication and transaction as well as that
among formal, informal, and semiformal organization. Without that
orientation I doubt that they could perform the suggested functions.
And by accepting leftovers for specialized study, sociology would provide our main information about social communication and about
informal (other than markets) and semiformal organization.

In economics the model of the firm has provided the logic of
decision theory. I now see the basics of that theory as the common
property of all social science, while economists pursue its special applications to firms, households, and other economic units. Refinements to
the core of market theory probably met diminishing returns some
decades ago, and I see the most important work for economists now as
lying in the areas that require cooperation with other disciplines—
economic development, the sheer power of economic units, the broader
(as contrasted to the marginal and homeostatic) aspects of enterprise
management, the interactions of economic and political systems, the
behavior and implications of multinational corporations, the mutual
interactions of market process with population and environment, and
the like. Having used the purist's retreat from political economy to
refine the economic discipline, practitioners in the subject should again
trade in the same intellectual market as other social scientists—if the

latter will also trade. My own background in economics has contributed the major methodology of this book, in its use of explicit models gradually modified to encompass more varied cases. Obviously I think the method should be exported to other areas of social science.

In recent decades political science has been moving in much the same directions I would prescribe for it. Numerous political scientists seem interested in formal organization, in and out of government. They are interested in the nature of power and its allocation within organizations, as well as in the processes for reaching decisions among conflicting interests that hold different amounts and types of power—in the communicational, transactional, and coalitional processes of deciding. Furthermore, Karl Deutsch, David Easton, J. David Singer, and many others are approaching the subject through system analysis. Perhaps the single most important topic that political science might study jointly with sociology and economics is the locus of power in various societies.

The main questions, of course, are not so much what jobs are done by which disciplines, but whether specialists can come to see themselves as parts of a larger venture and try to fit their own parts into the whole. If the spirit of this volume were followed, all social science disciplines would rest on the same central concepts, and all analytic concepts within each discipline would be derived from them. Conversation and cross-fertilization among disciplines should be considerably facilitated—in contrast to the mutual confusion often engendered by current interdisciplinary efforts. Furthermore the idiographic borders on and at least partially overlaps the humanistic. This is not to suggest that the poet adopt the language of this volume; the world faces enough tragedies already. It is to suggest that with additional cooperation from specialists in building C- and D-level conceptual structures on the A- and B-level ones developed here the clinician could adopt this language or intermediate levels of language based on it. And despite the differences in the way they go about things, there is much in common between clinician and poet. The poet, along with the artist, the choreographer, and the musician, may also be thought of as an experimenter in communication (Kuhn, 1963, pp. 192–199). In short, if we can manage to fill all the levels of Figure 4 we will have substantially bridged the gap between scientist and humanist, at least so far as the social sciences are concerned.

In the relation between the present and the proposed structuring of social knowledge, this section is intended to suggest avenues to explore, not straight-jackets to confine. If they do a job, no new concepts should be rejected because they do not fit. The sense of the framework is to try new concepts to see whether they *will* fit, and if not, to see whether the framework can be modified to accommodate them—which, after all, is what happened numberless times in getting the structure thus far. As I experienced it, the farther the structure advanced the more flexible, rather than rigid, it became, and additional concepts seemed easier, not harder, to incorporate. I am nevertheless aware that my own deep immersion may have blinded me to concepts that are intractable in this framework.

7.2   COOPERATION, CONFLICT, RESOLUTION

Although cooperation, conflict, and resolution have already been defined, they have not been explicitly analyzed as such because each is a "multiple" phenomenon—an aspect of other phenomena dealt with elsewhere but constituting no independent problem.

The definitions of cooperation and conflict are related to selector states: in the cooperative relation desired thing X also brings with it desired thing Y; in conflict desired thing X requires that desired thing Y be sacrificed. For purposes of the definition, it does not matter whether X and Y are desired by the same person (intrapersonal conflict or cooperation) or by different ones (interpersonal conflict or cooperation). Neither is concerned with detector states or communication. Although one might speak of conflicting perceptions when cues are ambiguous, or conflicting views of two geologists about continental drift, in present language these matters are said to be discrepant, uncertain, or inconsistent—but not conflicting. When such matters are construed as system phenomena, they are treated here as elements of a pattern system, to be judged solely for consistency within the criteria of some acting system.

We have also seen (2.50q) that people's goals can be in conflict, but not the people themselves. Further, conflict and cooperation are relations of goals or possibly of goal objects; they are not actions or behaviors of persons. That is, in a straightforward sense not only are cooperation and conflict not behaving systems but they are not even behaviors of systems. By contrast exchanges, threats, and integrative actions are behaviors of systems although they are not themselves behaving systems. It is nevertheless compatible with present definitions to view interacting parties whose goals conflict as a "conflict system" or to view those who employ threats on one another as a "threat system." This, I take it, is the sense in which Boulding (1962; 1968c, pp. 43ff) uses these terms. Because usage is by no means standardized, these terms have been discussed to clarify why and in what sense it is proper to refer to a conflict or a threat system in this framework, even though neither system states nor system interactions are accepted as themselves constituting systems. This section identifies the applicable analysis but does not carry it further than has already been done at other points in the volume.

Having indicated what conflict and cooperation are not, we now see where they are dealt with in this framework. Since they involve goals or satisfactions they can appear wherever selector states are involved—which in this context means in decisions (within or between persons), transactions, organizations, or variations or combinations of the three. Conflicts cannot appear in pure communications because no goals or motives are involved.

Decisions necessarily involve conflict. Conflict is identified with the presence of cost, and if there are no costs there is no need for a decision. All selfish and some nonselfish transactions involve both cooperation and conflict. They are cooperative in that both parties gain if a settlement is made within the overlap of EP's and conflicting in that the better the terms for one party the worse they are for the other. Organization is similarly cooperative in that the joint output per member is bigger or better than each could create separately, and increased benefit to one is achieved by the same act that also increases benefits to others. But organization also incorporates conflict in its transactions with members, since the greater the inputs from and the smaller the outputs to one member, the less and the greater, respectively, will be the inputs and outputs of other members. Transactions or organizations can, of course, be made in error, in which case the cooperative benefits might not in fact be realized.

By the same token, resolution of conflict is not a distinctive behavior but simply one aspect of other actions or interactions. As to decision by an individual, formulation of the opportunity and preference functions defines the conflict whereas selection of the preferred alternative resolves it. As to decision by a group, the differences in preference toward each perceived opportunity constitute the conflicts whereas the communicational, transactional, and coalitional processes that produce a decision constitute its resolution.

To move from the science to the policy of conflict resolution, according to 4.12.48 it is presumably most satisfactory to resolve such conflicts by communication, next best by transaction, and worst by dominant coalition, the last two without involving force. In transaction and dominant coalition, interactions in goods are better than in bads although the model provides no guide between nonviolent bads in transactions and coalitions using goods. Nor does it indicate whether the indicated priority of transaction over coalition remains valid if both involve force.

To move to transactions themselves, an overlap of EP's simultaneously constitutes the range of both cooperation and conflict —the total potential benefit and the area of disagreement over its distribution. Completion of the transaction brings the cooperative benefit and resolves the conflict.

Since cooperation and conflict appear in the basic concepts, they can also be found in almost every special case or combination of them. Competition, for example, is a conflictive transactional situation in which multiple B's seek to complete a transaction with A; this conflict is resolved when A selects the B he will deal with. An important aspect of freedom is a conflictive problem in interpersonal costs: the upstream farmer's freedom *to* dump sewage conflicts with the downstream farmer's freedom *from* pollution. If the latter persuades the former he will be better off using the effluent to fertilize his fields, the conflict is resolved by communication. If some payment is made to induce other disposal, it is resolved by transaction. And if government responds to a complaint by prohibiting the discharge, it is resolved by dominant coalition. The model does not indicate how to discover or implement these techniques in particular cases: this is a D-level problem of Figure 4 and beyond the scope of this volume.

Although this book identifies the conflicting or cooperative elements in various interactions, the analysis would not change if they were ignored. This means that cooperation and conflict are not themselves units of analysis, even though there may sometimes be great practical merit in trying to increase the former and decrease the latter.

To deal with matters labeled as conflict in traditional language, hostility in transactions reduces the power of both parties (3.26.12) whereas generosity increases the power of both (3.26.2). In fact the range of conflict—that is, of possible disagreement over terms— decreases with hostility and increases with generosity. Conflict disappears entirely when hostility becomes boycott, and the disadvantage of hostility lies not in conflict, since there is none, but in both parties' decline in power to the point where neither can get what he wants (3.26.131). If this terminology seems strange, one might ask whether the relation between shopkeeper and penurious window-shopper is conflictive when no transaction has even been contemplated by either. That both poverty and hostility are disadvantageous does not mean there is any analytic merit in construing either as conflict. In fact, to identify a situation as conflictive in either present or conventional meanings gives no clue in itself whether it is socially desirable or undesirable—and, if the latter, what action might be taken to improve it. All economic transactions and all competition are conflictive. But this does not mean they are socially disadvantageous. The starving mother comforting her starving child involves no conflict, but is hardly cause for rejoicing. With respect to transactions in goods, it is the high or low levels of power, not the absence or presence of conflict, that determine whether or not the situation is advantageous. Much the same could be said about organizations. There are also important questions about large discrepancies in power, but in themselves they are not necessarily more conflictive than equal power.

Similarly transactions in bads do not necessarily involve more conflict (present language) than those in goods, though ordinary language labels force as conflict. Their disadvantage lies in being zero- or negative-sum games (sometimes enormously negative), not in being conflictive. A more comprehensive list of the disadvantages of stress and threats is found in 3.25.4 and need not be repeated here.

It may be useful to distinguish between the cross-sectional and the developmental approaches to this problem. Cross-sectionally a cooperative relation between two parties is closely related to agreement by communication since both parties benefit from the same act. A conflict relation could possibly be settled by transaction but is less advantageous since each must accept a cost for his benefit. Yet over time it is entirely possible that the transactional relation with its conflict might, perhaps by facilitating division of labor, produce far greater benefit than the cooperative one without conflict. This approach contrasts somewhat with that which speaks of the "constructive role of conflict." Despite some relaxation from older views, the latter approach still seems to say: "Of course conflict is undesirable, but if you handle it right you may nevertheless get some incidental good out of it." The present approach says that conflict is neither advantageous nor disadvantageous in itself.

In summary, this approach includes no generic analytic concepts of conflict or cooperation—only conflictive or cooperative aspects of other concepts. But as soon as the cooperative or conflictive situation is identified within a decisional, transactional, or organizational context a conceptual base is there to analyze it.

7.3   PERSONALITY AND PSYCHIATRY

This section, like others in this chapter, does not suggest content for the study of personality and psychiatry but indicates how the concepts and vocabulary of this volume might be applied to this area. The basic model of man appeared in Section 2; the task here is to amplify it in light of the intervening materials.

Freudian terminology is absent from this model; there is no reference to id, ego, superego, or libido. Drives, needs, urges, and tension reduction are mentioned only incidentally. Personality, normal or abnormal, is viewed instead as an individual's total set of learned and unlearned DSE states, with emphasis on their durable aspects. Unconscious behavior (in the ordinary, not the Freudian, meaning) is construed as involving the main-level DSE systems alone whereas conscious behavior also involves their respective subsystems (Section 2.06). The spirit of the model is nevertheless distinctly Freudian in one respect—all motives trace directly or indirectly from pleasant and unpleasant sensations (Section 2.2).

Both positive and negative feedback are important to the content of DSE, particularly to states we would classify as normal or abnormal. Purely internal mental processes operate primarily, perhaps exclusively, by feedback that is positive with respect to environment. Given prior input, then without further input the brain can develop new concepts and motives internally by combining and restructuring existing concepts through a process known variously as pattern formation, theorizing, fantasizing, or daydreaming, depending on the circumstances. Without further input such internal operations would move farther and farther from reality. But there *are* continued inputs, which constitute the negative feedback that brings deviations from external reality back toward it. Motives as well as images can be "reality-tested" in this fashion—as when we observe whether we want things that are impossible to achieve or possibly fail to want things that are readily achievable. Although effector patterns are also subject to testing (2.41.23b), it is problematic whether they could be developed through internal positive feedback in parallel with detector and effector.

If negative feedback from the environment is inadequate, the deviation-amplifying positive feedback may produce behavior that is abnormal in the sense dealt with by the psychiatrist. Perhaps more often the feedback is adequate in itself but dissonance between it and the internal patterns is avoided or removed by ignoring the discrepancy as noted in 2.41.41. In that case behavior can remain out of touch with reality or become progressively more so. This observation is simple translated statement, of course, not explanation. We may also translate psychosis as detector malfunction and neurosis as selector malfunction; ataxia, aphasia, and other motor failures would similarly be translated as effector malfunction. "Translation" does not mean that the referents of the two sets of terms are identical but merely that the terms suggested here are apparently the closest system-language equivalent to the established terms. Much work would be required both to specify the differences and to evaluate the scientific and clinical merits of each set.

Internal feedback processes that are positive with respect to environment could simultaneously be negative with respect to self-image. That is, the disturbed person could homeostatically maintain his self-image by negative-feedback corrections of deviations from it while simultaneously getting farther and farther from reality by positive-feedback relations among inner images.

According to 2.10f, information about the environment is put into a hierarchical structure in which later learning consists of new combinations of patterns learned earlier. Later learning is thus emergent with respect to earlier learning. By contrast, decay of the higher levels of the hierarchy, which leaves only the lower levels extant, is a system-language description of regression. Since learned motives in the model are also hierarchically structured and are attached to concepts (2.23b and 2.24), motivational regression can be conceived as parallel to conceptual.

According to the model, all outputs of overt behavior are symbols of system states, and generalizations about the former really apply to the latter as well. The sense of this approach is to apply such generalizations forthrightly to thoughts and fantasies rather than to overt acts alone. Pleasant acts tend to be repeated and to reinforce behavior that leads to them, for example, while unpleasant ones tend to be discontinued and to reinforce avoidance. If we apply to thoughts generalizations derived from observing overt behavior we conclude that one tends to repeat pleasant thoughts and to discontinue or avoid unpleasant ones. At the same time, however, the model suggests that danger, and the associated emotion of fear, takes priority over pleasant emotions (2.41.1). No great problem arises when the valences or emotions are associated with clearly defined overt situations since the situations themselves determine which will dominate. Although the model cannot predict which will dominate in fantasizing, it does suggest that pleasant thoughts will take precedence over unpleasant ones if cortical valences only are involved. But if emotions are aroused fear will take precedence over the pleasant emotions and force fearful thoughts to dominate pleasant ones.

In this model unconscious behavior appears as follows, in analogy with overt behavior. One may learn a motor skill or he may not. If he does, he may perform it consciously, unconsciously, or some combination. Whether he performs the skill consciously or unconsciously, he may lose it through disuse or conditioned extinction or avoidance—though in either case a significant residual pattern remains in the nervous system and may at times consciously or unconsciously control behavior. By analogy, one may or may not learn an information concept, including concepts of his own sensations (2.10e), feelings, and actions, real or imagined. If he does, he may think about it consciously or he may not, and in either case it may or may not control overt behavior. Also, whether he holds the concept consciously or unconsciously, he may lose it through disuse or conditioned extinction or avoidance reaction. If he does, the pattern nevertheless remains residually and may sometimes control behavior. In this context, information concepts that never surface to the conscious level correspond to the Freudian preconscious. By contrast, if one learns a conditioned avoidance response to "thinking about" something unpleasant, the result corresponds to repression. This wording derives from ordinary learning theory whereas the Freudian way of putting it, by not tying unconscious behavior to learning theory, seems to presume additional, mildly mystical processes.

An individual's personality is strictly an idiographic phenomenon reflecting the totality of his inherited and learned patterns (2.10f, 3.13.1b, and 3.47). The number of alternative personalities is probably of an order of magnitude at least that of the interpersonal relations discussed in 3.47 and might approach that of the number of possible connections in the human brain—which I have estimated at about ten raised to the billionth power (Kuhn, 1963, p. 113). The complexity is so great that with bad luck the therapist may see a patient two hours a week for years and still have only a crude picture of his personality. A reasonably good picture, in present language, is a simulation model of that patient's personality, containing the main ingredients that seem to distinguish his major traits.

A simple system for categorizing personality types could be developed if it were found that successful simulation models for many different persons all focused on the same limited list of traits or dimensions. No such list seems to have appeared thus far, and perhaps no short list is possible. To think, by contrast, of a long list, which is another way of describing an unwieldy simulation, many definitions and propositions in this volume are possible dimensions along which personality might be described or measured. At the decision level, persons could be categorized according to their typical behaviors with respect to time preference, fixed cost, investment, sunk costs, and decision costs. As to decision costs, for example, does the individual carefully weigh alternatives or does he decide impulsively? And as to sunk costs, does he rationally ignore them or does he cry over spilt milk? Regarding other aspects of the model, what is his optimal level of exercise of faculties (Section 2.41.2), what are his priority levels among the emotions (Section 2.41.1), what are his dissonance and frustration overload levels (Section 2.41.4)—and so on?

Intersystem aspects of personality could be as detailed as the whole of Section 3. Here we can only sketch them roughly. As to communications, how does he view others as sources of information? How does he respond to confirming and disconfirming reports from others (3.4–.43)? To what extent does he modify his own communications to confirm the expectations of others (as in 3.41.51, .52, .53, .62, and .65)? To what extent does he augment the positive-feedback effects of those behaviors (other-directed?) as contrasted to seeking independent information (inner-directed?)? As to transactions, to what extent and with whom are his transactions generous, hostile, or indifferent? What expectations does he have about others in transactions with him? When transactional cues about others are ambiguous (3.44.8), how does he interpret them? To what extent does he apparently recognize and act on the usefulness of being liked (3.28.42)? Of considerable importance, to what extent does he use tactical misrepresentations to get good terms? To join the decisional and transactional, how strong a time preference does he show in using such misrepresentations for present benefit in contrast to investing in honesty, which often has higher present cost but greater long-run return?

Section 3.46 suggested ways in which transactional logic can be internalized with ensuing guilt, self-punishment, and the like. The degree to which a person behaves on the basis of such images, and does not test them overtly, is also a crucial dimension of his personality. So is the self-image he extracts from observing the responses of others to him. Regarding organization, one's concept of his role in family, firm, school, church, or other organization is a key aspect of his personality. So are his attitudes toward exercising authority and responsibility, delegating authority, participating in and accepting group decisions, producing jointly with others, and maintaining legitimacy. In short, in this model personality is categorized and described in terms of the system and its interactions with other systems. Workable categories of personality obviously cannot incorporate all these traits, even though a profile of many traits is more accurate than any named category (lethargic, compulsive) can be.

There are two main points here. First, many personality difficulties that require psychiatric help are laden with problems of interpersonal relationships. Hence a good social science ought to provide the conceptual apparatus the psychiatrist needs in dealing with these aspects of his problems. Similarly, good psychology ought to provide all the conceptual structure he needs for understanding internal processes. In that case there would be no *basic* theory that is distinctly psychiatric. Following Figure 4 the clinical psychiatrist would create a D-level simulation model of each of his patients from A, B, and C-level concepts of psychology and social science. Psychiatric theory, as such, would then presumably consist of superstructures that would categorize simulation models by major types, noting also the types that typically precede and follow each other, the psychological or social conditions that accompany them, and the stages of the transitions from one type to another.

In the same sense that the anthropologist would be a specialist in particular pattern systems of cultures, the psychiatrist (or clinical psychologist) would be a specialist in pattern systems of personalities. If culture and personality are fruitfully viewed as opposite phases of the same overall cycle, then the psychiatrist and the anthropologist should have considerable overlap in their analytic materials. And if we view the family as a small-scale society, then in family therapy the interest of the two should virtually coincide. Both should explore the nature of manageable and unmanageable inconsistencies among various parts of pattern systems.

The second main point concerns the usefulness of intrasystem and intersystem analysis for psychiatric purposes. I have sat in on a fair number of discussions of patients in connection with a part-time affiliation with the Department of Psychiatry at the University of Cincinnati. My intuitive and obviously biased conclusion is that the model can be clinically useful. But it is difficult to judge because no clinicians now think or report their "facts" with this model, and a test of it would require that some do so.

8   CONCLUSION

Traditional science deals with paired items in a cause-effect relation. If the science is deductive and well constructed, one can say "If A, then B." If the science is empirical one can say "If A, then pB," where p represents a probability, perhaps with its error of estimate. In some areas of physical science the model from which deductions are made is sufficiently like certain aspects of reality that one may go a long time without having to bother about whether he is dealing with a deductive or an empirical science. In an important shift of emphasis the science of systems has brought explicit recognition that the segments of complex reality interact with one another in many ways, including mutual causation and feedback loops in all sorts of odd places. That is, we now recognize explicitly in our science that real acting systems are complexly connected and we attempt to trace those connections. When the interacting units and the nature of the connections are identified, we have described the "structure" of the system as defined here.

Whatever the merit of the propositions generated in this volume, I think this approach has established one thing. Not only are the parts of real social systems interconnected in many and detailed ways but so are the parts of the analytic system we use to deal with them. Furthermore, to a degree I have scarcely even speculated about, some connections among parts of real social systems parallel the connections among parts of the analytic system. To compare this volume with two works mentioned earlier, the studies of March and Simon (1958) and Berelson and Steiner (1964) are collections of many empirical findings. The sheer number of findings testifies to the energy (and/or desperate need to publish) of our social and behavioral scientists whose experiments are collected there. Both volumes have a structure in that there is a logical rationale for their grouping and sequence of findings. But the overall body of science they reflect has little structure: most of its propositions have been reached independently and are not logically interrelated as they stand. Although the authors have to some extent synthesized groups of similar propositions to produce generalizations somewhat broader than the original empirical findings, with rare exception any of the propositions from the primary sources could be removed without affecting the validity or logical status of any other proposition.

It is conceivable that an empirical behavioral science could be constructed in which most findings could be connected in an interlocking set like the construction of the deduced findings here. At least two problems seem to preclude such a result. First, all researchers in a given area would have to accept common definitions and similar rules of identification. Otherwise the one who collected and related the findings would have to superimpose his own definitions on those reported by the researchers in a procrustean exercise that would certainly distort many of the findings. Second, many research findings are reported in terms of coefficients of correlation or some other statistical relation. Nothing short of a miraculous new statistical science would enable us to assess the credibility of a conclusion reached by deductively combining three other conclusions stated respectively as a coefficient of correlation, a chi square test, and a simple average—especially when the error of estimate of each was included. The probability is high that the confidence level in any such "deduced empirical" conclusion would be low, and I am sure there are other reasons that would strengthen this conclusion.

If we agree that propositions about social behavior are inter-related, that they ought to be related into a coherent structure, and that my dismal conclusions about comprehensively relating empirical findings are correct, we must conclude that a serious move in the direction of a deductive analytic structure is in order and perhaps long overdue. Whether this particular structure is sensible is one question, and I am quite aware that certain aspects might advantageously be changed. But whether social and behavioral scientists ought to shift toward building and refining this kind of structure is an entirely different question. One must assume that several different structures are possible—I can shudder already for the student of the future as he is introduced to five or six. For the moment, however, I will confine my worries to the acceptability of one framework, not to the prospect of a library of them.

Along with putting much social science on a deductive basis, the purpose of this volume is to unify, not to produce new knowledge. The following quotations from DiRenzo (1966) help clarify my intent. Although some ellipses represent large spatial gaps in his work, I think the quotes retain his meaning.

> A science is said to be mature if it has developed a solid and valid foundation for the construction of theory. . . . The behavioral sciences . . . remain characterized by much word play . . . with many concepts that seem to have a certain vagueness, not only about their substantial referent, but also with regard to their methodological utility. . . . The lack of precision . . . in many such cases is the result of unclear thought [p. ix].
>
> The several disciplines, in many instances, deal with the same phenomena in formally different ways. . . . Postulating the oneness of reality, these individual perspectives must be congruent with one another and, at an "ultimate" point, must converge into one comprehensive explanation. . . . Another specific thing that must be done [is to develop] a common scientific language for the behavioral sciences—and, wherever possible (such as in methodology), for all scientific disciplines. . . . It is most difficult to make scientific progress with the persistent use of common, everyday language whose ambiguous vocabulary with its multiple connotations is the chief impediment. The physical and biological sciences . . . were able to avoid such problems, partly because no generally accepted terminology existed for much of their phenomena. The behavioral sciences have not been so fortunate, but . . . need a standardization of scientific concepts and scientific language. . . . It is time that the "underground" of

the behavioral sciences—those that refuse to admit that the behavioral disciplines are, or can be, methodologically scientific—face up to scientific tenability, even though they themselves refuse to subscribe to it [pp. 285–288].

To counter conceptual vagueness I have tried to give precise definitions to all terms used and to employ their technical definitions consistently without reverting to everyday meanings. In particular I think the distinctions among power, bargaining power, and aggregate power—and such definitions as those of culture, communication, transaction, authority, delegation of authority, role, and organizational structure—help bring precision to concepts that have been vague. A prerequisite for undertaking this task is to sharpen the crucial difference between the roles of scientist and lexicographer. The lexicographer tries to find how a term is used, without regard to its scientific utility. The scientist does the reverse. In fact, when the scientist defines a term it is irrelevant to him, as scientist, whether the meaning he gives it corresponds at all with ordinary meanings of the same word—though the degree of correspondence may determine how widely he is understood and liked.

At this stage I am more interested in underscoring the necessity of precise definition than in discovering whether my own definitions are either precise or useful. Another major concern is to emphasize that referents of most scientific terms must be analytic constructs, not segments of reality. That such a statement will alienate many specialists only underscores the urgency of making it. It is only when we posit and define an unreal lever without weight, an unreal vacuum surrounding the earth, and unreal liquids without viscosity flowing through unreal pipes without friction that we can make precise statements about levers, falling bodies, and hydraulics. Recognition of this methodological principle is more imperative for the behavioral than for the physical scientist since the latter can discover or construct real cases close enough to his hypothetical models to keep him out of serious trouble. The behavioral scientist has not inherited this good fortune and can less afford to squander his resources pursuing the Grail of real referents for his main concepts. Although some will accept the desirability of conceptual simplification without argument, I urge on others the desirability of seeking it before we get too used to computers. As Boulding once noted in a lecture, if Ptolemy had had computers, Copernicus might not have seemed necessary.

For those who tend to brush off taxonomy as not significant I note that only by using precise definitions and numerous distinctions not previously used (power versus bargaining power, transactional and pretransactional stages, authority as a transactional relation, the separation of interactions into communications and transactions) was it possible to construct the deductive core of this volume. At least in this case the science (statements of relationships) is utterly dependent on the taxonomy.

As a pragmatic problem, the behavioral scientist's constituency blocks his path to real science. If the physicist is asked for a precise answer to the "mushy ice" problem cited in the Preface he can respond that it is essentially unanswerable—not really a question in "science"—and get away with it. But the social scientist is apt to be scorned by the public if he says he cannot give answers to mushy social questions, and he is thereby under pressure to avoid unreal concepts. For decades economists were taunted for their unreality. But I am sure they now have better science, and can deal with reality better, than if they had not first made the retreat from reality. The rest of social science must do the same, and I have pushed in this direction with such concepts as pure decision, perfectly transmitted culture, pure communication and transaction, and unlimited-purpose government. I see no danger to social policy in such a move. There are now enough of us to man both the front lines dealing with real problems with existing concepts and the strategic retreat for regrouping.

In DiRenzo's language, the central point of this volume is that the separate behavioral and social disciplines "must converge into one comprehensive explanation." This book is an experiment in constructing one such explanation, expressed in "a common scientific language for the behavioral sciences." Because it is based on system analysis—which from its inception was designed to cut across physical, biological, behavioral, and social lines—its language should hold better chance of applying to all scientific disciplines than one laid on a basis initially unique to social science. Were the latter partial unification to be done first it would then require an additional step to reach common ground with physical and biological sciences. And if the first step were widely accepted, inertia might preclude or long delay the second. Hence there is merit in taking both steps in one if we can manage it—and if the result is good social science.

The aim of this book is not to say "This is it," but to advance the conviction that such a unification is possible. And I hope that at least some of the structure proposed here proves relatively durable and helps provide a continuum from the most global generalizations to the most particular, as outlined in Chapter One and Figure 4. To the extent that this approach succeeds, it has several immediate consequences. First, it transfers a substantial batch of material now classified as social psychology from the realm of low-level empirical findings to that of conclusions deducible from several central models. Second, it ties certain central concepts of sociology directly to concepts of political science and organization theory by suggesting that the general principles in the first are the same as those in the last two, except that they operate more loosely in semiformal organization. Third, in doing so it accepts and attempts to expand upon Homans' (1961) view of "elemental forms" of social behavior, which are here taken to be communication, transaction, and organization along with their many combinations. Even if these consequences produce no immediate generalizations not already known to the specialists, the method should interest them.

The alleged unifying properties of this conceptual set have already been stated and need no recapitulation here. The unification, however, lies in suggesting a common substructure. There still remains much work for specialists in broad species of systems (political, economic, social), in individual members of species (the political system of ancient Greece, the economic system of contemporary Yugoslavia), in generic or individual subsystems (legislatures, the British Parliament, the family structure of Appalachian whites, the American copper industry), in particular patterns or instances of social interactions (culture as a special case of communication, the transactional patterns specific to the American political party, the authority structure in the Hopi family). As intrasystem behavior the decision process, in general or particular, is similarly an appropriate area for specialized study as it occurs under varied circumstances.

Given continued specialization, widespread use of unifying concepts could lead all specialists to share a common base and a largely common scientific language. Hence they could, and actually might, read more of one another's literature. They could also test hypotheses more readily; today specialists in one discipline are often unaware that replications of their work are currently reported in other fields but in language so different that their similarity goes unnoticed.

Feeling that a unified structure ought to get along with few terms, I am mildly disconcerted by the many I have had to define. Part of this problem is that many terms from system theory, psychology, communications, and decisions are here incorporated in a book about social science. But even with these additions I feel that the number of technical terms I use is much smaller than the sum of those used elsewhere to cover the same conceptual territory. In any case the Glossary overstates the number of terms needed since it includes many terms that are not required in this framwork, merely to indicate that they can be translated.

Similarly disconcerting is the fact that it has taken so many pages to state what seems to me a simple idea. Part of this difficulty is noted in 3.13.1 and 3.14.2—that since source and receiver of this message do not yet share a common set of signs and referents it often takes paragraphs or pages to convey what a single word might eventually do. Part of the wordage is also argument for the proposal, though I have tried to keep that argument short. A substantial part is sheer elaboration of applications of the model, partly to illustrate its capacity and partly to spell out the content of this kind of social science.

The assumed preference for the advantageous that runs through much of the model does not imply that people are presumed always to have it in fact. It merely means that the assumption seems to provide a better first-approximation model than does any visible alternative. The alternative of making *no* assumptions about goals does not seem sensible for analyzing real systems that do have goals. Where the informal organization is concerned we do without goals of the whole system but must still assume goals for the subsystems. In any case the advantage-preferring model seems to remain useful as we move, by relaxing rigid assumptions, from the many pure cases to the mixed and impure ones that more nearly approximate much of reality.

Let me put the question about the whole approach in its own language: Is good unified social science possible (opportunity function)? And, if so, do we want it (preference function)? And to deal with mutual interrelations of the two: To what extent is our desire for it conditioned by whether we perceive it to be possible (preference affected by perceived opportunity) and to what degree is our conclusion whether it is possible conditioned by whether we want it (perceived opportunity affected by preference)? Whether this book has really shifted the burden of proof from those who think unified social science possible to those who think it is not, as suggested in the Preface, is for each reader to decide. Even those who think it possible may still not consider it desirable. Some think it demeaning to suggest that human behavior can be explained scientifically at all, much less with a relatively simple set of central conceptual tools—Man is a Higher Creature than is amenable to such diagnosis. Although I understand the feeling, I do not share it.

One reservation I do share concerns the very real dangers of "premature closure" (Kaplan, 1964, pp. 70f). But I would add, as before, that there are also dangers of senile openness. Although I strongly insist that "unified" is not necessarily "closed," Kaplan makes clear why precise definitions are themselves a closure and T. S. Kuhn (1970, p. 64) makes clear why the more comprehensive a paradigm the more resistant it will be to change. Moreover, as can readily be appreciated by anyone who visualizes the processes of writing this book, I am painfully aware that the more widely interlocking a set of definitions, the larger the repercussions and the greater the difficulties of changing any of them. Although the difficulties are smaller for later "superstructure" concepts, I would be less than candid if I failed to report a feeling of tension at the mere thought of significantly redefining any of the basics. On both logical and emotional grounds (detector and selector) a tight unified conceptual structure is more closed than a loose one. But changes, even as extensive as designing complete alternative systems, *are* possible. And perhaps someday someone somewhere will make them. In any case, unless the social sciences are to remain perpetually adolescent the risks attendant on closure must someday be taken. And I do not think the time is any longer premature.

I close with a mild plea to specialists. Even if you foresee no direct advantages of this approach within your own field, unless you are reasonably sure of distinct disadvantages other than the initial retooling costs, which might be considerable, it seems to me there is an experimental obligation for enough specialists to acquire enough facility with the unified structure to give it a serious test in every field. To the question posed by Landauer (1971) whether integrated social science will stand as just one more discipline, the answer seems to be: If specialists adopt it into their workaday world, No. If they do not, Yes.

Glossary

The terms in this glossary fall into three main groups. First are all terms formally defined in the book. Most of these definitions are taken verbatim from the text, but some have been modified to compress into the glossary definition ideas that are handled in the text as amplification of the definition. The location of these definitions in the text is indicated by the coded paragraph number. Second are terms used more or less formally in the volume but not formally defined. Such terms are distinguished by the word *see* preceding the relevant paragraph number. Third are terms not formally used in the volume and mentioned only in passing if at all. These terms are marked with an asterisk. Terms in this category indicate how certain items might be defined if ever needed. There is no systematic basis for inclusions and exclusions in this category. We included terms that seemed likely to arise in the reader's mind in connection with topics discussed here. We excluded simple descriptive or identificatory terms like *matrilineal,* names of social patterns of particular times or places like *feudalism,* and most of the vocabulary of methodology.

abstraction: the formation of higher-level, emergent patterns that consist of similar elements among lower-level patterns. 2.10p.
abstract system: a pattern system whose elements consist of signs or concepts. 1.20n
acceptance (of B by A): detector and selector states of *A* about *B* such that *A* foresees an overlap of EP's in a sequence of future transactions with *B,* including the transactional aspects of communications and organizations. If *A* is a group and *B* an individual, acceptance of *B* by *A* is measured in

the number and magnitude of communications and transactions between A and B initiated by A. Acceptance is the obverse of *participation*.

accommodation (by A*):* a form of resolution of *conflict* in which A modifies his detector and selector (and possibly his effector) states in such ways as to produce interactions with B that are more satisfactory to A.

accountability: same as *responsibility*.

acculturation: the process of *culture* assisted by formal or semiformal sanctions of authority to motivate the learning of cultural patterns.

acting (behaving) system: a pattern, two or more elements of which interact. 1.20g.

action (social): communication, transaction, or organization in any of their pure, mixed, or variant forms or details.

actual aggregate power: see *power, actual aggregate*. 3.35.0.

adaptation (adaptive behavior): behavior that in some way changes the relation of a system to its environment, whether by altering itself, the environment, or both. 2.02.

  successful adaptation: adaptation that to some degree increases the likelihood of achieving some goal of the system. 2.02.

adjustment: adaptation, especially to the behavior of others.

advantage (in bargaining): see *bargaining advantage*. 3.22.0e.

advantageous: The more advantageous of two things is the one that will provide the greater satisfaction. Given rational behavior the more advantageous coincides with the preferred. 2.501.

agent: a third party C who acts on behalf of A in a transaction between A and B. 3.27.2.

aggression: counterattack against something perceived as threatening, with the aim of injuring or immobilizing the source of the threat. Also the system states that produce such counterattack. The definition allows, but does not require, that the counterattack employ force. Any interaction that apparently produces the desired result is admissible under the definition. See 2.41.12.

alienation: a set of system states of an individual in which he sees himself as having no role, or no role acceptable to himself, in a formal or semiformal organization. Occurrences of alienation must be approached clinically as an essentially idiographic problem, even though numerous individual cases may occur for similar, relatively widespread causes.

analytic system: see *abstract system*. 1.20n.

anomie: a partial failure of the process of *culture* in an individual, particularly with reference to his selector states toward roles he might be expected to fill and their specifications.

anxiety: a state of fear; usually applied to conditions in which the object of fear is indefinite, unknown, or imaginary.

approach: any behavior, including the acquisition or processing of information, directed toward some desired (positively valenced) outcome. 2.2.

arbitrage: A general principle of transactions is that their terms and consummation are determined by four variables—the value of each of two things to each of two parties or in each of two different times, places, markets, and so forth. Arbitrage is an application of this principle to the relation between two markets, particularly in the case of two curriencies that have different relative values in two different places. See 3.31.41.

*_assimilation:_ the result of successful _induction,_ with special reference to the cultural _fusion_ of a smaller into a larger cultural group, typically with physical dispersion of the former among the latter.

*_association:_ 1. a formal organization. 2. continued or repeated interaction.

*_attitude:_ one's detector and selector states regarding a particular thing. Where relevant, the detector may encompass information about one's effector.

*_attraction:_ a somewhat amorphous concept of a multiple phenomenon. Not an interaction but a set of system states regarding interaction, actual or potential. See 4.12.5. Overtly attraction consists of approach activities leading to _participation._ Internally attraction is an estimate of the value of participation, which evaluation might be made before or after participation.

authority: the ability to grant or withhold rewards or punishments (sanctions) for the performance or nonperformance of instructions. 4.12.10f.

delegation of authority: Among three successive levels of systems _A, B,_ and _C_ (with _A_ highest), the delegation of authority is a transaction between _A_ and _B_ in which _A_ agrees to provide sanctions to _C,_ directly or through _B_ when requested by _B,_ in return for which _B_ will request such release only as _C_ follows instructions from _B_ designed to effectuate goals specified by _A._ 4.12.10j.

legitimate: see _legitimate authority._ 4.12.10n.

illegitimate: see illegitimate authority. 4.12.10o.

effective: a situation in which instructions are in fact followed. 4.12.10p.

*_autonomy:_ the position of a system that is not a subsystem in a formal organization or, if it is such a subsystem, whose role specification is relatively vague.

avoidance: the opposite of _approach._

_AX: A_'s desire for _X._ 3.21.2a.

_AY: A_'s desire for _Y._ 3.21.2b.

bad (negative good): any object or event that reinforces an avoidance response —that is, has a negative valence. 3.25.0a.

bargaining: see _negotiations._ 3.22.0f.

bargaining advantage: bargaining power forces that enable one party to get terms more favorable than some reference terms to which they are compared. 3.22.0e.

bargaining coalition: see _coalition._ 3.27.50.

bargaining power: see _power, bargaining._ 3.22.0d.

behavioral system (human): the system of the human that selects behavioral outputs and guides their execution. It consists essentially of the nervous system—afferent, central, and efferent—and the associated organs such as eyes and ears that recode information from the environment into a form that will actuate sensory nerves. 2.03c. The control, as contrasted to the maintenance, system.

*_belonging (sense or feeling of):_ a loose, multiple phenomenon incorporating aspects of _identification, attraction, involvement,_ and _participation._

benefits: the goods or satisfactions received or receivable from an alternative in a response selection. 2.50o.

*_bindingness:_ the _stake_ one has in fulfilling a promise; or any other loss one may suffer by breaking off a relationship with another. In many cases bindingness coincides with responsibility. A decision may be binding on one who lacks the power to change or avoid it.

biological system (human): the biological entity of the individual human, bounded by skin, hair, nails, and the like. 2.03a. The maintenance, as contrasted to the control system.

bit: the amount of information required to select unequivocally between two mutually exclusive and equally probable items. 3.13.1a.

blockage: an action by *A* that prevents *B* from executing overt behavior—as by death, injury, physical constraint, or carrying *B* away. 3.25.0c.

boundaries (of an interaction): An interaction is bounded analytically by specifying the system (parties) involved, the type of interaction (such as transaction or movement of matter), and the time duration. 3.10f.

boundaries (of a system): Logically the boundaries of a system are defined by listing all the components of the system; any elements not listed are construed as falling outside the system. 1.20r.

boycott: a hostile shrinkage of EP to the point where there is no overlap and thus no transaction is consummated. 3.26.13.

bride price: a payment made by the man for a wife as part of the major bargain of a marriage. See 4.36.1.

*bureaucracy: the staff of a formal organization and their behavior within the organization.

BX: *B*'s desire for *X*. 3.21.2c.

BY: *B*'s desire for *Y* (his desire to retain *Y* in a transaction). 3.21.2d.

*capitalism: For general purposes we can construe capitalism as another name for a pure *market system*.

*caste: a *class* for which the semiformal and formal organization of the society has set rigid role prescriptions and terms of entering or leaving membership —typically by birth and death.

causative (imperfections in equilibrium): situations that actively push away from an equilibrium position as contrasted to frictional imperfections. See 3.18.64.

change: see the discussion of social change and developmental change in 5.2. See also 3.10b.

charisma: the ability of a person, usually in the face of some deep, widespread, but unmet desire, to generate long EP's in many people for the thing he can allegedly provide. See 4.12.6.

citizens: that portion of the public who are members of the government organization with specified rights and obligations, whether or not they belong to the government staff. 4.13.10g.

civil liberty: limitations on the power of government staff to interfere through the staff-recipient relation with certain specified actions of citizens. 4.13.516g.

class: the collection of people in a society who have approximately equal aggregate power. 3.35.0.

*class consciousness: awareness by individuals that they belong to a particular class. If action is taken by members to enhance the position of their class, the class becomes a semiformal organization. See 4.23.3.

*clique: a coalition or an informal organization within a formal organization.

closed system: a system in which interactions occur only among components of the system. That is, there are no inputs from or outputs to the environment. 1.20ee.

coalition (or bargaining coalition): coordinated action among actual or potential competitors (*B*'s) to improve their power position in dealing with *A*'s. 3.27.50. See *dominant coalition.* 4.12.42. For the equilibrium size of a coalition see 4.13.5143.

code: a set of categories (and hence a pattern system) of coded inforamtion. 3.10.0g.

coded (semantic) information: information that has been separated into distinguishable patterns—that is, categorized or grouped by similarity of pattern. 2.10b, e, f.

coercion (of B *by* A*):* a situation in a stress or threat transaction in which *A* is able to make *BX* so large that *B*'s likelihood of not giving *Y* to *A* is small. 3.25.0e. See also *consent.*

*cognitive style: the ways in which an individual typically processes information.

cohesion: a term applied to a complex multiple phenomenon to refer to the "ties that bind" people together—"the resultant of *all forces* [emphasis added] acting on members to remain in the group" (Cartwright, 1968, p. 91). The ties can be any of an almost limitless number of combinations of communicational, transactional, and organizational elements, which can presumably be analyzed in a particular case only through a simulation model. Directly or indirectly, virtually the whole of Chapters Seven through Fifteen, as well as parts of Sixteen and Seventeen, might be construed to deal with cohesive processes. See 4.12.5.

collective bargaining: coordinated action among *B*'s who conduct transactions with *A*, to improve their power position in dealing with *A*, but whose transactions with *A* stand in the relation of complements rather than substitutes (or competitors, as with coalition). 3.27.6.

collective concept formation: see *concept formation, collective.* 3.18.5b.

*collectivity: a simple multiplicity of persons who are not interacting.

*commitment: an explicit or implicit major bargain that some kind of relationship will continue. More broadly, a firm decision to pursue a particular course of action even at high costs; one such decision is to continue a relationship.

communication: a transfer of information (pattern) between systems or any movement of anything between systems analyzed with reference to its information content. 3.10.0a.

communication versus perception: see 3.10.0j.

social communication: mediated transfer of information coded at both *A* and *B* with the addition of contingency. 3.11.

*community (biological sense): an uncontrolled system of controlled subsystems; an ecological relation or informal organization. 4.21.3.

*community power structure: a class, coalition, or organization of persons of relatively large aggregate power who together exert considerably more than an average amount of power in decisions affecting some local area.

*comparative advantage: A general principle of transactions is that their terms and consummation are determined by four variables—the value of two things to two parties or in two different times, places, markets, and so forth. Comparative advantage is an application of general principle of trans-

actions (see *arbitrage*) to differences in relative values of two commodities in two different countries. See Section 3.2.

compatible decisions: When goods X and Y show scarcity costs and there exists sufficient aggregate power to acquire both, decisions in favor of both are compatible. 4.13.515.

competition: a conflict situation in which two or more parties seek to complete the same transaction. 3.27.30.

*compliance: behavior of a person on whom authority has been effectively exercised.

component: any interacting element in an acting system. 1.20h.

concept: a class pattern of coded information in a brain. 2.10e and f.

    concept-learning: the process of forming relatively durable concepts. Concepts of external objects correspond directly with recurrent patterns of sensations and indirectly with external patterns that actuated the sensations. 2.10h.

    concept formation: concept-learning in which the individual sorts reality into his own patterns. 2.10i.

    concept attainment: concept-learning in which the individual adopts patterns learned from others. 2.10j.

concept formation, collective: consensus by communication but with emphasis on developmental change. 3.18.5b.

conditioning: the main process by which the behavioral system forms non-inherited internal patterns that are significantly correlated in some way with patterns in the environment. 2.09.5.

conflict: the presence of costs as viewed before the decision is made. 2.50q. See also 7.2.

*confusion (in a system): a state of overload or dissonance or both.

conscious(ness): a primitive term whose meaning must be assumed in the reader. But see 2.06.3.

consensus: the process by which two or more persons reach similar detector, selector, or effector states or processes.

    consensus by communication: the process or reaching consensus by successive and mutual transfer of pattern. See 3.18.5a.

    consensus by transaction: the process of reaching consensus by transfer of valued things. A single exchange ratio between X and Y among a multiplicity of mutually accessible and competing A's and B's, developed through the external availability of alternative transactions. 3.27.4.

consent: Any conscious action originating inside a person's head and executed under his own neural controls is said to be done with his consent. If coercion is involved, we say that one's consent has been coerced but not that he has been coerced into acting without his consent. See 3.25.0e.

consistency (or interdependence of elements of a pattern system): Elements A and B are said to be consistent or interdependent if some change in element A is perceived by some acting system as requiring some change in A, B, or both to restore a "proper" relation between them. 1.201.

*conspicuous consumption: the deliberate acquisition, use, nonuse, or destruction of goods in such a way as to communicate to others awareness of one's large aggregate power. 3.36.7.

*conspiracy: in present terminology, either a coalition or a formal organization—though customary usage connotes disapproved action as well.

constitution: the content of a major bargain between sponsors and government about decision rules, broadly construed. 4.13.36e.

contingency: An action by A is contingent on B if it is designed or modified to affect B or A's relation to B. 3.10d.

> *mutual contingency:* a situation in which B's response to A's contingent behavior is in turn contingent on A. 3.10e.

contract: a promise to make delivery as agreed in a transaction. 3.24.3c.

*control: see *power,* especially 3.22.8.

control system (human): see *behavioral system.* 2.03c.

controlled system (cybernetic system): any acting system whose components and their interactions maintain at least one system variable within some specified range or return it to within that range if the variable goes beyond it, despite changes in forces that influence the state or level of that variable. A controlled system can also be defined as a goal-oriented system. 1.20s. Pragmatically, it must have identifiable DSE subsystems.

cooperation (cooperative relation): goods X and Y are said to stand in a cooperative relation when the same action produces both. 2.50p.

cooperative organization: a formal organization whose sponsors are the recipients of its outputs and whose sponsor goal is their own welfare as recipients. 4.12.72a.

costs: the satisfaction denied in the course of achieving other satisfaction. 2.50n. Costs can be intra- or interpersonal.

> *opportunity costs:* the dissatisfactions of having to avoid what one would like to approach. 2.50n.

> *disutility costs:* the dissatisfactions of having to approach what one would like to avoid. 2.50n.

*coupling: This term is not used at all here. But so far as humans are concerned, all couplings of two or more persons are interactions between them; hence in this volume all couplings are communicational, transactional, organizational, or some combination.

cues: current receipts of information via the senses that assist an organism in identifying a current pattern in its environment. 2.10n.

> *sufficient cues:* cues that permit identification of a pattern with a degree of confidence satisfactory for the purpose at hand. 2.10o.

culture: communicated, learned patterns. 3.17.0a.

> *cultural content:* the total set of cultural patterns common to some collectivity of people. 3.17.0b.

> *cultural process:* the communication of cultural content to and among that collectivity—particularly to its new members. 3.17.0c. See also *communication, social.*

> *perfectly transmitted culture:* a situation in which all members of a society possess an identical body of culture that is accurately communicated to all new members by the cultural process. 3.17.0e.

> *cultural fusion:* the merger of multiple cultures and their respective societies. 3.17.0g.

> *cultural diffusion:* a merging of cultural content of multiple societies by cultural communication but without merger of the societies. 3.17.0h.

currently: Since costs and benefits are in most cases thought of as flows, not stocks, some lapse of time must be allowed within which they are measured. This period could range from a few seconds (for the runner who conserves his strength during the first half of a dash to have more at the finish) to a lifetime (for the person who lives this life in such a way as to affect his next reincarnation). 2.50aa.

custom: culturally transmitted patterns of behavior.

cybernetics: the study of controlled systems. 1.20s.

cybernetic system: see *controlled system.* 1.20s.

decay: the developmental breakup of a system, after which only its subsystems or nonsystem components remain. Chapter One and 5.21.1. Decay can occur in both acting and pattern systems.

decision: response selection under conditions of complexity. 2.50c. See *judgment.*

decision costs and benefits: the costs or benefits entailed in the process of making the decision, without regard to the costs or benefits of any alternatives. 2.50n.

decision rule: some criterion for determining whose desires will constitute the group decision in the event of disagreement. 4.12.42.

acceptance of decision rule: an agreement (a) to accept as legitimate any decision made in accordance with those rules and to accept sanctions for violating them and (b) to abide by those rules in one's attempt to influence decisions. 4.13.33.

decision theory: the study of the decision process in humans with respect to the conscious content of all three DSE subsystems; content in this context includes a state of uncertainty. 2.50d. The study of objective rationality. 2.50u.

decoding: the selective recognition or identification of a pattern. 2.10m and 3.12.

defense mechanism: see *ego defense.*

deference: behavior by B that communicates (symbolizes) his acknowledgment of status inferior to A's. 3.35.0.

delegation of authority: see *authority, delegation of.* 4.12.10j.

delegation of responsibility: see *responsibility, delegation of.* 4.12.10k.

delivery: the act of giving in a transaction, as defined broadly in 3.21.2e. 3.24.3b.

democracy: a cooperative type of government organization in which all citizens are sponsors. 4.13.516a.

dependence (of A on B): With respect to any Y that B can provide to A, A's dependence is the same as AY. A's aggregate dependence is then $AY_1 + AY_2 + \cdots + AY_n$. Stated more broadly (but without changing the definition), A is dependent on B to the extent that communicational, transactional, or organizational relations between the two are advantageous to A.

destruction: any transformation that decreases utility. 2.50ff.

detection (as a stage in information transfer): the uncoded, essentially isomorphic transfer of a pattern from a medium to its receiver. 3.12. Detection is only one of several kinds of information-processing within the detector.

detector (of a controlled system): the function by which a system acquires information about its environment. 2.05a.

deviant: behavior that differs markedly from that of the role prescription. Also a person who engages in deviant behavior. See 5.1.

dictatorship: a profit or, much less likely, pressure organization of government that uses force to keep its coalition dominant—in contrast to using force merely to enforce decisions. 4.13.526.

diffusion (of cultures): see *cultural diffusion.* 3.17.0h.

discipline: 1. as field of study: a substantial complex of concepts and their interrelations. See Chapter One. 2. as organizational phenomenon: the exercise of authority with special reference to the application of particular acts of sanction to particular violations of instructions. 4.12.10g.

*disintegration: same as *decay.*

disinvestment: the reduction of the future cost/benefit ratio by withdrawing investment, not incurring fixed costs, or accident. 2.50z.

dissonance: pattern mismatch (in the human system). 2.25.

*division of labor: in this framework, a negentropic, emergent rise in differentiation that necessarily coordinates the separate functions thus differentiated into a higher level of system. This negentropic differentiation is accompanied by entropic trade. 4.22.

*divorce: legal termination of the legally sanctioned major bargain between husband and wife. Subsequent interactions between them are not subordinate to that major bargain.

dominance (of A*) and submission (of* B*):* a condition in a continuing relation between *A* and *B,* or in some aspect of that relation, in which *A* generally exerts the greater power. 3.37.

dominant coalition: that subgroup of an organization which holds the greatest power relevant to that organization or its decision rules and which as a unit supports a particular decision. 4.12.42.

*dowry: Same as bride price but with reversal of sexes. See 4.36.1.

DSE: a convenient reference to detector, selector, and effector. 2.04.

*dyad: two persons who interact through communication, transaction, or simple organization.

dynamic (steady-state) equilibrium: a situation in which components of an acting system or their states continue to move or change but are balanced so that at least one variable remains within some specified range. 1.20x.

ecological system: an uncontrolled system, some of whose component subsystems are controlled systems—namely, organisms. 1.4 and 4.21.3. The ecological system is defined precisely by the set of assumptions that constitute the model of it in 4.21.1.

*education: a subset of the *cultural process* (an identification, not a definition).

effective preference (EP): the selector and effector states of a party toward a transaction. 3.21.2f.

effector (of a controlled system): the function of executing the behavior selected by the selector. 2.05c.

*efficiency of knowledge: see Chapter One.

*ego defense: the attempt to see oneself, or to have oneself seen by others, as a person of power or at least as not without power. Since being without power is validly frightening, ego defense may easily involve strong emotions. See *ego image.*

*ego image: one's sense of his own status—in his own eyes, in the eyes of others, or both. Since status is perceived aggregate power, which depends on DSE states plus communicational, transactional, and organizational relations, all these elements may be involved in the ego image.

element: any identifiable entity—concrete or abstract, object or event, individual or collective. 1.20a.

*elite: 1. persons of large power within a particular category. See 3.3

(especially 3.35.0) and 4.23.3. 2. the *dominant coalition* (4.12.42) who control the decisions within a particular type or level of organization. For a whole society this dominant coalition might be thought of as the ruling class.

emergent (emergence): the development of higher-level systems by combining existing lower-level ones—typically, but not necessarily, with some disposal or rearrangement of parts. Once the higher-level system has emerged, new components may be added. Emergence can occur in both acting and pattern systems. Chapter One and 5.21.2.

emotion: in this framework, the biological state that characterizes a particular emotion (endocrine secretions, pulse changes, sweating) and the accompanying neural input to the *behavioral system* from those biological states. 2.08.0d.

employee point of view: the question as seen by an individual whether or not he should belong to an organization, in light of its costs and benefits to him. 4.11.3.

employer point of view: the question of whether or not the organization should have a particular person as a member, in light of the costs and benefits of his membership to the organization. 4.11.3.

encoding: the initial transfer of a pattern from its source to a medium. 3.12.

*encounter (between A *and* B*):* any situation mutually perceived by *A* and *B* in which actions by *A* have some actual or potential perceived effect on *B* with a resultant potential interaction. The encounter can apparently be analyzed as almost solely transactional, though with communicational adjuncts and consequences, so that the whole of 3.2 through 3.4 potentially applies.

*engagement (social): the initiation of a communicational, transactional, or organizational interaction with some expectation that it will be consummated.

environment: anything outside the boundaries of a system. See 1.20r.

environment of the individual human system: Since our attention is focused on the *behavioral system,* the environment of a given human system for our purposes includes the biological system of that person. 2.03d.

equifinality: the principle that in a closed system the final state of the system is unequivocally determined by the initial conditions of the system and that in an open system a given final state may be reached from different initial conditions and by different ways. 1.33.1 and .2.

equilibrium, cross-sectional: a situation in which one or more variables of a system remains constant or varies within only a limited range. Chapter One.

equilibrium, developmental: the continuance of a given level of system with neither *emergence* nor *decay.*

equilibrium, dynamic: see *dynamic equilibrium.* 1.20x.

equilibrium, static: see *static equilibrium.* 1.20w.

*esteem: 1. *status.* 2. approval of one's DSE states or communicational, transactional, organizational interactions by others.

*ethnocentrism: a detector state of humans in which the images of one's own society and culture are stronger, better defined, and more determinant of behavior than are the images of other societies and cultures; typically accompanied by selector states in which those images are strongly approved.

event: a change in pattern over time. 1.20d.

evolution: the operation of some selective criterion among essentially *random* variations. 2.02 and 5.21.2

executive function (of government): the process by which government staff effectuates the legislated instructions. Much, but by no means all, of the executive function consists of staff-recipient relations. 4.13.516e.

expectation (anticipation): an unperceived but activated portion of an information concept of object or event. Or the unexecuted but partially activated portion of a motor concept. (This is description, not definition.) See 2.25.3. As an integrative factor, see also 4.12.51.

*expressive (versus adaptive or instrumental) behavior: the *symbolizing* of selector states, or preference functions, as contrasted to the symbolizing of detector-effector states, or opportunity functions.

external power factors: see *power factors, external.* 3.32.

face: credibility of one's position in a transactional relation. See 3.23.23.

factor suppliers (of an organization): those parties who provide the organization's inputs. 4.12.10a.

family: a small, complex (if children are involved), multipurpose, cooperative organization with such dominant functions as procreation and sexual relationships, child-rearing, mutual assistance, and care of basic biological needs of its members. See 4.36 for simulation model of family.

feedback: a mutual interaction between a system A and some element in its environment B such that an action by A on B produces a return action by B on A. 1.20y and z. In connection with adaptive behavior, feedback may be construed more narrowly as the detection by A of the consequences of his prior action. See 2.07.2.

*femininity (sense of): a subset of a sense of *identity* dealing with the role of the female.

fixed cost: any cost or portion of cost currently incurred to prevent deterioration of the cost/benefit ratio whether or not that ratio was achieved by past investment. 2.50y.

flow: the movement of matter-energy or information across the boundaries of a system or across any arbitrary boundary within it. See 4.23. Compare with *stock.* See also 3.22.8 and 3.3.

*fool: see *knave.*

force: a subcategory of bads consisting of physical blockage, destruction, or biological pain or injury. 3.25.0d. Generalizing from the use of the term in the physical sciences, a force may be considered as anything with which A varies directly—as contrasted to a resistance, which is anything with which A varies inversely.

formal organization: see *organization, formal.* 4.10b.

freedom: the inverse, or relative absence, of cost. 3.29.

    sense of freedom: a condition in which some desired thing is perceived as having less cost than under some other condition to which it is compared. 3.29.

    freedom to: absence of cost in carrying out one's own actions. 3.29.

    freedom from: absence of cost imposed on one without his consent by actions of others. 3.29.

free good: any good or satisfaction achievable without cost. 3.29.

frictional (imperfections in equilibrium): situations that slow or block movement toward an equilibrium in contrast to causative imperfections. See 3.18.64.

friendship: a continuing relationship between two parties characterized by generous transactional relationships based on liking one another.

frustration: dissonance of a degree that arouses a negatively valenced emotion. 2.25.

functionalism: the study of a system with special reference to the role it plays as a subsystem of some larger system. Chapter One. Functionalist analysis is essentially cross-sectional. Being related to hierarchy, it can be applied either to acting or to pattern systems.

fusion (of cultures): see *cultural fusion.* 3.17.0g.

future: see *present.* 2.50bb.

general-purpose analytic tool: a scientific concept or proposition of wide generality. The term is used here solely for the widest degree of generality dealt with in this framework—that is, the most completely nomothetic concept or proposition. An *A*-level analytic tool. Chapter One.

general system: see 1.1.

generosity (by A toward B): a desire by *A*, not necessarily conscious, to help *B* (that is, to increase the level of *B*'s satisfaction or at least to increase the amount of goods he has available). The term is used only operationally in that it relates to the effect of the desire on overt behavior without regard to *A*'s "true" feelings or "ultimate" intent about *B*. See 3.26.0 and 3.26.2.

generous transaction: a transaction in which, in addition to the desires of the parties concerning the things to be exchanged, at least one party wishes to increase the satisfactions of the other via the transaction at hand—or at least to increase the amount of goods received by him. 3.26.0.

gift (by A to B): any valued thing *X* given by *A* to *B* without explicit contingency that *B* will give any *Y* in exchange. 3.36.

good: any external that has utility. 2.50i.

government: the formal and sovereign organization of a whole society. 4.13.10b.
   government staff: those persons who do the government's work. 4.13.10f.
   sponsors of government: see 4.13.10i.
   recipients of government: see 4.13.10j.
   stable government: see *stable government.* 4.13.26.

Gresham's law: A general principle of transactions is that their terms and consummation are determined by four variables—the value of two things to two parties or in two different times, places, markets, and so forth. Gresham's law is an application of the general principle of transactions (see *arbitrage*) to two kinds of money for two different uses, particularly as money and as nonmoney. See Section 3.2.

group: an informal or semiformal organization consisting of few enough persons that each can take explicit cognizance of each of the others. Should they take on an explicit joint task they become a formal organization. Without the joint goal the analysis remains that of communications, transactions, and their many combinations and interrelations. The group can also be viewed as a scale model of informal and semiformal society—which is perhaps what small-group research should be construed to deal with.

guilt: a form of fear—mainly fear of disapproval by others, fear of failure by one's own standards, or fear of punishment.

hegemony: the position of a dominant coalition, with special reference to international politics.

hierarchy of political party: see 4.13.514.

hierarchy of systems: any relation between systems in which one is a subsystem or supersystem relative to another system. The term *hierarchy* is not used here to designate different quantities or levels of essentially the same thing. This book does not speak, for example, of a hierarchy of wealth or of status. 1.20dd.

higher-order conditioning: the process by which A acquires a conditioned connection with C by virtue of no other connection than that A and C have each independently been conditioned to B. By extension, A becomes conditioned to N by virtue of a series of independent paired connections between A and N. 2.09.7.

holist (holism): the study of a system as a whole, functioning unit (in contrast to either functionalist or reductionist approaches). Chapter One. Holist analysis is essentially cross-sectional and can apply either to acting or to pattern systems. It sometimes means *idiographic* but not used thus here.

homeostasis: a condition in which a controlled system successfully maintains a steady-state equilibrium of one or more system variables. 1.20aa.

hostile transaction: a transaction in which, in addition to the desires of the parties concerning the things to be exchanged, at least one party wishes to reduce the satisfactions of the other via the transaction at hand. 3.26.0.

hostility: (by A toward B): a desire by A, not necessarily conscious, to hurt (reduce the level of satisfaction) of B. See 3.26.0 and 3.26.1.

ideal type (from Weber): an analytic model, particularly of the type referred to repeatedly here as the pure case.

identity (sense of): one's concept of his own role or roles. See 4.12.10b and 4.21.2. Viewed overtly, identification might be construed as *participation*.

ideology: a concept of desired goals and structure of a whole society. The concept may include processes and techniques for a transition to the desired state and often attends to levels of social system for which autonomy is desired.

idiographic: analysis of a situation that attends to its unique aspects, the multiplicity of details that make it up, or both. The second aspect is really a variation of the first since it is typically the particular concatenation of details that makes a situation unique. Contrasts with *nomothetic*. See also 3.47 and 6.

illegitimate authority: a situation in which instructions or sanctions from A to B fall outside the categories or terms specified in the major bargain of affiliation, 4.12.10o.

image: another term for an information concept. 2.10k.

impulsive (versus calculated) behavior: see 2.55.2 and *rationality*, 2.6.

incest taboo: a portion of the role specification of certain members of the same family, proscribing sexual relations or marriage between them. See 4.36.2.

incompatibility: a kind of opportunity cost in which, for reasons other than scarcity, X and Y are not both possible—at least at the existing state of technology and under the circumstances of the case. 2.50n.

incompatible decisions: When goods X and Y are incompatible, so that sacrifice of one is an inescapable cost of the other, decisions in favor of both are incompatible. 4.13.515.

individualism: an aspect of an *ideology* which emphasizes maximum autonomy in the lowest subsystem of the social system—namely, the individual.

induction: the communicational, transactional, and suborganizational relationships by which a new member of an organization, who is also necessarily the occupant of at least one role, is brought to acquire system states and behaviors of his role that he does not already possess. 4.12.510.

induction ceremony (role-change ceremony, rite of passage): any collection of behaviors that provide a communication—mutually recognized by the new role occupant and those with whom he will probably interact—that a particular party has been installed in a particular role. 4.12.510.

influence: see also 3.22.8 and 3.33. Compare with *power.*

  intellectual influence: the ability, through communication, to alter the detector of others so that certain things are no longer conceived or perceived as before. 3.15.0.

  moral influence: the ability, through communication, to alter the selector of others—that is, to change their motivational set about certain things. 3.15.0.

informal organization: see *organization, informal.* 4.2.

information: some aspect of a pattern that enables an observer to make inferences about something other than the pattern itself—or at least other than the observed portion of the pattern. 2.04.

  information versus matter-energy: see 1.20ii.

  to contain or possess information: A system is said to contain or possess information about its environment when some state or pattern within it is functionally related, as a dependent variable, to some state or pattern in its environment. 2.04.

information concept: an idea, mental picture, or image; distinguished from a *motor concept.* 2.10k.

inherited: inherited structures or processes are those that uniformly appear in every normal human at birth or by maturation independent of experience. 2.08.0a.

initiation: an *induction ceremony* or some part of it.

input: any movement of matter-energy or information from the environment across the boundaries and into an acting system. 1.20u.

instinct: a genetically preprogramed pattern of explicit and relatively detailed overt behavior set off by an explicit environmental situation and essentially invariant with respect to it. 2.08.0c.

institution: a first-level acting subsystem of the whole society. 4.31.

insubordination: a subordinate's rejection of a legitimate instruction from a superior in an authority relation. Since acceptance of legitimate instructions is a part of the major bargain, such rejection is generally construed as a rejection of the major bargain itself and hence as a ground for terminating the subordinate's membership in the organization.

integrated (social science): see *unified* (science).

integrative processes (integration): The term is not formally used in this volume but might be defined as the total set of interactions between, and contingent behaviors of, the occupants of an organization's roles, whose joint result is to fulfill the organization's goals. Any mutually advantageous interaction. 4.12.5. See also *cohesion*.

intellectual influence: see *influence, intellectual.* 3.15.0.

interaction (of A *and* B*):* a situation in which some change in *A,* through a movement of matter-energy or information, induces some change in *B* or the reverse. 1.20i. See also 3.10b. Among humans interaction means communication, transaction, or both. 3.10b.

social interaction: any interaction that involves contingency. 3.10c.

*interdependence: mutual *dependence.*

interdisciplinary: the application of analytic tools from more than one discipline to the understanding of some problem. Contrast to *unified* (science). Figure 3.

internal power factors: see *power factors, internal.* 3.32.0a.

interrelated transactions: Two transactions are said to be interrelated if the terms of one are affected by the occurence or expectation of the other. 3.28.

intersection: the juncture of two or more patterns to produce a third. See 2.10b.

intuition: the use in the brain of information that is not consciously formulated, or of not-consciously formulated connections between pieces of information, whether or not the pieces are themselves conscious. See 2.50f.

investment: any cost or portion of cost currently incurred not for current benefit but to raise the ratio of benefits to costs at some future time. 2.50x. In psychological terms, the act of deferring gratification in the expectation of raising total eventual gratification as a result.

*involvement: This term must be divided into its overt and its inner aspects. Overtly it is the same as *participation.* Internally it is one's sense of the magnitude of his own EP for transactional interactions, including the transactional aspects of communications, with some identifiable other party. See also *commitment.*

*judgment: a term that might be used for the kind of decision made by a judge when or to the extent that his own values are not involved. His task here concerns the detector functions, first of ascertaining fact and second of categorizing it—for example, as violation or nonviolation of a law or contract. Even if values are involved, as in awards of damages, his conclusion remains a judgment so long as objective, external values are considered. His conclusion becomes a decision in the present sense only when or to the extent that his own values (other than a desire for fairness or justice) become determinants of the outcome.

judicial function: the process of deciding or making a judgment about a dispute between two parties whether an agreement between them, or a decision that is legitimate under some agreement, is being executed according to its terms—the enforcement of legitimacy. 4.13.516h.

*knave: a person presumed to be in bad shape in his selector—as contrasted to a fool, who is in bad shape in his detector.

knowledge: coded information in a human brain. 2.10c.

leadership: the loose parallel in semiformal organization (and possibly in loose forms of simple formal organization) of the sponsor-management role in complex formal organization. 4.32.

legislative function (of government): the process of making decisions on behalf of sponsors for execution by the staff of government. The legislative function is the essence of the sponsor-staff relation. 4.13.516d.

legitimate authority (legitimacy): a situation in which instructions and sanctions from A to B fall within the categories and terms specified in the transaction of affiliation. 4.12.10n.

fully: uses accepted decision rules and rules for changing them. 4.13.36a.

conditionally: uses accepted decision rules but has no accepted rules for changing them. 4.13.36b.

tentatively: has no accepted decision rules but its decisions are accepted. 4.13.36c.

liberty: see *political liberty* and *civil liberty.* 4.13.516f and g.

*like: "A likes B" means that A's concept of B has a positive valence attached.

*linkage (between systems): any interaction between systems in which an output of one becomes an input of the other.

*loyalty: a generalized *commitment* by one party to assist another, especially when it is not in his immediate self-interest to do so.

maintenance system (human): see *biological system.* 2.03a.

main transaction (major bargain): an explicit or implicit agreement that there shall be a continued or repeated relationship of a particular sort. 3.28.5. Regarding government, see 4.13.36d.

major bargain: see *main transaction.* 3.28.5.

major bargain (or transaction of affiliation) in an organization: the agreement that a person will work with others in or as an organization toward some joint goal. 4.10f.

major bargain between sponsors and government: an explicit or implicit transaction of authority in which sponsors agree to accept decisions of government if made in accordance with specified decision rules and government agrees to abide by those rules. 4.13.36d. See also *constitution.*

management (of an organization): the sponsors or any of their agents or subordinates that are given the responsibility to see that actions by staff are directed toward the goals of the sponsors. 4.12.10l.

marginal decision-making: decision-making that attends to the benefits and costs of one additional unit of a good. 2.56.1.

market consensus: see *consensus by transaction.* 3.27.4.

market system: an informal organization for producing and exchanging exchangeable private goods. 4.22.

*marriage: the major bargain between two adults, presumably of different sex, to form the minimum core of a family.

matter-energy: a shortened form for matter, energy, or both. 1.20e.

meaning: 1. the referent of a sign. 3.10.0c. 2. the decoding that follows the detection of sensory inputs in a perception or the detection of signs in a message. See 3.10.0j, 3.11, and related materials. See also transaction as communication, 3.44.

mediated transfer of information: a transfer of pattern from A to B by virtue of its transfer first from A to M (medium) and then from M to B. 3.11.

medium: In a mediated transfer of pattern from A to B, the medium is the matter-energy on which the pattern is imposed by *A* and from which it is subsequently transferred to *B*. 3.12.

member: a person who has made a transaction of affiliation with an organization. 4.10d.

message: an item of new or potentially new information transmitted by signs in a communicational act. 3.10.0i.

mind: a nonexistent entity, as in "Never mind." The process, as contrasted to the structure, of the brain.

mob: a collection of people converted by a commonly perceived and accepted goal and technique into a temporary, loosely structured formal organization. Since productive actions normally require tighter structure and longer-term affiliation, the mob is typically able to perform only destructive actions. However, if the joint effectuation is the simple sum of individual actions, or of easily perceived instrumental interactions, a mob might nevertheless perform productive work like raising an earthen dike against a flood.

moral codes: a collection of concepts and propositions about roles and the interactions of the roles and their occupants as developed by the semiformal organization, with special reference to their approved or disapproved aspects. 4.35.

moral influence: see *influence, moral.* 3.15.0.

morale (that is, "good" morale): a condition in formal organization in which successful achievement of sponsor goals for the organization or suborganization is instrumentally or intrinsically rewarding to members of the staff. It may also be described as a condition in which goals of the informal organization within the formal (4.33) substantially coincide with, or at least do not conflict with, those of the formal organization.

motive: the state of the selector of a controlled system, particularly the human.
primary motive: an inborn valence that reinforces approach or avoidance behavior toward whatever sets it off. 2.21.0a.
secondary motive: a conscious or unconscious concept often but not necessarily corresponding to a pattern in the environment—which concept, by virtue of being conditioned directly or indirectly to a primary motive, is able to reinforce approach or avoidance behavior toward whatever sets *it* off. 2.21.0b. See also 2.21.0c and d regarding intrinsic and instrumental secondary motives.

motor concept: a neural pattern that corresponds to a pattern of overt behavior involving muscles and controls its execution. 2.10l.

mutual contingency: see *contingency, mutual.* 3.10e.

mutual interaction (of A *and* B*):* a situation in which a change induced in B by *A* in turn induces some change in *A*. 1.20j.

myth: a narrative used as part of the *induction* process to help communicate certain detector or selector states to members of a society. See 4.32.

nation: the whole of any society and culture—plus its territory—that includes a sovereign government. 4.13.10h.

nationalism: an aspect of *ideology* that puts strong emphasis on autonomy of the nation as a system, particularly against a supranational system.

negative (equilibrating) feedback: an oppositely paired mutual interaction. 1.20y.

negative good: a *bad.* 3.25.0a.

negotiations: the communications, tactics, and strategies by which *A* and *B* determine whether and on what terms they will conclude a transaction. 3.22.0f.

niche: the logical parallel in ecological systems of *role* in organizations. 4.21.2.

nomothetic: analysis that abstracts things from time and place by focusing on their common aspects. Contrasts with *idiographic.* Chapter One. See also 3.47 and 6.

nonsystem: any element that for purposes of a particular investigation does not change pattern—that is, does not consist of subelements that change relative to one another. 1.20o.

*normative process: the process of achieving consensus by communication. See 3.18.5.

*norms: a subset of the content of culture encompassing mainly attitudes and the performance of roles, with special emphasis on what is expected, approved, and made relatively uniform by the communicational *normative process* and the transactional use of *sanctions.* Norms are to the semiformal organization what role specifications, instructions, and rules are to the formal. See 4.32.

object: a pattern as it exists at a given moment in time. 1.20c.

objectivity: In a straightforward sense objectivity is a condition in which the detector function is not modified by the selector—though the problem is considerably complicated. Objectivity within the detector itself, in the sense of freedom from one's own conceptual set, is presumably impossible. See 2.43.7.

*office: an explicitly designated role in formal organization.

open system: a system that receives inputs from or produces outputs to its environment. 1.20ff.

opportunity function: the set of alternatives perceived to be possible and available in a given decision situation. 2.50t.

oppositely paired interactions: see *negative feedback.* 1.20y.

*organism: any living system that is not a subsystem of any other living system. An ecological system is not "biological" or "living" in this context, nor is an organization.

organization: any interaction of two or more parties in which attention is focused on the overall process and effect rather than on the interaction itself or on its effects on the separate parties. 4.10a.

formal organization: the consciously coordinated action of two or more parties toward the joint effectuation of some goal. 4.10b.

pure formal organization: a formal organization, whether or not it is a subsystem of some larger organization, all of whose detector, selector, and effector functions, including those of its subsystems, are directed toward the goals of the organization as an entity. 4.10c.

simple formal organization: an organization in which all details—such as contributions and withdrawals, goals, and techniques of making transformations—are settled in the major bargain. 4.10e.

complex formal organization: a formal organization characterized by a set of conditions stated in 4.12.0.

informal organization: an uncontrolled system of controlled human subsystems. 4.2. Informal organization is strictly defined by the model of 4.20.

semiformal organization: an organization which is informal in that it has no sponsors, goals, or decisions for the organization as a unit and does not interact as a unit, but formal in that some of its subsystems have an image of structures, roles, goals, and the like and take a sponsorlike interest to the extent of modifying their separate behaviors significantly in the interests of the whole organization. 4.3 and 4.32.

oscillation: the periodic movement of some variable above and below its equilibrium level. See 1.31.3.

output: any movement of information or matter-energy from any acting system across its boundaries to the environment. 1.20v.

overlap (of EP's): any situation in which at least one set of terms of a transaction is acceptable to both *A* and *B*. 3.21.2h.

*overload: a rate of input to a system that is larger than the system is able to handle. "Larger than . . . able to handle" means that a larger input leads to a decrease in efficiency in processing the input and possibly a decrease in the total amount processed. In extreme cases the processing stops completely. See 2.25.3.

*ownership (of X): the right to transform *X* or to give it in a transaction.

parameter (of a system): any trait of a system that is relevant to a particular analysis but does not change during the course of the analysis. 1.20gg.

parameter (of a system's environment): any trait of a system's environment that is relevant to a particular analysis but does not change during the course of the analysis. 1.20hh.

*parasitism: the ecological equivalent of *theft* between two organisms incapable of value-based transactions. When the term is used figuratively among humans it apparently refers to the continued receipt of gifts by one who is well able to get along without them.

*participation: Where *A* is the group and *B* an individual, participation by *B* consists of communications and transactions between *A* and *B* initiated by *B*. The obverse of *acceptance.*

party: a person, collectivity of persons, or organization engaging in an interaction. 3.10a. See also 3.20.

party (political): see *political party.* 4.13.514.

past: see *present.* 2.50bb.

pattern: any relationship of two or more elements. 1.20b.

pattern formation: see *concept-learning.* 2.10h.

*pattern maintenance: any of a variety of forces that lead to stability and equilibrium.

pattern recognition: see *perception.* 2.10m.

pattern system (nonacting or nonbehaving system): a pattern, two or more elements of which are mutually consistent or interdependent within some criterion of some acting system. 1.20k. The degree of systemness, according to some criterion of consistency.

*peer group (of A): *A*'s associates who have his general characteristics— especially those of about the same age during *A*'s youth. One's peers are those of about the same aggregate power; one's peer group is those members of one's own class with whom he associates.

perception: the process by which uncoded sensory inputs from some pattern in the environment selectively activate a particular coded pattern (concept) or group of patterns already learned and stored in the cortex and lead to an inference that an instance of that pattern currently exists in the environment. 2.10m.

perception versus communication: see 3.10.0j.

persistence (perceptual): Since perception resides in the identification and activation of an image already stored in the head, the same image may be activated regardless of the lighting, angle of view, distance, clothing on a person, and the like. Parallel statements could be made about other sensory modalities. See 2.11.

*personality: an idiographic pattern system encompassing the total set of system states and system interactions of the behavioral system of a particular individual—especially the portions of the total pattern that are both durable and unique to the individual.

personality types: see the discussion of simulation models of personality in 7.3.

physical blockage: see *blockage.* 3.25.0c.

plain power: When the distinction between power and bargaining power might not be clear, the adjective *plain* is added to the former. 3.22.0d.

political liberty: the freedom of citizens to exercise their sponsor function in the sponsor-staff relation of a cooperative government. 4.13.516f.

political party: a coalition of office-seekers and their supporters who wish to increase their power over government decisions. 4.13.514.

*politics: Depending on the emphasis one chooses, politics can be described variously in present terminology as follows: the exercise or study of interpersonal power; the process of achieving a group decision, with special emphasis on the dominant coalition (as in 4.12.42); the study of formal organization, with special attention to governments and their interactions; some combination of these definitions.

positive (nonequilibrating) feedback: a similarly (not oppositely) paired mutual interaction. 1.20z.

*poverty: a condition of small aggregate interpersonal power, particularly with reference to exchangeable goods.

power: the ability to bring about desired external states. 2.90a.

 actual aggregate power: To the extent that A's power factors exist independently of the images of them, his status may lie above or below his actual power. Actual aggregate power designates such independently existing power. 3.35.0.

 aggregate power: the ability to bring about some total quantity or level of such states, $Y_1 + Y_2 + \cdots Y_n$, considering both their number and individual magnitudes. 2.90e and 3.22.0c.

 bargaining power: the ability to get a particular good Y from others on good terms—that is, by giving relatively little in exchange. 3.22.0d.

 interpersonal power: the ability to induce others to bring about overt states one desires. 2.90c and 3.22.0a.

 intrapersonal power: the ability to bring about desired external states by oneself—to effectuate productive transformations. 2.90b.

 moral and intellectual power: the application of moral or intellectual *influence* toward one's interpersonal power. 3.22.8 and 3.33.

power factors: a convenient term for the EP's of both parties to a transaction, along with strategic modifications of them. 3.21.20i.

 internal power factors (of A*):* system states or characteristics of A (knowledge, skill, physical strength, willpower, beauty, friendliness, courage, ruthlessness) that give him power. 3.32.0a.

external power factors (of A*):* things or conditions outside *A* that give him power whether or not he is able to control them. These consist of material power factors and system states of others. 3.32.

*predisposition: a set of system states such that a given input is more likely to lead to one output than to another.

prefer: to prefer *X* to *Y* means that *X* is higher than *Y* in the relevant preference ordering. 2.50k.

preference function: the preference ordering of the decision-maker with respect to the alternatives he perceives to be available in a particular decision situation. 2.50s. A preference function is a selector state of the system.

preference ordering: a list of two or more items arranged in order of the strength of the valences attached to them with the strongest positive valance highest. 2.50j. A preference ordering is a selector state of the system.

*prejudice: 1. detector aspects: In the broadest sense prejudice is the perception of externals through the conceptual set already in the head and is thus inescapable (see discussion of objectivity in 2.43.7). More narrowly, a modification of detector by selector processes so that unfavorable traits of disliked persons or categories are attended to more reliably than favorable traits and vice versa. 2. selector aspects: a dislike of and hostility toward a category of persons without regard to the traits of the individual member.

present (in decision theory): a dimensionless line separating past from future. 2.50bb. See also *currently.*

pressure organization: a type of formal organization *C* whose outputs go directly or indirectly to recipient *B* for the purpose of improving the position of *A* in a transactional relation with B. 4.12.72d.

pressure transaction: a three-party relationship in which *C* employs tactics or strategies on *B* to improve the power or bargaining power of *A* in an *A-B* transactional relation. The full model includes a transactional relation between *A* and *C* and possibly between *B* and *C* as well. 3.27.1.

pretransactional: the unilateral, intrapersonal, transformational stage of creating goods or stress—in contrast to the subsequent interpersonal, or social, exchange of goods or relief of stress. 3.2 and 3.25.131.

private goods: goods that can be acquired and used by some parties whether or not others have the same good. 4.13.10c.

problem: any situation requiring a decision. 2.50e.

*process: the behaviors of the subsystems of a system, particularly their interactions and the resulting mutual modifications thereof. As contrasted to structural change, see 5.21.

production: any transformation that increases utility. 2.50ee.

profit organization: a type of formal organization run in the interests of its sponsors—who, except incidentally, are not the same people as the recipients and whose outputs go to recipients on the basis of selfish transactions. 4.12.72b.

*property: that which is owned by a party. See *ownership.*

*property rights: a set of rules about transformations or transactions of property along with rules of evidence of ownership.

public: all persons within the sovereign area of a government, but including only the nonstaff aspects of government staff. 4.13.10e.

public goods: goods whose acquisition is necessarily the same for all persons in a society or subdivision whether they individually want them or not. "Shared values" are logically public goods for small groups such as families. 4.13.10d.

random: in this framework, unplanned and undirected—not a mathematically prescribed situation like that of probability theory. 2.02.

rationality (rational behavior): the act or process of selecting the preferred alternative in an opportunity function. 2.50u.

  subjective rationality: the selection of the alternative that is rational given the dominant motive or impulse at the moment of decision, in light of whatever detector and selector functions then prevail. 2.50u.

  objective rationality: a selection that is rational given explicit assumptions about opportunity and preference functions, consciously related. 2.50u.

real system: an acting or pattern system whose elements consist of matter-energy. 1.20m.

receiver: the point or system to which a pattern is transferred. 3.12.

recipients (of an organization): those parties who receive the organization's outputs. 4.12.10a.

reductionist: analysis of a system with special reference to its components and their interactions. Reductionist analysis of a system may or may not require a different kind of science than *holist* analysis. Reductionist analysis is essentially cross-sectional and can be applied either to acting or to pattern systems. Chapter One.

*reference group: that society or subsociety whose culture one has adopted, which guides his behavior, and which may or may not be the same society of which he is a member.

referent: the pattern represented by a sign; the meaning of the sign. 3.10.0c. See also 3.10.0h and l.

reflex: an inborn specific overt response (involving muscles) to the excitation of specific sets of sensory nerves. 2.08.0b.

regression: decay of higher-level patterns in a hierarchy of neural patterns, leaving only lower-level patterns. 7.3.

regulatory powers (of government): the process by which a government C acts as a pressure organization on B in order to improve the power or bargaining power of A in transactional relations between A and B. The process may also be described as that of joining a coalition with A against B. A and B for this purpose may be particular parties or categories of parties. See 4.13.43.

reinforce: to strengthen, by conditioning, a connection between an environmental stimulus and a motivated or reflexive response. We extend the term to include single instances of cognitively guided behavior. 2.09.5.

residual (contributions to and withdrawals from a formal organization): contributions of inputs to or withdrawals of outputs from an organization that are not specified by prior agreement. 4.12.85.

resolution (of conflict): the making of a decision that accepts those costs which, before the decision, constituted the *conflict.* 2.50q. The discovery of an additional alternative response that reduces or eliminates costs is included within the meaning of resolution. See also 7.2.

*respect (of A for B): 1. B's status in A's eyes. 2. A's approval of B's behavior or accomplishment. Although some behavior is defined here only with reference to status, either type of respect might lead to deferential be-

havior. If so, respect is A's detector and selector states regarding B whereas deference is the effector process that reflects them.

response selection: the process of determining which of two or more alternative behavioral responses, including inaction, the system will perform. 2.50a. A simple response selection is any response selection that does not qualify as a *decision.* 2.50b.

responsibility: the position of a party on whom authority is exercised. 4.12.10i.
delegation of responsibility: from the definitions of authority, responsibility, and delegation of authority, it is clear that when authority is delegated from A to B, C is responsible to B to do work for whose performance B is responsible to A. 4.12.10k.

*revolt: sustained and repeated insubordination—particularly by subordinates who are not free to terminate their major bargain with an organization and leave it (citizens of a government or children in a family).

*right: the *freedom* under some *main transaction* to do or receive something desired on terms or conditions specified in the main transaction.

*rituals: acts that symbolize (and hence communicate) and affirm certain attitudes (intrasystem) or relationships (intersystem), whether with one's fellowmen or with his deities. See also *induction ceremony,* 4.12.510.

rite of passage: see *induction ceremony,* 4.12.510.

role: the set of system states and actions of a subsystem of an organization including its interactions with other systems or nonsystem elements. 4.12.10b. A role is a pattern system.
role specification (description): a statement or description of the content of a role. 4.12.10c.
role occupant: the person or other acting system that effectuates the behavior specified in the role. 4.12.10d.
role in informal organization: same as above. See 4.21.2.
*ascribed versus achieved role: see 4.21.2.

rules (of an organization): that portion of a role specification which deals with the kinds and terms of interactions to be engaged in by a role occupant—particularly those aspects that are common to all, or to a substantial category, of roles. 4.12.10b.

*ruling class: see *elite* (2).

sanctions: rewards or punishments used in the exercise of authority. 4.12.10h.

satisfaction: the positive-valence state of the selector from achieving a positively valenced result or avoiding or decreasing a negative one. 2.50g.

scale-model societies: see 4.34. For a discussion of the family as a scale-model society, see 4.36.

scarcity: a kind of *opportunity cost* in which X and Y are both possible (not incompatible), but the decision-maker lacks sufficient power to acquire both. 2.50n.

*security (sense of): a situation in which one perceives himself as having sufficient aggregate power to achieve or maintain those conditions that are important to him.

selector (of a controlled system): the function of selecting a response to a given environmental state. 2.05b.

*self-actualization (fulfillment, expression, and the like): the exercise of one's system faculties at or near optimal level. See 2.41.24.

selfish: short for selfish-indifferent. The attitude or stance of a party in which he tries to maximize his own benefit from a transaction—to give as little and to receive as much as possible—and in which he is indifferent to the position of the other, wishing neither to help nor to hurt the other. That is, neither party feels *hostility* or *generosity* toward the other. 3.21b.

semantic information: see *coded information.* 2.10b.

semantic sign: see *sign, semantic.* 3.10.0e.

semiformal organization: see *organization, semiformal.*

sensation (sensory inputs): actuation of sensory nerves of the external modalities or actuation of any nerves that convey information about the biological system into the central nervous system. 2.08.0f and 2.10d.

service organization: a type of formal organization whose outputs go to the recipients on the basis of generous transactions. 4.12.72c. For a discussion of government as service organization, see 4.13.53.

sign: a pattern transmitted in a communicatoin that stands for some other pattern, which is its meaning. 3.10.0b.

    semantic sign: a sign whose referent is an information concept. 3.10.0e.

    syntactic sign: a sign that represents a relationship among semantic signs and possibly the mode of intersection of their referents. 3.10.0f.

simulation (simulation model): a single-purpose analytic tool designed to analyze a particular situation for a particular purpose. Chapter One and Section 6.

*social: involving the interaction of two or more persons.

    social pressure (or control): the exercise of authority in semiformal organization. 4.32.

    *social engineering: To the extent that the social system is construed to be *informal* and *semiformal*, social engineering presupposes formalizing some aspects of it so that its structure, processes, and goals may be directed toward "improving" the society.

    social interaction: see *interaction, social.* 3.10c.

    *social mobility: the flow of persons among roles, classes, or status positions. See 4.23.

    social structure: see *structure* (of an organization). 4.12.10e and 4.21.2.

    social system: broadly, any organization—formal, informal, or intermediate. More explicitly, the semiformal organization of the whole society. 4.31.

*socialization: induction into informal or semiformal organization. 4.12.510. The name given to the total body of communications and transactional and organizational behaviors that communicate the body of a culture to a particular individual and by sanctions enforce some degree of conformity to it.

society: a collectivity of people having a common content and process of culture. 3.17.0d.

source: the point from which a communication originates. 3.12.

sovereignty: the authority position of a formal organization (and conceivably of an individual, such as a king) that is not a subsystem of any larger organization whose authority over it or its subsystems includes legitimate use of force, but whose own authority over its own subsystems does include legitimate force. 4.13.10a.

sponsors (of an organization): the persons whose motives concerning it constitute the goals of the organization as a unit. 4.12.10a.

stability: the continuance of a system with neither *emergence* nor *decay.* Stability presupposes *equilibrium.* Chapter One.

stable government: a condition in which neither the government nor the public makes any significant fraction of its actions contingent on the possibility that the government (as contrasted to particular staff members in it) will be replaced. 4.13.26.

staff (of an organization): those persons who effectuate the transformations that constitute the organization's function—that is, the transformations that are instrumental to achievement of the organization's goals. 4.12.10a.

stake: any valued thing S other than X or Y, access to which is contingent on performance of some act relative to a transaction. 3.24.3a.

*state: a nation, its people, and government.

state (of a system): a condition of some system variable. 1.20q. See also 2.03b and e.

static equilibrium: a situation in which the forces acting on or exerted by all components in an acting system have come to a state of rest and no further change takes place. 1.20w.

status: apparent or perceived aggregate power—the images of A's aggregate power in the heads of other persons. 3.35.0. See also *ego image.*

steady-state equilibrium: see *dynamic equilibrium.* 1.20x.

stock: the matter-energy or information within a system, or any arbitrary portion of it, at a given moment of time. See 4.23. Compare with *flow.* See also 3.22.8 and 3.3.

strategy (by A): any attempt by A to modify the EP's in a transaction—that is, to change the selector or effector states of either party. 3.24.0.

*stratification (social): a term that describes the fact that individuals have different amounts and types of power and that for convenience those of roughly similar amounts and types can be categorized into groups which can be given names. This book does not provide a system for categorizing these differences. It uses only the generic term *class* and in Sections 3.3 and 4.23 attempts to explain differences in class and give reasons why persons in a given class may find it advantageous to close it to upward mobility from lower classes. The cultural process would assist in maintaining class differences, as would the fact that certain kinds of power are transferable from one generation to the next.

stress: an act or situation that constitutes a bad when applied to someone in a stress transaction and generates a desire that it be removed. 3.25.0b.

stress transaction: a transaction in which a bad is applied unilaterally by A to B prior to negotiations (or to a particular state thereof) along with the express or implied promise that it will be relieved if, and only if, B does as A requests. 3.25.0f.

strong competitor: a B who has a strong likelihood of consummating a transaction with A as compared to some other B with a smaller likelihood (a weak competitor). 3.27.30.

structure (of an organization): the pattern of an organization described in terms of its subsystems and their roles. 4.12.10e.

structure of informal organization: same as above. See 4.21.2. That definition pertains also to semiformal organization.

submission: see *dominance.* 3.37.

subordinate: see *superior* and *subordinate.* 4.12.10m.

subsidiary transaction (or bargain): a transaction over some detail of a continuing relationship on the assumption that the *main transaction* will continue. 3.28.5.

subsystem: a system that is a component of some larger system. 1.20bb.

sunk costs and benefits: costs and benefits or parts thereof that cannot be changed by the decision at hand. 2.50v.

superior (and subordinate): a pair of persons *A* and *B* who hold authority and responsibility, respectively, in an organization under the same transaction of authority. 4.12.10m.

*superorganic system (in the sense of Spencer and Kroeber): a system consisting of multiple organisms—for humans, an organization as here defined.

supersystem (suprasystem): the larger system of which a given system is a component. 1.20cc.

*supportive: Behavior of *A* that is supportive of *B* consists of a generous stance by *A* in transactions with *B,* including the transactional aspects of his communications. The latter would presumably strengthen *B*'s ego image.

symbol: any pattern produced by a system (person) in a medium external to itself, which pattern corresponds in some way to an internal pattern. 3.10.0d.

*symbiosis: Among humans, symbiosis can be said to exist in any mutually advantageous transactional relation. Symbiosis includes the communications that facilitate transactions or that are themselves transactional by virtue of having valued content. For other species symbiosis is also transactional in a rudimentary way but notably lacks the conceptual and communicational skills of humans that are prerequisites for the negotiation of terms and hence for the application of transactional analysis.

*synergic (synergistic): a joint effect of multiple, independent actions but not intended or capable of being produced by any action taken separately. Among humans, informal organization epitomizes synergistic interaction.

syntactic sign: see *sign, syntactic.* 3.10.0f.

*syntactic system: see *pattern system.* 1.20k.

system: any pattern whose elements are related in a sufficiently regular way to justify attention. 1.20f.

>*system variable:* see *variable* of a system. 1.20p.

>*system state:* see *state* of a system. 1.20q.

tactics (by A*):* a communicational adjunct to transactions in which *A* seeks to modify perceptions of EP's—that is, to change the detector states of either party regarding them. 3.23.0.

tension: a state of conflict, and hence of dissonance, that has given rise to negatively valenced feelings. 2.50r.

*theft: a "gift" transaction initiated by the receiver without the consent of the "giver," presumably contrary to his wishes and often without his knowledge.

thermodynamics, second law of: Generalized for application to all kinds of systems, this law states that all closed systems are subject to loss of differentiation. 1.33 and 5.21.1.

threat (transaction): an express or implied promise made by *A* that a bad will be applied unilaterally to *B* if and only if *B* does not do as *A* requests. The threat is simultaneously a promise *not* to apply the bad if *B* does do as *A* requests. 3.25.0g.

*threatening (psychological sense): Any event that is perceived as likely to produce a significant loss of one's power or status may be said to be threatening. The loss presumably occurs through the deterioration of one's internal or external power factors (3.32.0) or one's beliefs about them. See also *ego image* and *ego defense.*

time preference: the relative strength of desires for present as contrasted to future satisfactions. 2.50cc.

*traditional society: a society that approximates a perfectly transmitted culture—though with differentiated roles.

transaction: any interaction between parties analyzed with reference to its value content to the parties; or a transfer of anything of value between parties. 3.20.

transaction costs and benefits: the costs or benefits of the transactional process —as contrasted to the costs of giving up an item in exchange for another or the benefits from the item received. 3.26.3–4.

transformation: any effector process that alters the environment or the system's relation to it. 2.50dd.

truthful: A message is considered truthful if it seeks to communicate the source's images correctly—without regard to the relation between those images and reality. See 3.17.6.

uncoded information: patterns in matter-energy which, by virtue of having been imposed by some other matter-energy, may be said to contain or constitute information about the latter. 2.10a.

*unconscious: Unless otherwise indicated, the term means simply "not conscious," with no Freudian implications of repression.

uncontrolled system: any acting system that does not fill the definition of a *controlled system.* 1.20t.

unified (science): the extraction of the general-purpose analytic tools that are common to more than one discipline. Conversely, the construction of multiple disciplines on a common analytic base. Figure 3. Contrast to *interdisciplinary.*

utility: the satisfaction that can be produced by some external when acquired or achieved; the ability of some external to create satisfaction. 2.50h.

valence: a property of the nervous system that reinforces approach or avoidance responses and typically, but not necessarily, produces what we call pleasant or unpleasant feelings. 2.08.0g.

value: the position of anything in a preference ordering. 2.50m. The preference ordering can be either individual or social.

variable (of a system): any element in an acting system that can take at least two distinguishably different states. 1.20p.

variable costs: any cost or portion of cost currently incurred to produce current benefit. 2.50w.

voluntary: not coerced. See 3.25.0e.

Bibliography

ACKOFF, R. L. "Systems, Organizations, and Interdisciplinary Research." *General Systems,* July 1960, *5,* 1–8.

ARROW, K. J. *Social Choice and Individual Values.* New York: Wiley, 1951.

ASHBY, W. R. *An Introduction to Cybernetics.* New York: Wiley, 1958.

BARRINGER, H. R., BLANKSTEN, G. I., and MACK, R. W. (Eds.) *Social Change in Developing Areas.* Cambridge, Mass.: Schenkman, 1965.

BERELSON, B., and STEINER, G. A. *Human Behavior: An Inventory of Scientific Findings,* New York: Harcourt Brace Jovanovich, 1964.

BERRIEN, F. K. *General and Social Systems.* New Brunswick, N.J.: Rutgers University Press, 1968.

BERTALANFFY, L. VON. *General System Theory.* New York: Braziller, 1968.

BLAU, P. M. *Exchange and Power in Social Life.* New York: Wiley, 1964.

BLAU, P. M., and SCOTT, W. R. *Formal Organizations.* New York: Intext, 1962.

BOORMAN, S. A. "Analogues in the Social Sciences." *Science,* Oct. 27, 1972, 391–394.

BOULDING, K. E. "Welfare Economics." In B. J. Haley (Ed.), *Survey of Contemporary Economics.* Homewood, Ill.: Irwin, 1952.

BOULDING, K. E. *Economic Analysis.* (3rd ed.) New York: Harper and Row, 1955.

BOULDING, K. E. *The Image.* Ann Arbor: University of Michigan Press, 1956.

BOULDING, K. E. *The Skills of the Economist.* Cleveland: Howard Allen, 1958.

BOULDING, K. E. *Conflict and Defense: A General Theory*. New York: Harper and Row, 1962.

BOULDING, K. E. "The Verifiability of Economic Images." In S. R. Krupp (Ed.), *The Structure of Economic Science*. Englewood Cliffs, N.J.: Prentice-Hall, 1966.

BOULDING, K. E. *Beyond Economics*. Ann Arbor: University of Michigan Press, 1968a.

BOULDING, K. E. "The Legitimation of the Market." *Nebraska Journal of Economics and Business*, Spring 1968b, 7 (1), 3–14.

BOULDING, K. E. "General Systems Theory—The Skeleton of Science" (1956). Reprinted in W. Buckley (Ed.), *Modern Systems Research for the Behavioral Scientist*. Chicago: Aldine, 1968c.

BOULDING, K. E. *Economics as a Science*. New York: McGraw-Hill, 1970.

BREGER, L. "Motivation, Energy, and Cognitive Structure in Psychoanalytic Theory." In J. Marmor (Ed.), *Modern Psychoanalysis: New Directions and Perspectives*. New York: Basic Books, 1968.

BROOM, L., and SELZNICK, P. *Sociology: A Text with Adapted Readings*. New York: Harper & Row, 1957.

BUCHANON, J. N. "An Individualistic Theory of Political Process." In D. Easton (Ed.), *Varieties of Political Theory*. Englewood Cliffs, N.J.: Prentice-Hall, 1966.

BUCKLEY, W. *Sociology and Modern Systems Theory*. Englewood Cliffs, N.J.: Prentice-Hall, 1967.

BUCKLEY, W. *Modern Systems Research for the Behavioral Scientist*. Chicago: Aldine, 1968.

CALHOUN, D. "History and Theory: The Need for Decadence." In M. Sherif and C. W. Sherif (Eds.), *Interdisciplinary Relationships in Social Sciences*. Chicago: Aldine, 1969.

CAMPBELL, D. T. "Blind Variation and Selective Retention in Creative Thought as in Other Knowledge Processes." *General Systems*, 1962, 7, 57–70.

CAMPBELL, D. T. "Variation and Selective Retention in Socio-Cultural Evolution." In H. R. Barringer, G. I. Blanksten, and R. W. Mack (Eds.), *Social Change in Developing Areas*. Cambridge, Mass.: Schenkman, 1965.

CARTTER, A. *Theory of Wages and Employment*. Homewood, Ill.: Irwin, 1959.

CARTWRIGHT, D. "The Nature of Group Cohesiveness." In D. Cartwright and A. Zander (Eds.), *Group Dynamics: Research and Theory*. (3rd ed.) New York: Harper and Row, 1968.

CARTWRIGHT, D., and ZANDER, A. (Eds.) *Group Dynamics: Research and Theory*. (3rd ed.) New York: Harper and Row, 1968.

CHAMBERLAIN, N. W. *A General Theory of Economic Process.* New York: Harper & Row, 1955.

CHARNES, A., and COOPER, W. W. "The Theory of Search: Optimum Distribution of Search Effort." *Management Science,* 1958, *5,* 450–458.

CHERRY, C. *On Human Communication.* New York: Wiley, 1957.

CHURCHMAN, C. W. *The Systems Approach.* New York: Dell, 1968.

COFER, C. N., and APPLEY, M. H. *Motivation: Theory and Research.* New York: Wiley, 1964.

COLINVAUX, P. A. *Introduction to Ecology.* New York: Wiley, 1973.

DAHL, R. A., and LINDBLOM, C. E. *Politics, Economics, and Welfare.* New York: Harper & Row, 1953.

DEUTSCH, K. W. *The Nerves of Government.* New York: Free Press, 1963.

DEUTSCH, K. W., PLATT, J., and SENGHAAS, D. "Conditions Favoring Major Advances in Social Science." *Science,* Feb. 5, 1971, 450–459.

DIRENZO, G. J. (Ed.) *Concepts, Theory, and Explanation in the Behavioral Sciences.* New York: Random House, 1966.

DOWNS, A. *An Economic Theory of Democracy.* New York: Harper & Row, 1957.

DUNN, E. S. *Economic and Social Development: A Process of Social Learning.* Baltimore: Johns Hopkins, 1971.

EASTON, D. *A Framework for Political Analysis.* Englewood Cliffs, N.J.: Prentice-Hall, 1965.

EASTON, D. "Categories for the Systems Analysis of Politics." In D. Easton (Ed.), *Varieties of Political Theory.* Englewood Cliffs, N.J.: Prentice-Hall, 1966a.

EASTON, D. (Ed.) *Varieties of Political Theory.* Englewood Cliffs, N.J.: Prentice-Hall, 1966b.

EMERSON, R. M. "Power-Dependence Relations." *American Sociological Review,* Feb. 1962, *27* (1), 31–41.

ETZIONI, A. *A Comparative Analysis of Complex Organizations.* New York: Free Press, 1961.

ETZIONI, A. *The Active Society: A Theory of Societal and Political Processes.* New York: Macmillan, 1968.

ETZIONI, A. *A Sociological Reader on Complex Organizations.* New York: Holt, 1969.

FESTINGER, L. *A Theory of Cognitive Dissonance.* Evanston, Ill.: Row, Peterson, 1957.

FIRTH, R. (Ed.) *Themes in Economic Anthropology.* London: Tavistock, 1967.

FISK, G. *Marketing Systems: An Introductory Analysis.* New York: Harper & Row, 1967.

GOFFMAN, E. *Encounters.* New York: Bobbs-Merrill, 1966.

GOLD, M. "Power in the Classroom." In D. Cartwright and A. Zander (Eds.), *Group Dynamics*. New York: Harper & Row, 1968.

GRINKER, R. R. "Conceptual Progress in Psychoanalysis." In J. Marmor (Ed.), *Modern Psychoanalysis: New Directions and Perspectives*. New York: Basic Books, 1968.

HARE, A. P. *Handbook of Small Group Research*. New York: Free Press, 1962.

HARRIS, M. *The Nature of Cultural Things*. New York: Random House, 1964.

∨ HAWLEY, A. H. *Human Ecology: A Theory of Community Structure*. New York: Ronald, 1950.

HOLSTI, K. J. *International Politics: A Framework for Analysis*. Englewood Cliffs, N.J.: Prentice-Hall, 1967.

HOMANS, G. C. *Social Behavior: Its Elementary Forms*. New York: Harcourt Brace Jovanovich, 1961.

HOMANS, G. C. *The Nature of Social Science*. New York: Harcourt Brace Jovanovich, 1967.

JAYNES, E. T. "Information Theory and Statistical Mechanics." *The Physical Review*, May 15, 1957, *106* (4), 620–630; Oct. 15, 1957, *108* (2), 171–190.

JORDON, N. *Themes in Speculative Psychology*. London: Tavistock, 1968.

KABRISKY, M. *A Proposed Model for Visual Information Processing in the Brain*. Chicago: University of Illinois Press, 1966.

KAPLAN, A. *The Conduct of Inquiry*. New York: Intext, 1964.

KUHN, A. "Toward a Uniform Language of Information and Knowledge." *Synthese*, June 1961, *8* (2), 127–153.

KUHN, A. *The Study of Society: A Unified Approach*. Homewood, Ill.: Irwin, 1963.

KUHN, A. *Labor: Institutions and Economics*. (2nd ed.) New York: Harcourt Brace Jovanovich, 1967.

KUHN, A. "Synthesis of Social Sciences in the Curriculum." In I. Morrissett and W. W. Stevens, Jr. (Eds.), *Social Science in the Schools: A Search for Rationale*. New York: Holt, 1971a.

KUHN, A. "Types of Social Systems and System Controls." In *Man in Systems* (1968 Proceedings of the Society for General Systems Research). New York: Gordon and Breach, 1971b.

KUHN, T. S. *The Structure of Scientific Revolutions*. (2nd. ed.) Chicago: University of Chicago Press, 1970.

LAING, R. D. *Knots*. New York: Pantheon, 1970.

LANDAUER, C. "Toward a Unified Social Science." *Political Science Quarterly*, Dec. 1971, *86* (4), 563–585.

LANGE, O. "The Scope and Method of Economics." *Review of Economic Studies,* June 1934, *1,* 19–32.

LASSWELL, H. D., and KAPLAN, A. *Power and Society.* New Haven: Yale University Press, 1950.

LENNARD, H. L., and BERNSTEIN, A. *The Anatomy of Psychotherapy.* New York: Columbia University Press, 1960.

LEVI-STRAUSS, C. *Structural Anthropology.* New York: Basic Books, 1963.

LEWIN, K. *Field Theory in Social Science.* New York: Harper & Row, 1951.

LINTON, R. *The Study of Man.* New York: Appleton-Century-Crofts, 1936.

LIPSET, S. M., TROW, M. A., and COLEMAN, J. S. *Union Democracy: The Internal Politics of the International Typographical Union.* Glencoe, Ill.: Free Press, 1956.

MACARTHUR, R. H. *Geographical Ecology: Pattern in the Distribution of Species.* New York: Harper & Row, 1972.

MACK, R. W. *"Theoretical and Substantive Biases in Sociological Research."* In M. Sherif and C. W. Sherif (Eds.), *Interdisciplinary Relationships in the Social Sciences.* Chicago: Aldine, 1969.

MAINE, H. *Ancient Law.* New York: Holt, 1906.

MARCH, J. G. "The Power of Power." In D. Easton (Ed.), *Varieties of Political Theory.* Englewood Cliffs, N.J.: Prentice-Hall, 1966.

MARCH, J. G., and SIMON, H. A. *Organizations.* New York: Wiley, 1958.

MARGENAU, H. "What Is a Theory?" In S. R. Krupp (Ed.), *The Structure of Economic Science.* Englewood Cliffs, N.J.: Prentice-Hall, 1966.

MARUYAMA, M. "The Second Cybernetics: Deviation-Amplifying Mutual Causal Processes." (1963) Reprinted in W. Buckley (Ed.), *Modern Systems Research for the Behavioral Scientist.* Chicago: Aldine, 1968.

MERRIAM, C. E. *Political Power.* New York: Whittlesey House, 1934.

MICHELS, R. *Political Parties.* (1911) Glencoe, Ill.: Free Press, 1949.

MILLER, G., GALANTER, E., and PRIBRAM, K. *Plans and the Structure of Behavior.* New York: Holt, 1960.

MILLER, J. G. "Living Systems." Three articles in *Behavioral Science,* 1965, *10* (3, 4).

MILLER, J. G. "Living Systems: The Organization." *Behavioral Science,* 1972, *17* (1).

MOORE, W. E. *Social Change.* Englewood Cliffs, N.J.: Prentice-Hall, 1963.

NADEL, S. F. "Social Control and Self Regulation." *Social Forces,* 1953, *31,* 265–273.

OLDS, J. "Physiological Mechanisms of Reward." *Nebraska Symposium of Motivation, 1955.* Lincoln: University of Nebraska Press, 1955.

PARSONS, T. "The Political Aspect of Social Structure and Process." In D. Easton (Ed.), *Varieties of Political Theory.* Englewood Cliffs, N.J.: Prentice-Hall, 1966.

PARSONS, T., and SMELSER, N. *Economy and Society.* New York: Free Press, 1956.

PEN, J. "A General Theory of Bargaining." *American Economic Review,* March 1952, *42* (1), 24–42.

PEPITONE, A. *Attraction and Hostility.* New York: Atherton, 1964.

PIMENTEL, D. "Complexity of Ecological Systems and Problems in Their Study and Management." In K. E. F. Watt (Ed.), *System Analysis in Ecology.* New York: Academic, 1966.

PINES, M. *The Brain Changers.* New York: Harcourt Brace Jovanovich, 1973.

POLANYI, M. *The Tacit Dimension.* New York: Doubleday, 1966.

POLSBY, N. W. *Community Power and Political Theory.* New Haven: Yale University Press, 1963.

POORE, M. E. D. "Integration in the Plant Community." In A. MacFadyen and P. J. Newbould (Eds.), *British Ecological Society Jubilee Symposium.* Supplement to the *Journal of Ecology,* 1964, *52.*

POTTER, V. R. *Productive Living in the Society.* Unpublished paper delivered at a symposium on "Changing Concepts of Productive Living." University of Wisconsin School of Education, June 1966.

PRINGLE, J. W. S. "On the Parallel Between Learning and Evolution." (1951) Reprinted in W. Buckley (Ed.), *Modern Systems Research for the Behavioral Scientist.* Chicago: Aldine, 1968.

RAPOPORT, A. "Some System Approaches to Political Theory." In D. Easton (Ed.), *Varieties of Political Theory.* Englewood Cliffs, N.J.: Prentice-Hall, 1966.

RIKER, W. *The Theory of Political Coalitions.* New Haven: Yale University Press, 1962.

ROBINSON, J. A., and MAJAK, R. R. "The Theory of Decision Making." In J. C. Charlesworth (Ed.), *Contemporary Political Analysis.* New York: Free Press, 1967.

ROTHENBERG, J. *The Measurement of Social Welfare.* Englewood Cliffs, N.J.: Prentice-Hall, 1961.

ROTHENBERG, J. "Values and Value Theory in Economics." In S. R. Krupp (Ed.), *The Structure of Economic Science.* Englewood Cliffs, N.J.: Prentice-Hall, 1966.

RUSSELL, B. *Power: A New Social Analysis.* New York: Norton, 1938.

RUSSETT, C. E. *The Concept of Equilibrium in American Social Thought.* New Haven, Conn.: Yale University Press, 1966.

SAHLENS, M. D. "On the Sociology of Primitive Exchange." In M. Banton (Ed.), *The Relevance of Models for Social Anthropology.* London: Tavistock, 1965.

SAMUELSON, P. A. *Foundations of Economic Analysis.* (1947) New York: Atheneum, 1965.

SAMUELSON, P. A. (8th ed.) *Economics.* New York: McGraw-Hill, 1970.

SCHELLING, T. C. "An Essay on Bargaining." *American Economic Review,* 1956, *46* (3), 281–306.

SCHRÖDINGER, E. "Order, Disorder, and Entropy." (1945) Reprinted in W. Buckley (Ed.), *Modern Systems Research for the Behavioral Scientist.* Chicago: Aldine, 1968.

SHANNON, C., and WEAVER, W. *The Mathematical Theory of Communication.* Urbana: University of Illinois Press, 1949.

SHERIF, M. *The Psychology of Group Norms.* New York: Harper & Row, 1936.

SHERMAN, M., and KEY, C. B. "The Intelligence of Isolated Mountain Children." *Child Development,* 1932, *3,* 279–290.

SIEGEL, S., and FOURAKER, L. E. *Bargaining and Group Decision Making.* New York: McGraw-Hill, 1960.

SIMON, H. "The Architecture of Complexity." *General Systems,* 1965, *10,* 63–76.

SIMON, H. "Political Research: The Decision-Making Framework." In D. Easton (Ed.), *Varieties of Political Theory.* Englewood Cliffs, N.J.: Prentice-Hall, 1966.

SINGER, J. D. *A General Taxonomy for Political Science.* New York: General Learning Press, 1971.

SMELSER, N. J. *The Sociology of Economic Life.* Englewood Cliffs, N.J.: Prentice-Hall, 1963.

SMITH, M. G. "A Structural Approach to Comparative Politics." In D. Easton (Ed.), *Varieties of Political Theory.* Englewood Cliffs, N.J.: Prentice-Hall, 1966.

SPITZ, R. "Hospitalism." In *The Psychoanalytic Study of the Child.* Vol. 1. New York: International Universities Press, 1945.

TAWNEY, R. H. *Equality.* London: G. Allen, 1931.

THIBAUT, J. W., and KELLEY, H. H. *The Social Psychology of Groups.* New York: Wiley, 1959.

THORSON, T. *Biopolitics.* New York: Holt, 1970.

TRIBUS, M. "Information Theory as the Basis for Thermostatics and Thermodynamics." *General Systems,* 1961, *6,* 127–138.

TRIBUS, M. *Rational Descriptions, Decisions and Designs.* New York: Pergamon, 1969.

VICKERS, G. "Motivation Theory—A Cybernetic Contribution." *Behavioral Science,* 1973, *18* (4), 242–249.

WALTON, R. E., and MCKERSIE, R. B. *A Behavioral Theory of Labor Negotiations.* New York: McGraw-Hill, 1965.

WARD, P. "The Awkward Social Science: History." I. Morrissett and W. W. Stevens, Jr. (Eds.), *Social Science in the Schools: A Search for Rationale.* New York: Holt, 1971.

WATSON, J. B., and MORGAN, J. J. B. "Emotional Reactions and Psychological Experimentation." *American Journal of Psychology,* 1917, *28* (2), 163–174.

WATT, A. S. "The Community and the Individual." In A. MacFayden and P. J. Newbould (Eds.), *British Ecological Society Jubilee Symposium.* Supplement to the *Journal of Ecology,* 1964, *52.*

WATT, K. E. F. (Ed.) *System Analysis in Ecology.* New York: Academic, 1966.

WEBER, M. *The Theory of Social and Economic Organization.* New York: Oxford University Press, 1947.

WHITE, L. A. *The Science of Culture.* New York: Grove Press, 1949.

WHITE, L. A. *The Evolution of Culture.* New York: McGraw-Hill, 1959.

WHYTE, L. L. *Internal Factors in Evolution.* London: Tavistock, 1965.

WILLIAMS, C. B. *Patterns in the Balance of Nature.* New York: Academic, 1964.

WILLIAMSON, M. A. "A Case for Aimless Browsing." *Research/Development,* May 1966, 38–39; Oct. 1966, 32–33.

Index